**Oxford Applied Mathematics
and Computing Science Series**

General Editors

J. Crank, H. G. Martin, D. M. Melluish

W. E. WILLIAMS
University of Surrey

Partial
differential
equations

CLARENDON PRESS · OXFORD
1980

Oxford University Press, Walton Street, Oxford OX2 6DP

OXFORD LONDON GLASGOW
NEW YORK TORONTO MELBOURNE WELLINGTON
KUALA LUMPUR SINGAPORE HONG KONG TOKYO
DELHI BOMBAY CALCUTTA MADRAS KARACHI
NAIROBI DAR ES SALAAM CAPE TOWN

Published in the United States
by Oxford University Press
New York

British Library Cataloguing in Publication Data

Williams, William Elwyn
    Partial differential equations.-(Oxford
    applied mathematics and computing science series).
    1. Differential equations, Partial
    I. Title II. Series
    515'.353   QA374   79-41197

    ISBN 0-19-859632-4
    ISBN 0-19-859633-2 Pbk

Typeset by The Universities Press (Belfast) Limited
Printed in Great Britain at the University Press, Oxford
by Eric Buckley, Printer to the University

In memory

of

Judy

Preface

This book is intended to serve as an introduction, primarily for mathematicians, to the theory of partial differential equations and, it is hoped, should be suitable for final-year undergraduate, and first-year postgraduate, students who are following a reasonably comprehensive first course on partial differential equations. Most of the material covered in the first nine chapters has been included, at various times, in final-year courses that I have given during the past twenty years but I would be hesitant about attempting to include all this material in a single course.

The approach is classical in the sense that the methods and notation of functional analysis are not used though relatively new concepts such as 'weak solutions', 'shocks', and Green's functions which are useful to applied mathematicians are discussed. There is also no use, in any systematic fashion, of the formal theories of generalized functions or of distributions except that the elementary concept of the delta-function is freely used to develop the theory of Green's functions. My intention has been to try and emphasize the relationship between a given (generally linear) partial differential equation and the type of problems for which solutions exist and to describe the properties of solutions of the canonical second-order linear equations. I have attempted not to over-emphasize the development of special methods of solution and this aspect of the subject is largely confined to Chapter 7 where most of the general techniques for solving linear equations are described.

A number of exercises are included in the text and these vary from routine applications of the basic theory to problems taken from recent examination papers set at various Universities. I should like to thank the Universities of Cambridge, East Anglia, Liverpool, and Manchester for permission to include questions from their examination papers. Questions from Oxford University examination papers have been included by permission of Oxford University Press and I am grateful to the latter body for giving me permission to include these questions. I am also greatly indebted to my friend and colleague, Dr. R. Shail, for his help in reading, and commenting, on the complete manuscript and in reading through the proofs.

Surrey   W.E.W.
November 1979

Contents

1 Introduction

THE basic definitions are given in §1.1 and the various types of problems that can occur for partial differential equations are illustrated by specific examples in §1.2. In §1.3 it is shown that the problem of determining a function which makes some integral involving it have a maximum or a minimum value is equivalent to finding a particular solution of a partial differential equation. This type of reduction is of particular relevance to mathematical physics as the behaviour of many physical systems can be formulated succinctly in terms of some maximum or minimum principle involving integrals and the methods described in §1.3 enable the governing partial differential equation to be obtained fairly easily from such maximum or minimum principles. It is important from the practical point of view that the solution of a partial differential equation varies continuously with any boundary data imposed and a problem in which this is the case is said to be 'well-posed'. This question of 'well-posedness' is considered briefly in §1.4 where a simple counter-example is used to illustrate the fact that a seemingly reasonable problem need not be well-posed. In §1.5 the basic results relating to the various Fourier series expansions of a given function are summarized, together with the corresponding results for expansion as a Fourier–Bessel series and as a series of Legendre polynomials. Particular cases of these results are required at various points in the subsequent chapters and it is convenient to collect them all together at this stage.

1.1. Basic definitions

A partial differential equation in two independent variables is a relationship of the form

$$F(x, y, u_x, u_y, u_{xx}, u_{yy}, u_{xy}, u_{xxx}, u_{xxy}, \ldots) = 0,$$

where $u_x = \partial u/\partial x$ etc; the *order of a partial differential equation* is the *order of the highest derivative (or derivatives)* occurring. This suffix notation will also be used, whenever appropriate and when no ambiguity can arise, to denote the total derivative with respect to x of a function of the single variable x.

A *linear* equation is one which is linear in the dependent variable u and all of its partial derivatives occurring in the equation. A linear equation is therefore of the form

$$Lu = g(x, y), \tag{1.1}$$

where Lu is a sum of terms each of which is a product of a function of x and y with u or one of its partial derivatives. For first- and second-order equations the respective general forms for Lu are

$$Lu = a(x, y)u_x + b(x, y)u_y + c(x, y)u,$$

$$Lu = a(x, y)u_{xx} + 2b(x, y)u_{xy} + c(x, y)u_{yy}$$
$$+ d(x, y)u_x + e(x, y)u_y + f(x, y)u.$$

A linear equation is said to be *homogeneous* when, in equation (1.1), $g \equiv 0$ (and therefore an inhomogeneous equation corresponds to $g \not\equiv 0$). The definition of L shows that the difference between two solutions of equation (1.1) is a solution of the corresponding homogeneous equation. Thus solutions of the inhomogeneous equation can be obtained by adding a solution of the homogeneous equation to any particular solution of the inhomogeneous equation. This is analogous to writing the solution of an ordinary linear differential equation as the sum of a particular integral and a complementary function. It also follows from the definition of L that

$$L(Au_1 + Bu_2) = ALu_1 + BLu_2,$$

where A and B are constants, showing that a sum of constant multiples of solutions of the homogeneous linear equation is also a solution of the equation. This is the *principle of superposition* and it forms the basis of most practical methods of solving linear equations.

A *quasi-linear equation* is an equation which is linear in the highest derivative (or derivatives) occurring; an example of a first-order quasi-linear equation is

$$(1 + u^2)u_x + u_y = x^2.$$

An *almost linear* or *half-linear* partial differential equation is a quasi-linear equation in which the coefficients of the highest order derivatives are functions only of the independent variables; an example of such an equation is

$$x^2 u_{xx} + 4xy u_{yy} + uu_x + u^2 = 0.$$

1.2. Typical problems

Practical problems in many fields of application can be reduced to the solution of a partial differential equation or equations. Such a reduction normally requires considerable background knowledge and it is inappropriate to present here the detailed reduction for particular cases. It seems worthwhile, however, to list some typical problems which can occur in practice. The simplest such problems for second-order linear equations occur in the classical areas of physics such as heat conduction, acoustics, fluid motion, and electromagnetic theory. Historically, a detailed physical understanding of the underlying phenomena has proved invaluable in establishing the mathematical theory of various types of second-order linear partial differential equations. Such an understanding however will not be presumed and most of the results and methods will not refer to particular applications, whether physical or otherwise, as detailed knowledge of such applications is likely to be non-uniform. It should however be borne in mind that, particularly in studying a new partial differential equation, an understanding of the underlying phenomena can be extremely useful in clarifying the mathematical structure of a partial differential equation.

(i) *Simple birth process*

The probability-generating function $G(x, y)$ for a simple birth (Yule–Furry) process satisfies

$$G_y + ax(1-x)G_x = 0,$$

with $G(x, 0) = x^j$, where a is a constant and j a positive integer. This is a first-order linear homogeneous equation and the dependent variable is prescribed on a curve (the x-axis) in (x, y)-space.

(ii) *Incoming calls at a telephone exchange*

With certain assumptions regarding the duration of calls the problem of determining the probable number of incoming calls at a telephone exchange reduces to solving

$$G_y + a(x-1)G_x = b(x-1)G,$$

with $G(x, 0) = x^j$, where a and b are constants and j is a positive integer. This is another example of a first-order equation with the dependent variable prescribed on a curve.

(iii) *Temperature in a metallic lamina*

The temperature $u(x, y)$ in a metallic lamina, whose boundary is a closed curve C kept at a constant temperature u_0, satisfies

$$u_{xx} + u_{yy} = 0, \tag{1.2}$$

and $u = u_0$ on C, where x and y are Cartesian coordinates in the plane of the lamina. Equation (1.2) is *Laplace's equation* in two dimensions. The problem of finding the solution of a partial differential equation taking prescribed values on a closed curve (or surface in three dimensions) is termed a *Dirichlet* problem.

When C is not kept at a steady temperature but a steady known flow of heat is maintained round C then it can be shown from the theory of heat conduction that the derivative of u normal to C is known. A problem of this kind, where the normal derivative of the dependent variable is prescribed on a closed curve (or surface), is referred to as a *Neumann problem.*

If a steady source of heat such as a flame is applied within C then equation (1.2) has to be replaced by

$$u_{xx} + u_{yy} = g, \tag{1.3}$$

where g is a known function (related to the applied heat source) and this equation is the two-dimensional *Poisson's equation.*

(iv) *Transverse vibrations of a string*

The displacement u in the small transverse vibrations of an infinitely long taut string, which extends along the x-axis when in equilibrium, satisfies the equation

$$c^2 u_{xx} - u_{tt} = 0, \tag{1.4}$$

where c is a constant, and the conditions

$$u(x, 0) = f(x), \qquad u_t(x, 0) = g(x), \tag{1.5}$$

where f and g are given functions. In equation (1.4), which is known as the *one-dimensional wave equation*, t is a time variable and equations (1.5) prescribe the displacement f and the velocity g at the instant $t = 0$. In (x, t)-space the conditions (1.5) are applied on the line $t = 0$ and can be interpreted as prescribing u and its normal derivative on this line. For a second-order equation, a problem where the dependent variable and its normal derivative are prescribed on a given curve is termed a *Cauchy problem*. The problem posed by equations (1.4) and (1.5) is, by analogy with the underlying physical problem, also referred to as an *initial-value problem*. Prescribing a function along a curve means its derivative along the curve can be found and therefore, if the normal derivative is also known, both first derivatives are known on the curve. Therefore for nth order equations a Cauchy problem is one where the dependent variable and all its derivatives of all orders up to and including the $(n-1)$th are prescribed on a curve.

For a string of finite length a and held fixed at the end points $x = 0$ and $x = a$ (as, for example, for a violin string) equations (1.4) and (1.5) have to be supplemented by

$$u(0, t) = u(a, t) = 0,$$

and the resulting problem is called a mixed *initial-value–boundary-value problem*.

Equations (1.2) and (1.4) would appear to be of a similar type and yet the typical boundary conditions posed in practical problems are very dissimilar. It will transpire however that the two equations are very different in nature and that it is not possible in general to solve a Dirichlet problem for equation (1.4), and solving Cauchy problems for equation (1.2) poses serious problems. In these cases the physical context provides an excellent guide for developing an appropriate mathematical theory.

(v) *Heat conduction in a thin rod*

The temperature u in a thin straight rod with insulated sides satisfies

$$u_{xx} = ku_t, \tag{1.6}$$

where k is a constant, x is a Cartesian coordinate along the rod, and t is a time variable. A typical problem is that of determining the temperature at any time, given the temperature distribution at $t = 0$ and the temperatures at the two ends $x = 0$ and $x = a$, say. Thus the conditions are

$$u(0, t), \ u(a, t), \ u(x, 0) \quad \text{given.} \tag{1.7}$$

The conditions (1.7) pose a mixed *initial-value–boundary-value* problem for u similar to that for the transverse displacement of a string fixed at two ends. The main difference in the present case is that only u is prescribed at $t = 0$; this is essentially because equation (1.6) only involves the first derivative with respect to t. It would not be possible to prescribe u and u_t independently as $k u_t(x, 0) = u_{xx}(x, 0)$.

It is shown in Chapter 3 that there are three different classes of second-order linear (and half-linear) equations and that equations (1.2), (1.4), and (1.6) are, respectively, typical (and in fact canonical) members of each class.

(vi) *Electrostatics*

The static electric field $\mathbf{E}$ due to a time-independent charge distribution can be shown to be of the form $\mathbf{E} = -\text{grad } u$ where

$$\nabla^2 u = u_{xx} + u_{yy} + u_{zz} = f, \tag{1.8}$$

and f is related to the charge density of the given distribution. Equation (1.8) is the three-dimensional *Poisson's equation* and reduces, when $f \equiv 0$, to the three-dimensional Laplace's equation. In a typical electrostatic problem u will be prescribed on some given surface and in unbounded regions will tend to zero at infinity; hence electrostatic problems are generally Dirichlet ones.

(vii) *Electromagnetic wave propagation*

Each Cartesian component of the electric and magnetic field vectors in a homogeneous medium satisfies

$$c^2 \nabla^2 u = u_{tt}, \tag{1.9}$$

where c is a constant and t is a time variable. Equation (1.9) is the *three-dimensional wave equation* and it is also satisfied by the velocity potential of small-amplitude sound waves. In physical problems u and u_t are generally known at $t = 0$ at all points of space and u or $\partial u/\partial n$ are known on any bounding surface.

(viii) *Plateau's problem*

The problem of finding the surface of minimum area passing through a given plane curve C (Plateau's problem) reduces to solving

$$(1 + u_y^2)u_{xx} + (1 + u_x^2)u_{yy} - 2u_x u_y u_{xy} = 0, \qquad (1.10)$$

when u is given on C.

In the following section it will be shown that the problem of solving Dirichlet and Neumann problems for some partial differential equations is equivalent to finding a function such that a given integral involving the function and its derivatives (such an integral is termed a functional) has a stationary value (often a minimum). A problem of this type is said to be a problem in *variational calculus* or a *variational problem* and use of variational calculus provides an alternative approach to the solution of partial differential equations. Many of the equations of mathematical physics can also be derived directly from some minimum principle such as one of minimum energy or Hamilton's principle of least action in mechanics.

1.3. Variational formulation of partial differential equations

The variational approach is probably best illustrated by considering a particular example before proceeding to the general case. We therefore attempt to find a function u taking specified values on the boundary C of a region D and such that the integral (or functional) $I(u)$, defined by

$$I(u) = \int_D (u_x^2 + u_y^2 + 2gu) \, dx \, dy, \qquad (1.11)$$

where g is a known function of x and y, has a minimum value. It will be assumed that u and its first and second derivatives are continuous within D. The problem is more complicated than the normal minimum problems of differential calculus as there is clearly an infinity of possible functions that could be substituted into the right-hand side of equation (1.11). The correct approach to the problem can be found by careful consideration of what is meant by stating that u produces a minimum value of I. This means that substituting into the right-hand side of equation (1.11) any function other than the correct one produces a greater value for I. In particular then

$$I(u + \varepsilon f) \geq I(u),$$

where u is the correct function and ε a constant, for any function f vanishing on C (this is necessary in order that u and $u + \varepsilon f$ take the same values on C). Hence,

$$\int_D \{u_x^2 + u_y^2 + 2\varepsilon(u_x f_x + u_y f_y) + \varepsilon^2(f_x^2 + f_y^2) + 2g(u + \varepsilon f)\} \, dx \, dy$$
$$\geq \int_D (u_x^2 + u_y^2 + 2gu) \, dx \, dy$$

or

$$2\varepsilon \int_D (u_x f_x + u_y f_y + gf) \, dx \, dy + \varepsilon^2 \int_D (f_x^2 + f_y^2) \, dx \, dy \geq 0.$$

$$(1.12)$$

For small values of ε the first term on the left-hand side of equation (1.12) will dominate and hence, as ε can be positive or negative, the coefficient of ε must vanish so that

$$\int_D (u_x f_x + u_y f_y + gf) \, dx \, dy = 0,$$

Hence,

$$\int_D (\text{grad } u \cdot \text{grad } f + gf) \, dx \, dy = 0$$

or

$$\int_D (\text{div } (f \text{ grad } u) - f\nabla^2 u + gf) \, dx \, dy = 0. \qquad (1.13)$$

Equation (1.13) can be rewritten, on using the divergence

theorem, as

$$\int_C f\frac{\partial u}{\partial n}\,\mathrm{d}s - \int_D f(\nabla^2 u - g)\,\mathrm{d}x\,\mathrm{d}y = 0, \qquad (1.14)$$

where $\partial u/\partial n$ is the normal derivative of u on C. The condition $f \equiv 0$ on C gives

$$\int_D f(\nabla^2 u - g)\,\mathrm{d}x\,\mathrm{d}y = 0, \qquad (1.15)$$

for all f such that $f \equiv 0$ on C.

If it is assumed that there exists a point P in D such that $\nabla^2 u - g \neq 0$ at P then, by continuity, $\nabla^2 u - g$ will be non-zero and one-signed in a neighbourhood of P. It is also possible to construct a suitable function f which is one-signed and non-zero within such a neighbourhood and vanishes outside it (e.g. $f = (a^2 - x^2)^3(b^2 - y^2)^3$ within $|x| \leq a$, $|y| \leq b$ and zero outside the rectangle satisfies all the conditions). Hence $f(\nabla^2 u - g)$ will be one-signed in some neighbourhood of P and zero outside and the left-hand side of equation (1.15) will be non-zero. This is a contradiction and hence the original assumption was false and therefore

$$\nabla^2 u = g, \qquad (1.16)$$

with u known on C. Hence the problem of solving equation (1.16) with u taking prescribed values on C is equivalent to finding u taking known values on C and such that the functional I defined in equation (1.11) has a minimum value.

The above analysis can be extended to the case when I is defined by

$$I(u) = \int_D F(x, y, u, u_x, u_y)\,\mathrm{d}x\,\mathrm{d}y, \qquad (1.17)$$

where F is a given function. For a minimum (or stationary) value to be attained it is necessary, by analogy with the arguments applied to equation (1.11) that the coefficient of ε in $I(u + \varepsilon f) - I(u)$ must vanish for all $f \equiv 0$ on C. Taylor's theorem gives this coefficient to be

$$\int_D (fF_u + f_x F_{u_x} + f_y F_{u_y})\,\mathrm{d}x\,\mathrm{d}y$$

or, equivalently,

$$\int_D f \left\{ F_u - \frac{\partial}{\partial x}(F_{u_x}) - \frac{\partial}{\partial y}(F_{u_y}) \right\} dx\, dy$$
$$+ \int_D \left\{ \frac{\partial}{\partial x}(fF_{u_x}) + \frac{\partial}{\partial y}(fF_{u_y}) \right\} dx\, dy.$$

It follows by applying the divergence theorem that the second integral vanishes, since $f \equiv 0$ on C, and the condition for a stationary value is thus

$$\int_D f \left(F_u - \frac{\partial}{\partial x}(F_{u_x}) - \frac{\partial}{\partial y}(F_{u_y}) \right) dx\, dy = 0.$$

Repeating the argument following equation (1.15) now gives

$$F_u = \frac{\partial}{\partial x}(F_{u_x}) + \frac{\partial}{\partial y}(F_{u_y}) \tag{1.18}$$

and equation (1.18) is known as the Euler–Lagrange equation.

Plateau's problem discussed in the previous section is an example of a problem of the general type considered above. The direction cosine of the angle between the z-axis and the normal to the surface $z = u(x, y)$ is $(1 + u_x^2 + u_y^2)^{-\frac{1}{2}}$ and projecting the element dS of area of the surface onto the (x, y)-plane gives

$$(1 + u_x^2 + u_y^2)^{-\frac{1}{2}}\, dS = dx\, dy.$$

The area of the surface is

$$\int_D (1 + u_x^2 + u_y^2)^{\frac{1}{2}}\, dx\, dy,$$

where D is the area in the (x, y)-plane bounded by the curve C. Equation (1.18), with $F = (1 + u_x^2 + u_y^2)^{\frac{1}{2}}$, then gives equation (1.10).

The only boundary conditions considered have been Dirichlet ones and we consider briefly, for the particular case of I defined by equation (1.11), the possibility of imposing other boundary conditions. Removal of the restriction that u is prescribed on C means that f has no longer to vanish on C and that equation (1.14) must hold for all f. In particular, of course, it must still hold for $f \equiv 0$ on C and this, as before, yields equation (1.16).

Equations (1.14) and (1.16) then give

$$\int_C f\frac{\partial u}{\partial n}\,\mathrm{d}S = 0$$

for all f and hence $\partial u/\partial n$ must be identically zero on C. Thus, consideration of the minimum problem of I for all u gives a Neumann condition. This condition arose naturally from the variational problem and it is therefore termed a natural boundary condition. If a term $-2\int_C hu\,\mathrm{d}s$, where h is known, is included on the right-hand side of equation (1.11) then it is easy to see that, for the new functional to be stationary, equation (1.16) must still hold and $\partial u/\partial n$ will have to be equal to h on C.

1.4 Well-posed problems

It is necessary, when considering the solution of partial differential equations, to introduce a third concept, complementary to that of existence and uniqueness, namely that of a well-posed problem. A problem with a unique solution is said to be well-posed if any small change in the data produces a small change in the solution. It would clearly be ill-advised in general to apply numerical methods to problems not satisfying this condition and the practical validity of a formulation which leads to a non-well-posed problem is doubtful. It is comparatively easy to pose an apparently reasonable problem for a partial differential equation and yet find that it is not well-posed. It is therefore important to attempt to classify problems which are well-posed.

The simplest example of a non-well-posed (or ill-posed) problem is one first proposed by Hadamard, viz. the solution of the two-dimensional Laplace's equation (equation (1.2)) in $y \geq 0$ with

$$u(x, 0) = 0, \qquad u_y(x, 0) = n^{-1}\sin nx.$$

This problem has the solution $u \equiv 0$ when $u_y(x, 0) = 0$ but for finite values of n the solution is $n^{-2}\sin nx \sinh ny$. When n is large the solution differs considerably from zero thus showing that even a very small change from the boundary data $u_y = 0$ can produce a large change in the solution.

1.5. Fourier series and their generalizations

An important and very useful method for solving boundary-value problems for partial differential equations is that known as the 'method of separation of variables' (cf. §§3.4, 4.3, and Chapter 7). It is essential, for the successful application of this method, to be familiar with the elements of the theory of the expansion of a given function as an infinite series of simpler functions. The simplest such expansions are the various Fourier series associated with a given function and it is convenient to include at this point a summary of the basic theorems associated with Fourier series and also some of the generalizations of those theorems to cover expansions involving more complicated functions than the trigonometric ones. Detailed proofs may be found in (Ritt 1970).

For any function $f(x)$ continuous in $0 \le x \le a$ we define five series $F(x)$, $C(x)$, $S(x)$, $C^*(x)$, $S^*(x)$ by

$$F(x) = \tfrac{1}{2}a_0 + \sum_{n=1}^{\infty} \left(a_n \cos \frac{2n\pi x}{a} + b_n \sin \frac{2n\pi x}{a} \right), \qquad (1.19)$$

$$C(x) = \tfrac{1}{2}c_0 + \sum_{n=1}^{\infty} c_n \cos \frac{n\pi x}{a}, \qquad (1.20)$$

$$S(x) = \sum_{n=1}^{\infty} d_n \sin \frac{n\pi x}{a}, \qquad (1.21)$$

$$C^*(x) = \sum_{n=1}^{\infty} e_{2n-1} \cos(2n-1) \frac{\pi x}{2a}, \qquad (1.22)$$

$$S^*(x) = \sum_{n=1}^{\infty} f_{2n-1} \sin(2n-1) \frac{\pi x}{2a}, \qquad (1.23)$$

where the coefficients are given by

$$a_n = \frac{2}{a} \int_0^a f(x) \cos \frac{2n\pi x}{a} \, dx, \qquad (1.24)$$

$$b_n = \frac{2}{a} \int_0^a f(x) \sin \frac{2n\pi x}{a} \, dx, \qquad (1.25)$$

$$c_n = \frac{2}{a} \int_0^a f(x) \cos \frac{n\pi x}{a} \, dx, \qquad (1.26)$$

$$d_n = \frac{2}{a} \int_0^a f(x) \sin \frac{n\pi x}{a} \, dx, \qquad (1.27)$$

$$e_n = \frac{2}{a} \int_0^a f(x) \cos(2n-1) \frac{\pi x}{2a} \, dx, \qquad (1.28)$$

$$f_n = \frac{2}{a} \int_0^a f(x) \sin(2n-1) \frac{\pi x}{2a} \, dx. \qquad (1.29)$$

A basic result in the theory of Fourier series is that each of the series F, C, S, C^*, S^* is equal to $f(x)$ in $0 < x < a$. The situation at the end points is slightly more complicated and we have that

$$F(0) = F(a) = \tfrac{1}{2}[f(0) + f(a)],$$
$$C(0) = f(0), \qquad C(a) = f(a),$$
$$S(0) = S(a) = 0$$
$$S^*(0) = 0, \qquad S^*(a) = f(a), \qquad C^*(0) = f(0), \qquad C^*(a) = 0.$$

One of the reasons for attempting to represent the solution of a partial differential equation as a Fourier series is that it is possible, by choosing the appropriate Fourier series, to satisfy some types of boundary conditions automatically. For example the series $S(x)$ represents a function which vanishes when $x = 0$ and $x = a$, whilst $C^*(x)$ represents a function which has a zero derivative at $x = 0$ and vanishes at $x = a$. The series $F(x)$ on the other hand represents a function such that it and its first derivative at $x = 0$ are equal to the corresponding values at $x = a$. Clearly the trigonometric functions are not the only sets of functions which could be used to make certain that various boundary conditions are satisfied automatically, though they are particularly suitable for solving certain classes of problems for partial differential equations (this is discussed in Chapter 7). The simplest method of generating other suitable sets of functions is by choosing them to be the eigenfunctions of the differential equation

$$\frac{\mathrm{d}}{\mathrm{d}x}\left(p\frac{\mathrm{d}y}{\mathrm{d}x}\right) + (q + \lambda r)y = 0, \quad a < x < b \text{ (}a \text{ and } b \text{ both being finite)}$$

$$(1.30)$$

with

$$\alpha y(a) + \beta y_x(a) = 0 = \gamma y(b) + \delta y_x(b) \qquad (1.31)$$

where α, β, γ, δ, λ, are constants and p, q, r are continuous functions of x with $p \neq 0$ in $a \leq x \leq b$. The problem posed by equations (1.30) and (1.31) only possesses non-trivial solutions for particular values of λ and is known as a Sturm–Liouville eigenvalue problem. It is shown in (Ritt 1970) that there exist non-trivial solutions of this problem for an infinitely denumerable set of values of $\lambda(\lambda_1, \lambda_2, \ldots)$ and that the corresponding solutions (i.e. the eigenfunctions) $\phi_1, \phi_2, \ldots$ are such that, for any continuous function $f(x)$,

$$f(x) = \sum_{n=1}^{\infty} g_n \phi_n(x), \qquad a < x < b, \qquad (1.32)$$

where

$$g_n = \frac{\displaystyle\int_a^b rf\phi_n \, dx}{\displaystyle\int_a^b r\phi_n^2 \, dx}.$$

The functions ϕ_n are orthogonal in the sense that

$$\int_a^b r\phi_n \phi_m \, dx = 0, \qquad n \neq m, \qquad (1.33)$$

and the equation for g_n is obtained formally by multiplying both sides of equation (1.32) by $r\phi_m$, integrating with respect to x from a to b and using equation (1.33). It should be noted that the trigonometric functions of equations (1.20)–(1.23) are eigenfunctions of the Sturm–Liouville problem for $p \equiv r = 1$ and $q = 0$, with one member of each of the pairs (α, β), (γ, δ) equal to zero. The trigonometric functions of equation (1.19) are also the solution of a Sturm–Liouville eigenvalue problem; in this case the boundary conditions of (1.31) are replaced by the conditions

$$y(a) = y(b), \qquad y_x(a) = y_x(b).$$

The class of sets of functions which can be used to represent a given function can be further extended by removing one or both of the restrictions $p(a) \neq 0$ and $p(b) \neq 0$. In this case the problem is said to be a singular Sturm–Liouville problem and one or both of the boundary conditions in equation (1.31) have to be replaced by a finiteness condition at the end (or ends) at which p vanishes. Two particular singular problems, namely those associated with Bessel's equation and Legendre's equation, are of some practical significance and we therefore summarize the relevant expansion theorems. The basic results are

(i) If $f(x)$ is continuous in $0 < x < a$ then

$$f(x) = \sum_{n=1}^{\infty} h_n J_m\left(j_{nm} \frac{x}{a}\right), \qquad 0 < x < a, \qquad (1.34)$$

where $J_m(z)$ is Bessel's function of order m, j_{nm} $(n = 0, 1, \ldots)$ are the positive zeros, arranged in ascending order of magnitude, of J_m, and

$$a^2 h_n = \frac{2}{J_{m+1}^2(j_{nm})} \int_0^a x f(x) J_m\left(j_{nm} \frac{x}{a}\right) \mathrm{d}x. \qquad (1.35)$$

An expansion of this type is known as a Fourier–Bessel expansion and the properties of such expansion are described in Ritt (1970).

(ii) If $f(x)$ is continuous in $|x| \leq 1$ then

$$f(x) = \sum_{n=0}^{\infty} k_n P_n(x), \qquad (1.36)$$

where $P_n(x)$ is the Legendre polynomial of degree n and

$$k_n = \frac{2n+1}{2} \int_{-1}^{1} f(x) P_n(x)\, \mathrm{d}x. \qquad (1.37)$$

The theory of expansions in terms of Legendre polynomials is discussed in (Ritt 1970).

Bibliography

Ritt, R. K. (1970). *Fourier series*. McGraw-Hill, New York.

2 First-order equations in two independent variables

THIS chapter treats various aspects of the solution of first-order partial differential equations. Linear equations are discussed 'semi-qualitatively' in §2.1 and methods of solving such equations are described in §2.2 and illustrated by means of several worked examples. The discussion is extended in §2.3 to include quasi-linear equations and methods of solving these equations are described in §2.4. Linear and quasi-linear equations are, of course, just particular cases of the general first-order equation but, in view of the complications arising in the general case, it is preferable to look at these two classes of the general equation independently. In §2.5 some aspects of the solution of the general equation are described and solutions given for some particular cases. Finally in §2.6 the notion of a solution of a partial differential equation is extended to cover functions which are not differentiable everywhere and are possibly not even continuous and the concept of a 'weak' solution of a partial differential equation is introduced, together with that of a 'shock'. The occurrence of 'shocks' in a practical context is illustrated by applying the theory of 'weak' solutions to solve an idealized model of a traffic-flow problem.

2.1. Basic properties of the linear equation

The general first-order linear equation is

$$au_x + bu_y + cu = d, \qquad (2.1)$$

where a, b, c, d are functions of x and y. The nature of this equation can be made slightly more explicit by introducing the vector $\mathbf{v}$ defined by $\mathbf{v} = a\mathbf{i} + b\mathbf{j}$, where $\mathbf{i}$ and $\mathbf{j}$ are the usual Cartesian unit vectors. The equation then becomes

$$\mathbf{v} \cdot \operatorname{grad} u + cu = d,$$

or

$$vu_s + cu = d, \qquad (2.2)$$

where u_s denotes the derivative of u in the direction of $\mathbf{v}$. At any given point P the vector $\mathbf{v}$ makes an angle $\tan^{-1} b/a$ with the x-axis and therefore u_s at P is the rate of change, at P, of u in the direction of the tangent to the particular member, through P, of the family of curves defined by

$$\frac{\mathrm{d}y}{\mathrm{d}x} = \frac{b}{a}. \tag{2.3}$$

The partial differential equation therefore reduces, on these curves, to an ordinary differential equation.

The general solution of equation (2.3) will involve an arbitrary constant and therefore the family is often referred to as a 'one-parameter' family of curves. The curves defined by equation (2.3) are known as the *characteristic traces* (*or base curves*) (some authors call them the characteristics) associated with equation (2.1). The problem of finding the characteristic traces is one in the theory of ordinary differential equations and for most 'reasonably well-behaved' functions there is one, and only one, characteristic trace through any given point. (A more precise condition is that there exists a solution of equation (2.3) in some neighbourhood D of a point (x_0, y_0), with $y(x_0) = y_0$, provided that there exists a constant K such that $|b(x, y_2)/a(x, y_2) - b(x, y_1)/a(x, y_1)| \leq K |x_2 - x_1|$ for all points (x_1, y), (x_2, y) in D. This is equivalent to requiring that b/a satisfies a Lipschitz† condition of order 1 in x for all points of D.) A point through which there passes one, and only one, characteristic trace will be referred to as a non-singular point. It should be noted that particular difficulties occur at points where $a = b = 0$.

The theory of ordinary differential equations suggests that, as equation (2.1) reduces to an ordinary differential equation on a trace, prescribing u at a non-singular point of a characteristic trace C should enable u to be found at all points of C. Therefore prescribing u on an arc Γ, all points of which are non-singular, and which intersects no trace twice, should enable u to be found in a region covered by the characteristic traces through Γ. The typical problem for the first-order linear equation is therefore the

---

† The function $F(x, y)$ satisfies a Lipschitz condition of order α in x within a region D if, for all (x_1, y) and (x_2, y) in D,

$$|F(x_1, y) - F(x_2, y)| \leq K |x_2 - x_1|^{\alpha} \quad \text{for some constant } K.$$

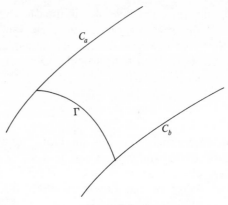

Fig. 2.1.

Cauchy problem and, for the arc Γ shown in Fig. 2.1, we would expect such a problem to have a solution defined in the region bounded by the traces C_a and C_b through the end points of Γ. It should be noted that it is not possible to state that u will be defined everywhere in the region spanned by the characteristic traces. (The theory of ordinary differential equations is a local one in that existence theorems are generally only valid in a neighbourhood of the initial point and u may not therefore be defined everywhere on a given trace). We would also not expect u to be defined at points of intersection of characteristic traces (these would be singular points of the traces), since integrating the values of u on Γ along two different traces would usually give two different values of u at the point of intersection. Difficulties would also arise if, as shown in Fig. 2.2, the curve Γ intersected a given trace twice. There is no *ab initio* reason for expecting that the value of u at P_2 obtained by integrating equation (2.2) along C from P_1 to P_2 is equal to the prescribed value of u at P_2. Even the tangency at a point of a given initial curve and a characteristic trace can give rise to difficulties because the prescribed derivative of u along Γ (and hence along the tangent trace) at the point of tangency need not satisfy equation (2.2) and u would not be differentiable at such points. In the exceptional case when Γ is tangential at each point to a different characteristic trace, u would not generally be differentiable on Γ and would not exist as a solution of the differential equation. Such a curve is said to be the envelope of the family of traces.

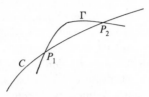

F IG. 2.2.

It is clearly not possible to prescribe u arbitrarily on a trace as no solution would exist if the prescribed data did not satisfy equation (2.1). Even data consistent with equation (2.1) would not, however, be sufficient to determine u off the trace as equation (2.1) provides no information about the rate of change of u in any direction oblique to the trace. A characteristic trace is therefore a curve such that knowledge of u on it is insufficient to determine u uniquely at points off it. The lack of uniqueness stems from the impossibility of determining, from values of u on a trace, the derivatives of u in any direction oblique to the trace; this prevents a Taylor series being found at any point of a trace. Some of the difficulties associated with Cauchy problems posed on traces can be seen by attempting to solve $u_x = 1$ for the two separate cases when, on the trace $y = 1$, (i) $u = 2x$ and (ii) $u = x$. In the first case the condition on the trace is inconsistent with the equation whereas in the second case there is consistency but a unique solution cannot be found as $x + c(y - 1)$, for all constants c, satisfies all the conditions.

The absence of information about the derivative normal to a trace suggests that there could be a discontinuity in this derivative across a trace. This can be illustrated by considering the problem $u_x = 1$, $u(0, y) = y$, $y > 0$, $u(0, y) = -y$, $y < 0$. The solution is $u = x + |y|$ and the equation is satisfied for all x and y. u_y is discontinuous across $y = 0$ (and is not defined at $y = 0$), which is the trace through the point of discontinuity of u_y on $x = 0$, suggesting that the discontinuity in a derivative normal to a characteristic trace at a point is propagated along the trace through that point.

The problem of solving equation (2.1) with u prescribed on a given plane curve is equivalent to determining the surface $u = u(x, y)$ containing the three-dimensional curve C defined by the values of u on the given plane curve (this being the projection on $u = 0$ of C). The determination of u along a characteristic trace is equivalent to finding a curve in the surface $u = u(x, y)$ and intersecting C at one point. The process of finding u in a given plane region is thus that of assembling the surface $u = u(x, y)$ (often called the *integral* or *solution surface*) from the curves constructed through each point of C. These space curves are the curves which are normally defined as the characteristics and their projections on $u = 0$ are the characteristic traces, (hence the nomenclature). The non-uniqueness of solutions when u is prescribed on a trace shows that characteristics in space are curves which are contained in more than one integral surface and are therefore the curves formed by the intersection of such surfaces.

A method of solving equation (2.1) is described in §2.2 and the various examples solved there illustrate the points made above.

We shall not attempt to formulate general conditions to specify the domain of existence of the solution of a first-order equation taking prescribed values on a curve which is not at any point tangent to a characteristic trace. The precise extent of such a domain is very dependent on the nature of the exact form of equation (2.1), but, in general, the solution will be defined in the region covered by the characteristic traces through the initial curve and in which these traces do not intersect. The region of existence must also, as explained above, exclude points where the traces intersect an initial curve a second time, etc. It is possible to make a general assertion of this kind for linear equations as there are global existence theorems for equations of the form of equation (2.2) with 'reasonably well-behaved' coefficients (e.g. if $c/(a^2 + b^2)^{\frac{1}{2}}$, $d/(a^2 + b^2)^{\frac{1}{2}}$, satisfy suitable Lipschitz conditions). The same is not true of quasi-linear equations (when c and d can be functions of u) as in general there are no global existence theorems for non-linear equations.

2.2. Solution of linear equations

The most direct method of solving the Cauchy problem for a linear partial differential equation is one based on a parametric representation of the traces. If x and y are functions of a parameter t then the characteristic traces are defined by

$$\frac{\mathrm{d}x}{\mathrm{d}t} = a, \qquad \frac{\mathrm{d}y}{\mathrm{d}t} = b, \tag{2.4}$$

and, on the traces,

$$\frac{du}{dt} = u_x \frac{dx}{dt} + u_y \frac{dy}{dt} = au_x + bu_y.$$

Therefore, along the traces, equation (2.1) simplifies to

$$\frac{du}{dt} + cu = d, \tag{2.5}$$

which is a more precisely formulated form of equation (2.2).

Equations (2.4) and (2.5) have to be solved with x, y, and u taking prescribed values $x_0(s)$, $y_0(s)$, $u_0(s)$ on a curve Γ, s being a parameter on Γ. The problem is therefore to find solutions $x(s, t)$, $y(s, t)$, $u(s, t)$ such that

$$x(s, 0) = x_0(s), \qquad y(s, 0) = y_0(s), \qquad u(s, 0) = u_0(s), \tag{2.6}$$

where it has been assumed that $t = 0$ corresponds to a point on Γ. The solution of equation (2.1) is then found by inverting the relations $x \equiv x(s, t)$ and $y \equiv y(s, t)$ to give s and t as functions of x and y; the resulting expressions are then substituted into $u \equiv u(s, t)$ so as to obtain u in terms of x and y. In order that the inversion can be carried out at a given point it is necessary, by the implicit function theorem, that the Jacobian $x_s y_t - y_s x_t$ be non-zero. In particular it is necessary that inversion is possible in a neighbourhood of Γ (i.e. for $t = 0$) and this gives, on using equation (2.4), that, on Γ,

$$x_{0s} b - y_{0s} a \neq 0, \tag{2.7}$$

which is the condition for Γ not to be a characteristic trace or tangential to one.

The breakdown of the solution at curves tangent to a trace, or at the point of intersection of traces, will be reflected in the vanishing of the Jacobian for particular values of t. The nature of the breakdown is often seen more clearly from an investigation of the geometry of the family of traces than from an analytic investigation of the vanishing of the Jacobian.

In the above analysis, parametrization provided a direct method of showing that equation (2.1) becomes an ordinary differential equation on the traces. An alternative way of making this property explicit would be to introduce two new independent variables with one of them (ϕ, say) being constant on the traces. We would then expect that the partial differential equation, when rewritten in terms of the new variables, would not involve a derivative with respect to ϕ as such a derivative would not represent a rate of change of u along a trace. The absence of one derivative from the equation also means that the equation can, in principle, be integrated by the normal methods used for ordinary differential equations. The choice of the second independent variable is entirely arbitrary as long as it is independent of ϕ and for simplicity we take it to be x. The function ϕ is obtained by writing the solution of the equation defining the traces, viz.

$$\frac{dy}{dx} = \frac{b}{a},$$

in the form $\phi(x, y) = $ constant. It follows that, as $d\phi = 0$ on the traces, ϕ has to satisfy

$$a\phi_x + b\phi_y = 0.$$

The chain rule for partial derivatives gives

$$u_x = \left(\frac{\partial u}{\partial x}\right)_\phi + \left(\frac{\partial u}{\partial \phi}\right)_x \phi_x, \qquad u_y = \left(\frac{\partial u}{\partial \phi}\right)_x \phi_y,$$

where $\left(\dfrac{\partial u}{\partial x}\right)_\phi$ and $\left(\dfrac{\partial u}{\partial \phi}\right)_x$ denote the respective partial derivatives of u as a function of the independent variables x and ϕ. Equation (2.1) then becomes

$$a\left(\frac{\partial u}{\partial x}\right)_\phi + \left(\frac{\partial u}{\partial \phi}\right)_x \{a\phi_x + b\phi_y\} + cu = d,$$

which, in view of the condition satisfied by ϕ, simplifies to

$$a\left(\frac{\partial u}{\partial x}\right)_\phi + cu = d. \tag{2.8}$$

Equation (2.8), as anticipated, does not involve the derivative with respect to ϕ and can, in principle, be integrated (y in a, c, and d has to be expressed in terms of x and ϕ). The general solution of equation (2.8) (and therefore of equation (2.1)) will involve an arbitrary function of ϕ (corresponding to the arbitrary constant occurring in the general solution of a first-order ordinary differential equation). It is possible to use this general solution to determine the solution to specific Cauchy problems but these are generally solved more directly by the previous parametric approach.

If only the general solution is required then the above method, using new variables x and ϕ, involves slightly less algebra than the parametric method. To obtain the general solution via the latter approach it is necessary first to solve the differential equations of (2.4) for some convenient initial values. The resulting expressions have then to be inverted to give s and t as functions of x and y. The general solution of equation (2.5) will involve an arbitrary function of s and substituting into this general solution the expressions obtained for s and t will give the same general solution as that obtained by direct integration of equation (2.8).

If u satisfies equation (2.1) with $d = 0$, then substituting $v = F(\phi)u$, for any function F, into equation (2.1) gives

$$av_x + bv_y + cv = F(\phi)[au_x + bu_y + cu] + uF_\phi[a\phi_x + b\phi_y]$$
$$= 0.$$

Therefore, when u is a particular solution of the Cauchy problem with conditions prescribed on a trace, a second solution is $F(\phi)u$ where $F = 1$ on the given trace (so that both solutions take the same value on the trace). This demonstrates explicitly, for the particular case $d = 0$, the non-uniqueness of the solution of the Cauchy problem on a trace.

Example 2.1

Solve

$$u_x + u_y + u = 1$$

with $u = \sin x$ on $y = x + x^2$, $x > 0$.

Equations (2.4) and (2.5) become

$$\frac{\mathrm{d}x}{\mathrm{d}t} = 1, \qquad \frac{\mathrm{d}y}{\mathrm{d}t} = 1, \qquad \frac{\mathrm{d}u}{\mathrm{d}t} + u = 1 \qquad (2.9)$$

and the boundary conditions in parametric form are

$$x(s, 0) = s, \qquad y(s, 0) = s + s^2, \qquad u = \sin s, \qquad s > 0.$$

The appropriate solutions of equations (2.9) are

$$x = s + t, \qquad y = s + s^2 + t, \qquad u = 1 - (1 - \sin s)e^{-t}.$$

It follows from the first two of these equations, or directly by integrating $dy/dx = 1$, that the traces are the family of lines $y - x = s^2 = \text{constant}$ (see Fig. 2.3).

The initial conditions are prescribed on $y = x + x^2$ which is a parabola going through the origin O and the trace $y = x$ through 0 is tangent, at O to the parabola. The parabola only intersects the trace in the region $y > x$ and, when $x > 0$, intersects each trace in this region once and only once. The solution is therefore defined in the region $y > x$.

The relationship $y - x = s^2$ confirms that the solution is only defined for $y > x$ and the third of the equations can be written as

$$u = 1 - e^s (1 - \sin s)e^{-(s+t)}.$$

Replacing $s + t$ by x and substituting $(y - x)^{\frac{1}{2}}$ for s gives

$$u = 1 + e^{-x}(\sin(y - x)^{\frac{1}{2}} - 1)e^{(y-x)^{\frac{1}{2}}}.$$

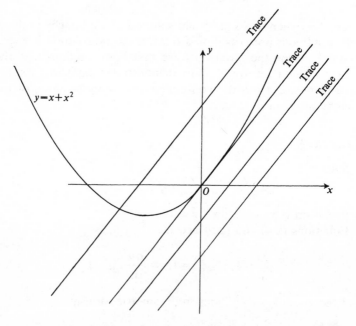

FIG. 2.3.

Differentiating the given condition $u(x, x+x^2) = \sin x$ with respect to x gives

$$u_x + (1+2x)u_y = \cos x \quad \text{for} \quad y = x + x^2,$$

and setting $x = 0$ shows that the given condition satisfies the partial differential equation at $x = 0$. This is the point of tangency of the trace and the curve $y = x + x^2$ and, as the given conditions are consistent with the partial differential equation, we would expect that the solution is valid even on the tangent trace i.e. $y = x$. This is confirmed by the fact that u as defined above is a differentiable function of both x and y separately, even at $x = y$. Changing the condition on the parabola to $u = 1 + \sin x$ $(u(x, x+x^2) = 1 + \sin x$ is no longer consistent with the equation at $x = 0$) can be shown to give

$$u = 1 + e^{-x}\sin(y-x)^{\frac{1}{2}}e^{(y-x)^{\frac{1}{2}}},$$

which is not differentiable with respect to y and x separately when $y = x$. This lack of differentiability (and hence of the existence of the solution at $y = x$) reflects the inconsistency between the new condition on the parabola and the governing differential equation.

The general solution, and the particular solution just found, can also be obtained by the alternative method described above. The traces are the family of lines $y - x = \text{constant}$ and therefore an appropriate choice for ϕ is $y - x$ and equation (2.8) becomes

$$\left(\frac{\partial u}{\partial x}\right)_\phi + u = 1.$$

The general solution of this equation is

$$u = 1 + e^{-x}f(\phi) = 1 + e^{-x}f(y-x),$$

where f is an arbitrary function of ϕ. The condition $u = \sin x$ when $y = x + x^2$ gives

$$\sin x = 1 + e^{-x}f(x^2),$$

and therefore

$$f(\phi) = (\sin \phi^{\frac{1}{2}} - 1)e^{\phi^{\frac{1}{2}}}, \quad \phi > 0.$$

Substitution of this form for f into the general solution gives the solution found previously. f is only determined from the given conditions for $\phi > 0$, confirming that the solution can only be found in the region $y > x$.

Example 2.2

  Solve

$$xu_x + (x^2 + y)u_y + \left(\frac{y}{x} - x\right)u = 1$$

with $u = 0$ on $x = 1$.
  Equations (2.4) and (2.5) give

$$\frac{dx}{dt} = x, \qquad \frac{dy}{dt} = x^2 + y, \qquad \frac{du}{dt} + \left(\frac{y}{x} - x\right)u = 1$$

and the boundary conditions are

$$x(s, 0) = 1, \qquad y(s, 0) = s, \qquad u(s, 0) = 0.$$

(The initial curve is parallel to the y-axis so that y is an appropriate choice for s.) The equations for x and y may be integrated to give

$$x = e^t, \qquad y = (s - 1)e^t + e^{2t}$$

and substituting these into the equation for u gives

$$\frac{du}{dt} + (s - 1)u = 1,$$

whose solution is

$$u = \frac{1}{s - 1}\{1 - e^{-(s-1)t}\}.$$

Eliminating t between x and y gives $s - 1 = (y - x^2)/x$ so that

$$u = \frac{x}{y - x^2}\left(1 - x^{-(y - x^2)/x}\right).$$

Eliminating t between the expressions for x and y shows that the traces are the parabolae $y/x = x + s - 1$ and these are shown for various values of s in Fig. 2.4. The traces intersecting $x = 1$ cover the whole plane excluding the line $x = 0$. Thus the solution will not be defined on $x = 0$ except at $x = 0$, $y = 0$ but, as all the traces

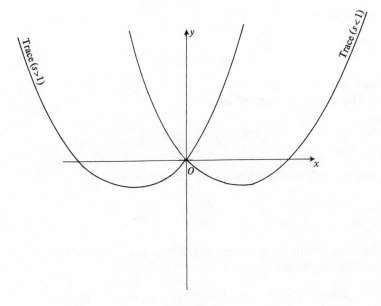

Fig. 2.4.

intersect at the origin, there will be a singularity there. The analytic solution confirms these observations.

The traces are the curves

$$\frac{y}{x} - x = \text{constant}$$

and an appropriate choice for ϕ in equation (2.8) is $(y/x) - x$. Equation (2.8), when rewritten in terms of ϕ and x, becomes

$$x\left(\frac{\partial u}{\partial x}\right)_\phi + \phi u = 1,$$

and the general solution of this is

$$u = \phi^{-1} + x^{-\phi} f(\phi),$$

where f is an arbitrary function. In the Cauchy problem solved above the function f can be found from the given conditions and the previous solution again obtained.

Example 2.3

Solve

$$u_x + 3y^{\frac{2}{3}}u_y = 2$$

with $u(x, 1) = 1 + x$.

The traces are defined by

$$\frac{dx}{dt} = 1, \qquad \frac{dy}{dt} = 3y^{\frac{2}{3}}$$

and the differential equation on the traces is

$$\frac{du}{dt} = 2.$$

The conditions at $t = 0$ are $x(s, 0) = s$, $y(s, 0) = 1$, $u(s, 0) = 1 + s$ and the equations integrate immediately to give

$$x(s, t) = s + t, \qquad y(s, t) = (t + 1)^3, \qquad u(s, t) = 2t + 1 + s.$$

Eliminating s and t gives

$$u = x + y^{\frac{1}{3}}.$$

The characteristic traces are defined by

$$\frac{dy}{dx} = 3y^{\frac{2}{3}}$$

and are the family of cubic curves $y = (x + s)^3$; the line $y = 1$ intersects each trace only once and thus, as the curves cover all the plane, we would expect the solution to be defined everywhere but it turns out not to be differentiable at $y = 0$. The reason for this apparent contradiction is that the line $y = 0$ is also a characteristic trace ($y = 0$ satisfies $dy/dx = 3y^{\frac{2}{3}}$), and also all the traces are tangent to $y = 0$ at their point of intersection with it. Thus, from previous arguments we would expect non-differentiability at $y = 0$.

The Jacobian $x_s y_t - y_s x_t$ is equal to $3(t + 1)^2$ and vanishes at $t = -1$ (i.e. $y = 0$) again showing that the solution breaks down when $y = 0$.

Exercises 2.1

1. In each of the following problems find the equation, in parametric form, of the characteristic trace through the point (x_0, y_0). Hence, in each case, find the function $u(x, y)$ satisfying the given conditions.

(i) $u_x - 2u_y = u$, $u = s$ when $x = 0$, $y = s$.

(ii) $xu_x + yu_y = 2u$, $u = s^2$ when $x = s$, $y = 1$.

(iii) $yu_x - xu_y = 2xyu$, $u = s^2$ when $x = y = s$, $s > 0$.

2. Obtain, for each of the following partial differential equations, the equations of the characteristic traces in the form $\phi(x, y) = $ constant. Hence determine the general solution of each equation.

(i) $(x + y)(u_x - u_y) = u$.

(ii) $yu_x + xu_y = 1$.

(iii) $x(x + y)u_x + y(x + y)u_y = -(x - y)(2x + 2y + u)$.

3. Find the general solution of

$$-x^2 y u_x + x^3 u_y + xyu = 2y(x^2 + y^2)$$

and the particular solution which takes the value $(a^2 + y^2)^{\frac{1}{2}}$ on $x = a \cos \alpha$, $0 \le y \le a \sin \alpha$. Find the values of x for which these conditions determine u on $y = 0$.

4. Solve

$$-2xyu_x + 4xu_y + yu = 4xy, \qquad x > 0,$$

with $u = 2x^{\frac{1}{2}}$ on $y = 0$, $x > 0$. In what part of the half-plane $x > 0$ is u determined if its value on the x-axis is only known for $1 \le x \le 3$?

5. Find the solution of

$$yu_x + u_y = x$$

that takes the value $u = \frac{2}{3}s^3$ when $x = s^2$, $y = s$. Given that s varies in the range $1 < s < 2$, determine the region of the (x, y)-plane in which the solution is determined.

6. Obtain for

$$x^2 u_x - xyu_y + y^2 = 0,$$

the equation, in parametric form, of the trace passing through (x_0, y_0). Hence find the solution of the equation such that $u = 2s^3$

when $x = s$, $y = s^2$, $1 \le s \le 2$. In what region of the (x, y)-plane is the solution defined?

7. Show that, for an arbitrary differentiable function $f(x)$, the equation

$$x^2 u_x + 2xy u_y = xu$$

cannot have a solution such that $u = f(x)$, when $y = 4x^2$. Find the general form of f such that a solution can be found and for such f verify, by obtaining two independent solutions, that the problem does not have a unique solution.

8. Find u satisfying

$$xu_x - yu_y = u$$

and such that $u = s^2$ when $x = y = s$, $1 \le s \le 2$, stating the region in which your solution is defined.

Show that, for an arbitrary differentiable function $f(s)$, the equation has no solution equal to $f(s)$ on the curve $x = s$, $y = 1/s$, $s > 0$. Determine the general form of f when a solution exists and show that, in this case, there is no unique solution.

2.3. Quasi-linear equations

The general first-order quasi-linear equation is

$$au_x + bu_y = c, \tag{2.10}$$

where a, b, c are functions of x, y, and u, and u will normally be prescribed on some plane curve Γ. The problem is, as has been mentioned earlier, equivalent to finding a surface (an integral surface) in (x, y, u)-space, satisfying equation (2.10) and containing the three-dimensional curve C defined by the values of u on Γ. (x and y will be functions of some parameter s on Γ and hence so will u and the three coordinates $x(s)$, $y(s)$, $u(s)$ define the curve C and Γ is the projection of C on the xy-plane.)

Guided by the analysis for the linear equation we now try to find a family of plane curves such that prescribing u at one point of a given curve of the family determines u on at least a small part of that curve. This is equivalent to looking for space curves lying in the surface $u = u(x, y)$ and passing through specified points. If such curves exist then the surface formed by the members of the family through C will be the required integral surface.

The direction cosines of the normal to the surface $u = u(x, y)$ are proportional to u_x, u_y, and -1 (the surface is defined by $\phi = u - u(x, y) = 0$ so that the direction cosines of the normal are proportional to the components of grad ϕ in (x, y, u)-space) and thus any curve satisfying

$$\frac{dx}{a} = \frac{dy}{b} = \frac{du}{c} \qquad (2.11)$$

will be perpendicular to the normal and, therefore, in the surface. The parametric form of equations (2.11) is

$$\frac{dx}{dt} = a, \qquad \frac{dy}{dt} = b, \qquad \frac{du}{dt} = c, \qquad (2.12)$$

and these equations define a family of curves in (x, y, u)-space; these curves are the characteristics of equation (2.10). The solutions of equation (2.12) will, like those of equations (2.4) and (2.5), be functions of two parameters s and t with varying values of s corresponding to different points of intersection with the initial curve C. The integral surface is therefore constructed by assembling the characteristics through C. This is illustrated in Fig. 2.5, where the dotted lines denote the characteristics.

Each characteristic can be drawn independently of neighbouring ones, which suggests that characteristics defined by equation (2.11) again have the property that knowledge of the solution on them does not determine it uniquely off them. Curves with this property also occur in the theory of second-order equations and

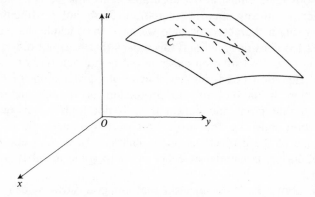

FIG. 2.5.

of systems of first-order equations and the word 'characteristic' is used to describe all such curves. Characteristics for quasi-linear equations were not derived as curves on which prescribing boundary data led to non-uniqueness as it seems preferable in this instance to emphasize the way they arise naturally in constructing the solution. In more general situations their precise role in constructing the solution is not as obvious however. It will now be shown that the alternative approach of seeking curves which do not enable u to be determined uniquely off them leads directly to equation (2.11) and hence, by reversing the above arguments, to a method of constructing the solution.

If, on some curve C, $x = x(s)$, $y = y(s)$, $u = u(s) = u(x(s), y(s))$ then

$$u_s = \frac{du}{ds} = x_s u_x + y_s u_y$$

is known on C and it now follows by using equation (2.10), that

$$\left.\begin{array}{l} u_y(bx_s - ay_s) = cx_s - au_s, \\ u_x(ay_s - bx_s) = cy_s - bu_s. \end{array}\right\} \tag{2.13}$$

u_x and u_y are not determinate, and hence u cannot be found uniquely off C, when

$$bx_s - ay_s = 0. \tag{2.14}$$

It also follows that, for u differentiable, the right-hand sides of equations (2.13) must also vanish and hence that C is a characteristic. If, on any curve C, equation (2.14) holds without the curve being a characteristic then we have to conclude that equation (2.13) does not hold and this means that u_s is not differentiable on C and no solution exists in the neighbourhood of such a curve. Equation (2.14) implies that at any given point of C its projection is parallel to the projection of the characteristic through that point and hence, if C is not a characteristic, its projection must be the projection of a characteristic or the envelope of the projections of the characteristics. (An envelope of a family is tangential at every point to some member of the family.)

The above analysis suggests that integral surfaces can only intersect in characteristics and this can be shown directly by considering two particular intersecting surfaces $u = U$ and $u = V$. On the curve of intersection

$$du = U_x \, dx + U_y \, dy \quad \text{and} \quad du = V_x \, dx + V_y \, dy$$

so that, for consistency,

$$\frac{dx}{U_y - V_y} = \frac{dy}{V_x - U_x} = \frac{du}{V_x U_y - V_y U_x}. \tag{2.15}$$

U and V satisfy equation (2.10) so that

$$aU_x + bU_y = c, \qquad aV_x + bV_y = c$$

and hence

$$\frac{a}{U_y - V_y} = \frac{b}{V_x - U_x} = \frac{c}{V_x U_y - V_y U_x}. \tag{2.16}$$

Eliminating U_x, V_x, etc. between equation (2.15) and (2.16) gives equation (2.11), thus showing that the curve of intersection is a characteristic.

2.4 Solution of quasi-linear equations

Equations (2.12) can be solved in exactly the same way as equations (2.4) and (2.5); a and b, however, now vary with u so that integration is likely to be more difficult and we therefore describe an alternative approach which does not involve the direct determination of u, x, and y as functions of s and t. The approach is due to Lagrange who deduced that a solution of equation (2.10) would be provided by

$$F(\phi, \psi) = 0 \quad \text{(for any } F)$$

or equivalently

$$\phi = f(\psi) \quad \text{(for any } f),$$

where $\phi(x, y, u)$ and $\psi(x, y, u)$ are independent functions such that

$$a\phi_x + b\phi_y + c\phi_u = a\psi_x + b\psi_y + c\psi_u = 0. \tag{2.17}$$

(In this context independence means that the normals to the surfaces $\phi = $ constant, $\psi = $ constant are not parallel at any point of intersection.) The functional relationship $F(\phi, \psi) = 0$ provides an implicit relationship between u, x and y, and in practice this can often be inverted to give u in terms of x and y. Comparison of equations (2.17) and (2.11) shows that a solution of equation (2.17) will be provided by writing a solution of equation (2.11) in the form ϕ = constant. Thus ϕ and ψ satisfying equation (2.17) are integrals of equation (2.11) (equation (2.11) represents a curve at the intersection of surfaces ϕ = constant and ψ = constant) and are most easily obtained by direct integration of this equation. If the ratio a/b, say, is independent of u, then solving

$$\frac{\mathrm{d}y}{\mathrm{d}x} = \frac{b}{a}$$

in the form $\phi(x, y)$ = constant still gives a solution of equation (2.17) (ϕ_u now being zero).

The intersection of the surfaces $\phi = c_1$ and $\psi = c_2$ forms a two-parameter system of curves satisfying equations (2.12) and imposing the condition $F(c_1, c_2) = 0$ thus produces a one-parameter family of characteristics. Thus Lagrange's method is effectively equivalent to constructing an integral surface from the characteristics through some arbitrary curve with varying values of the parameter corresponding to different points on the initial curve. Lagrange's result can also be proved fairly easily analytically as follows.

Equation (2.17) gives

$$\frac{a}{\dfrac{\partial(\phi, \psi)}{\partial(y, u)}} = \frac{b}{\dfrac{\partial(\phi, \psi)}{\partial(u, x)}} = \frac{c}{\dfrac{\partial(\phi, \psi)}{\partial(x, y)}} \qquad (2.18)$$

and differentiating the identity $F(\phi, \psi) = 0$ gives

$$F_\phi(\phi_x + \phi_u u_x) + F_\psi(\psi_x + \psi_u u_x) = 0,$$
$$F_\phi(\phi_y + \phi_u u_y) + F_\psi(\psi_y + \psi_u u_y) = 0.$$

Hence, on eliminating F_ϕ and F_ψ we obtain

$$\frac{\partial(\phi, \psi)}{\partial(y, u)} u_x + \frac{\partial(\phi, \psi)}{\partial(u, x)} u_y = \frac{\partial(\phi, \psi)}{\partial(x, y)},$$

and it now follows immediately on using equation (2.18) that u is a solution of equation (2.10).

Normally, the most difficult part of the solution is finding ϕ and ψ; no general method is available but there are two particular techniques which can often prove useful.

(i) If one of the pairs of equations in (2.11) has an integral (ϕ say) depending on only two of the variables (e.g. x and y) then substituting for y, say, in terms of x and ϕ in another pair will give an ordinary differential equation involving u and y (ϕ being regarded as constant) and it may be possible to obtain a second integral from this equation.

For the linear equation (2.1), $\phi =$ constant defines the solutions of equation (2.3) (i.e. the traces) and substituting for y in terms of x and ϕ in the appropriate form of equation (2.11) gives

$$\frac{\mathrm{d}x}{a} = \frac{\mathrm{d}u}{d - cu},$$

which, since a, c, and d are now to be regarded as functions of x and ϕ, is just equation (2.8). The approach leading to equation (2.8) is therefore a particular instance of Lagrange's method.

(ii) We have that for all p, q, r,

$$\frac{\mathrm{d}x}{a} = \frac{\mathrm{d}y}{b} = \frac{\mathrm{d}u}{c} = \frac{p\,\mathrm{d}x + q\,\mathrm{d}y + r\,\mathrm{d}u}{pa + qb + rc},$$

and it may be possible to determine p, q, r so that a simple solution can be obtained. The particular choice $pa + qb + rc = 0$ gives $p\,\mathrm{d}x + q\,\mathrm{d}y + r\,\mathrm{d}u = 0$ and this can often be integrated directly.

Both the parametric approach and Lagrange's method are illustrated in the following examples. It is to a certain extent a matter of taste as to which method should be used, the parametric approach has the advantage of giving a clearer picture of the nature of the characteristics than the Lagrangian one but it can involve heavy manipulation. It should be remembered, when using the parametric approach, that multiplying each of a, b, and c by the same arbitrary function still gives a set of three equations which define the characteristics (the multiplication is equivalent to re-defining the parameter t). In some cases multiplication by some appropriate factor is helpful in producing a more manageable set of equations. An instance of this occurs in Example 2.5 where the direct integration of equations (2.12) leads to complicated special functions whereas multiplication by a suitable factor gives an easily soluble set of equations.

It should also be remembered that, as for linear equations, the inversion process in the parametric solution breaks down whenever the Jacobian $x_s y_t - y_s x_t$ vanishes and hence singularities are to be expected whenever this Jacobian vanishes.

Example 2.4

Find the general solution of

$$(y+u)u_x + yu_y = x - y$$

and also that integral surface containing the curve $y = 1$, $u = 1 + x$, $-\infty < x < \infty$.

The characteristics are defined by

$$\frac{dx}{y+u} = \frac{dy}{y} = \frac{du}{x-y}$$

and hence

$$\frac{d(x+u)}{x+u} = \frac{dy}{y},$$

giving one integral to be $\phi = (x+u)/y$. Also

$$\frac{d(x-y)}{u} = \frac{du}{x-y}$$

so that a second integral is

$$\psi = (x-y)^2 - u^2.$$

Thus the general solution is

$$(x-y)^2 - u^2 = F\left(\frac{x+u}{y}\right)$$

and in order that the surface contains the given curve we must have that

$$(x-1)^2 - (1+x)^2 = F(1+2x),$$

giving $F(u) = -2(u-1)$ and hence

$$(x-y)^2 - u^2 = -\frac{2}{y}(x+u-y).$$

This is a quadratic for u with solutions

$$u = \frac{1}{y} \pm \left[(x - y) + \frac{1}{y} \right],$$

the positive sign has to be chosen so that the given curve is contained in the surface and hence

$$u = \frac{2}{y} + (x - y).$$

The solution is singular on the plane $y = 0$ and hence is only defined for $y > 0$.

The parametric form of the characteristic equation is

$$\frac{dx}{dt} = y + u, \qquad \frac{dy}{dt} = y, \qquad \frac{du}{dt} = x - y$$

and the conditions on $t = 0$ are

$$x(s, 0) = s, \qquad y(s, 0) = 1, \qquad u(s, 0) = 1 + s, \qquad -\infty < s < \infty.$$

The equation for y integrates immediately to give

$$y = e^t,$$

and eliminating x between the first and third equations gives

$$\frac{d^2 u}{dt^2} = u$$

with the general solution

$$u = A e^t + B e^{-t}.$$

This gives

$$x = (A + 1) e^t - B e^{-t} + C$$

and substituting for x and u in the third equation shows that $C = 0$. The conditions on $t = 0$ then give

$$x = (s + 1) e^t - e^{-t}, \qquad u = s e^t + e^{-t}.$$

Eliminating s and t gives the previous result, and the existence of the singularity at $y = 0$ can be seen by calculating the Jacobian which is found to be e^{2t} and therefore vanishes as $t \to -\infty$. The

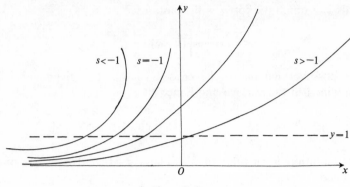

F IG . 2.6.

projections, on $u = 0$, of the characteristics are defined by $x = (s+1)e^t - e^{-t}$, $y = e^t$, and these are sketched, for various values of s, in Fig. 2.6. All have $y = 0$ as an asymptote.

Example 2.5

Find the integral surface of

$$(y^2 - u^2)u_x - xyu_y = xu$$

containing $x = y = u$, $x > 0$.

The characteristics are defined by

$$\frac{dx}{(y^2 - u^2)} = \frac{dy}{-xy} = \frac{du}{xu},$$

the second pair can be integrated to give $\phi = yu$, and also each of the above ratios is equal to

$$\frac{x\,dx + y\,dy + u\,du}{0},$$

so that a second integral is $\psi = x^2 + y^2 + u^2$. The general solution is $uy = f(x^2 + y^2 + z^2)$ and the condition that the given line is contained in the surface gives $x^2 = 3f(x^2)$ i.e. $3f(z) = z$. The solution is thus defined by

$$3uy = x^2 + y^2 + u^2;$$

solving this quadratic for u gives the appropriate solution to be

$$u = \frac{3y - (5y^2 - 4x^2)^{\frac{1}{2}}}{2}.$$

The non-differentiability of u when $5y^2 = 4x^2$ reflects the fact that there are no longer available any global existence theorems and that it is not possible to obtain domains of existence with any precision.

In parametric form the problem reduces to solving

$$\frac{dx}{dt} = y^2 - u^2, \qquad \frac{dy}{dt} = -xy, \qquad \frac{du}{dt} = xu,$$

with $x(s, 0) = y(s, 0) = u(s, 0) = s > 0$. The problem posed by these equations is an extremely complicated one and their solution can only be found in terms of elliptic functions. We therefore divide the right-hand sides of each equation by xy to give the equivalent, but simpler, set

$$\frac{dx}{dt} = \frac{y^2 - u^2}{xy}, \qquad \frac{dy}{dt} = -1, \qquad \frac{du}{dt} = \frac{u}{y}.$$

These equations can be integrated directly to give

$$x^2 = -(s - t)^2 + 3s^2 - \frac{s^4}{(s - t)^2}, \qquad y = s - t, \qquad u = \frac{s^2}{s - t},$$

and eliminating s and t gives $x^2 + y^2 + u^2 = 3uy$, as before.

Example 2.6

Find the integral surface of

$$yu_x - xu_y = 0$$

containing the curve $x^2 + y^2 = 1$, $u = y$.

The characteristics are defined by

$$\frac{dx}{dt} = y, \qquad \frac{dy}{dt} = -x, \qquad \frac{du}{dt} = 0,$$

and two integrals of these equations are $\phi = u$ and $\psi = x^2 + y^2$. The characteristics are thus circles parallel to the (x, y)-plane. The general solution is $u = f(x^2 + y^2)$ and f has to satisfy the condition $y = f(1)$ which is impossible. Hence no solution exists.

The curve $x^2 + y^2 = 1$, $u = y$ is an ellipse whose projection on the (x, y)-plane is a circle, and the projection of the characteristic through any given point of the ellipse coincides with the projection of the ellipse. Thus equation (2.14) holds and hence no solution exists.

Example 2.7

Find u satisfying

$$uu_x + u_y = 1$$

and $u(2s^2, 2s) = 0$, $s > 0$.
   Equation (2.12) gives

$$\frac{dx}{dt} = u, \qquad \frac{dy}{dt} = 1, \qquad \frac{du}{dt} = 1,$$

and the conditions at $t = 0$ are

$$x(s, 0) = 2s^2, \qquad y(s, 0) = 2s, \qquad u(s, 0) = 0, \qquad s > 0.$$

Hence

$$x = \tfrac{1}{2}t^2 + 2s^2, \qquad y = 2s + t, \qquad u = t.$$

Eliminating s and t gives

$$(u - \tfrac{1}{2}y)^2 = x - \tfrac{1}{4}y^2$$

and the appropriate solution for u is thus

$$u = \tfrac{1}{2}y - (x - \tfrac{1}{4}y^2)^{\frac{1}{2}}.$$

The solution is not defined for $x < \tfrac{1}{4}y^2$ and is not differentiable when $x = \tfrac{1}{4}y^2$, this non-differentiability being due to the fact that $x = \tfrac{1}{4}y^2$ is the envelope of the projection of the characteristics, i.e. it is tangent at each point to the characteristic trace at that point. This can be seen as follows: Eliminating t between x and y gives $x = 2s^2 + \tfrac{1}{2}(y - 2s)^2$ and dx/dy at $(4s^2, 4s)$ is $2s$. The point $(4s^2, 4s)$ also lies on $x = \tfrac{1}{4}y^2$ and dx/dy at this point is also $2s$, thus $x = \tfrac{1}{4}y^2$ is the envelope of the characteristics. (It can in fact be proved from the geometrical definition of an envelope that the envelope of a two-parameter family of curves is obtained by

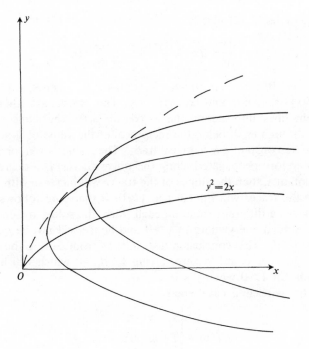

$$y^2 = 2x$$

F<small>IG</small>. 2.7.

eliminating s, t from the three equations $x = x(s, t)$, $y = y(s, t)$, $x_s y_t - y_s x_t = 0$.)

The situation is illustrated in Fig. 2.7 where the solid arcs denote the projections of the characteristics through the points of the initial curve and the envelope is shown as a dotted curve. The initial curve, the envelope, and the characteristic trace are coincident at the origin so u will not be differentiable there.

Use of the Jacobian shows that the solution does not exist when $t = 2s$ and this again gives $x = \frac{1}{4}y^2$.

Example 2.8

Solve

$$uu_x + u_y = 0, \qquad y \geq 0,$$

with $u(x, 0) = f(x)$, where $f(x)$ is a continuous differentiable function of x.

Equations (2.12) give

$$\frac{dx}{dt} = u, \qquad \frac{dy}{dt} = 1, \qquad \frac{du}{dt} = 0,$$

also $x(s, 0) = s$, $y(s, 0) = 0$, $u(s, 0) = f(s)$, hence $u = f(s)$, $x = tf(s) + s$, $y = t$, and $x = yf(s) + s$. The traces are therefore straight lines intersecting the x-axis at $(s, 0)$, the slope of the trace through $(s, 0)$ being the reciprocal of the value of u at $(s, 0)$ and u is constant on a given trace. The values of u on $y = 0$ are therefore propagated along the traces. If $f(s)$ is a decreasing function of s, then the slopes of the traces will increase with s and hence the traces will intersect in $y > 0$. It therefore follows that, as u takes a different value on each trace, the solution cannot be single-valued everywhere in $y > 0$ and hence will not be defined for all $y > 0$. This conclusion also follows from setting the Jacobian $x_s y_t - y_s x_t$ equal to zero, giving $1 + tf_s = 0$ which will have a solution for $t > 0$ when $f_s < 0$.

In the particular case when

$$f(x) = \begin{cases} 1, & x \le 0, \\ 1 - x, & 0 \le x \le 1, \\ 0, & x \ge 1. \end{cases}$$

the solution is given by

$$x(s, t) = \begin{cases} s + t, & s \le 0, \\ (1-s)t + s, & 0 \le s \le 1, \\ s, & s \ge 1, \end{cases} \qquad u(s, t) = \begin{cases} 1, & s \le 0, \\ 1 - s, & 0 \le s \le 1, \\ 0, & s \ge 1, \end{cases}$$

and is illustrated graphically in Fig. 2.8, where the straight lines are the characteristic traces in the various regions. The value of u on a trace emanating from a point $(s, 0)$ $(0 \le s \le 1)$ is $1 - s$ and, as these traces all intersect at $(1, 1)$, this implies that u is multi-valued at $(1, 1)$ and hence the solution breaks down at this point. It is also seen in Fig. 2.8 that the traces emanating from $(s, 0)$ with $s < 0$ intersect those emanating from $(s, 0)$, $s > 1$, and, as u has different values on these traces, it follows that it is not defined in the quadrant $x \ge 1$, $y \ge 1$.

The multi-valuedness of u in $y > 0$ can also be deduced analytically as follows. For a general function $f(s)$ we have that $x_s = 0$

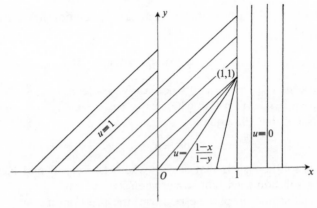

FIG. 2.8.

when $1 + t f_s = 0$ and, when f_s is a decreasing function of s, this will have, for some values of t, a solution $s = s_0$. Thus x, as a function of s, has a stationary value at $s = s_0$ and in a neighbourhood of this stationary value there will be, corresponding to any given x, two possible values of s. Therefore s is a multi-valued function of x and hence so is u. In the particular case when $f(s) = \sin s$ we have that $x_s = 0$ when $\cos s = -1/t$ which has solutions for all $t \geq 1$ and hence a single-valued solution does not exist for $y \geq 1$. It should also be noted that $x_s = 0$ implies u_x is infinite.

We discuss in §2.6 some of the problems raised by this example.

Exercises 2.2

1. Find the general solutions of

   (i) $(mu - ny)u_x + (nx - lu)u_y = ly - mx$; l, m, n constant.
   (ii) $(x + u)u_x + (y + u)u_y = 0$.
   (iii) $(x^2 + 3y^2 + 3u^2)u_x - 2xyu_y + 2xu = 0$.

2. Obtain a relationship between u, x, and y in each of the following problems.

   (i) $(y - u)u_x + (u - x)u_y = x - y$; $u = 0$ when $y = 2x$.
   (ii) $(2xy + 2y^2 + u)u_x - (2x^2 + 2xy + u)u_y = 2u(x - y)$; $\quad u = 2x^2$
when $y = 0$.
   (iii) $(3y - 2u)u_x + (u - 3x)u_y = 2x - y$; $u = 0$ when $x = y$.

3. Find, in parametric form, the solutions of the following

problems and hence obtain, in each case, a relationship between u, x, and y.

(i) $xuu_x + y^2u_y = u^2$; $x, y > 0$; $u = s$ when $x = \dfrac{1}{s}$, $y = 2s$.

(ii) $x^2u_x + uu_y = 1$; $u = 0$ when $x = s$, $y = 1 - s$.

(iii) $uu_x + u_y = 1$; $u = \frac{1}{2}s$ when $x = y = s$, $0 \le s \le 1$.

4. Show that the equation

$$uu_x + u_y = 1$$

has no solution such that $u = \frac{1}{2}y$ when $4x - y^2 = 0$ and has no unique solution such that $u = y$ when $2x - y^2 = 0$.

5. Determine, in parametric form, the solution of

$$u_y + uu_x + \tfrac{1}{2}u = 0$$

with $u(x, 0) = \sin x$. Show that there will be a single-valued differentiable solution in $y < 2 \log 2$ but that no such solution can exist for $y > 2 \log 2$.

2.5. The general first-order non-linear equation

We now consider the general problem of finding u satisfying

$$F(x, y, u, p, q) = 0, \tag{2.19}$$

where $u_x = p$, $u_y = q$, and F is a given function, and taking given values on a plane curve Γ.

The methods used for solving linear and quasi-linear equations suggest that a fruitful method of obtaining an integral surface would be to construct it from characteristics. In this context the most convenient definition of characteristics is that of a family of curves such that conditions on them do not determine u uniquely off them. We therefore introduce new coordinates ϕ and ψ and try to find ϕ, say, such that equation (2.19) provides no information about u_ϕ. This derivative will be in a direction oblique to the curve, $\phi = $ constant, and, if it is not determined by equation (2.19), u cannot be continued off the curve by using Taylor series. Hence equation (2.19) becomes

$$F(x(\phi, \psi), y(\phi, \psi), u(\phi, \psi), u_\phi\phi_x + u_\psi\psi_x, u_\phi\phi_y + u_\psi\psi_y) = 0, \tag{2.20}$$

and the condition that u_ϕ does not occur explicitly is

$$F_{u_\phi} \equiv 0 = \phi_x F_p + \phi_y F_q. \tag{2.21}$$

The gradient of the tangent to the curve $\phi = \text{constant}$ is $-\phi_x/\phi_y$ so that

$$\frac{dy}{dx} = \frac{F_q}{F_p}$$

and hence the parametric equations of the curve C', say, are,

$$\frac{dx}{dt} = F_p, \qquad \frac{dy}{dt} = F_q. \tag{2.22}$$

Also, on C',

$$\frac{du}{dt} = u_x \frac{dx}{dt} + u_y \frac{dy}{dt} = pF_p + qF_q, \tag{2.23}$$

and, for the particular case of the quasi-linear equation, equation (2.23) reduces to the third relation of equation (2.12). Equations (2.22) and (2.23) are no longer sufficient to determine a curve in (x, y, u)-space as p and q are not known and hence equations have to be obtained for these.

Differentiating equation (2.19) with respect to x and y gives

$$F_x + pF_u + p_x F_p + q_x F_q = 0, \tag{2.24}$$

$$F_y + qF_u + p_y F_p + q_y F_q = 0. \tag{2.25}$$

Using $q_x = p_y$ and equations (2.22) reduces equations (2.24), (2.25) to

$$\frac{dp}{dt} = -F_x - pF_u, \tag{2.26}$$

$$\frac{dq}{dt} = -F_y - qF_u. \tag{2.27}$$

Equations (2.22), (2.23), (2.26), and (2.27) form a set of five ordinary differential equations for x, y, u, p, and q. These can, in principle, be solved provided that all the variables are known at $t = 0$. Normal Cauchy conditions are enough to specify x, y, u at $t = 0$ in terms of some parameter s on an initial curve, but specifying the values of p and q at $t = 0$ is not as straightforward.

Knowledge of u, and hence of its tangential derivative, on a curve gives one relation between u_x and u_y and, for consistency, a second one is obtained from equation (2.19). The values so found for u_x and u_y are the values of p and q at $t = 0$. The given values at $t = 0$ can therefore be used to solve the above differential equations and express all five variables as functions of s and t. These latter relations can then be inverted to give u, p, and q as functions of x and y. It is however necessary to prove that these quantities are such that $p = u_x$ and $q = u_y$ and that equation (2.19) holds. Direct differentiation shows that, when equations (2.22), (2.23), (2.26), and (2.27) hold, $dF/dt = 0$. The conditions at $t = 0$ are chosen so that $F = 0$ at $t = 0$ and therefore $F \equiv 0$.

It now remains to show that $p = u_x$ and $q = u_y$ and in order to do this we first form the function $V(s, t)$ defined by

$$V = u_s - px_s - qy_s.$$

Initially, $u(s, 0) = u(x(s, 0), y(s, 0))$ so that

$$u_s(s, 0) = x_s p(s, 0) + y_s q(s, 0),$$

giving $V(s, 0) = 0$. Also

$$
\begin{aligned}
V_t &= u_{st} - p_t x_s - q_t y_s - p x_{st} - q y_{st} \\
&= u_{ts} - p x_{ts} - q y_{ts} - p_t x_s - q_t y_s \\
&= (u_t - px_t - qy_t)_s + p_s x_t + q_s y_t - p_t x_s - q_t y_s. \quad (2.28)
\end{aligned}
$$

Substitution from equations (2.22) and (2.23) shows that the term in brackets on the right-hand side of equation (2.28) is zero, and further substitution for x_t, y_t, p_t, q_t from the appropriate equations gives

$$
\begin{aligned}
V_t &= p_s F_p + q_s F_q + x_s F_x + y_s F_y + F_u(px_s + qy_s) \\
&= \frac{dF}{ds} - F_u(u_s - px_s - qy_s).
\end{aligned}
$$

Thus, as $F \equiv 0$,

$$V_t + V F_u = 0$$

and, as V satisfies a first-order linear partial differential equation, it now follows from the condition $V(s, 0) = 0$ that $V \equiv 0$ so that

$$u_s = px_s + qy_s. \quad (2.29)$$

Equations (2.22) and (2.23) show that

$$u_t = px_t + qy_t. \qquad (2.30)$$

Equations (2.29) and (2.30) have a unique solution for p and q provided that

$$x_s y_t - y_s x_t \neq 0. \qquad (2.31)$$

Equation (2.31) is, as mentioned earlier, the condition necessary in order to be able to invert the expressions for x and y to give s and t (and hence u) as functions of x and y. Hence equation (2.31) is the condition for $u(x, y)$ to exist and u_s and u_t can be calculated from it to give

$$u_s = u_x x_s + u_y y_s,$$
$$u_t = u_x x_t + u_y y_t.$$

Solving the above equations for u_x and u_y yields the same expressions as those obtained from equations (2.29) and (2.30) for p and q so that $p \equiv u_x$, $q \equiv u_y$. Hence, provided equation (2.31) is satisfied in a neighbourhood of the initial curve, equations (2.22), (2.23), (2.26), (2.27) provide a solution of equation (2.19). These equations define a curve in (x, y, u)-space, which is the characteristic, and they also define at each point of the characteristic the derivatives u_x and u_y and hence define the orientation of some tangent plane defined by the direction ratios p, q, -1. This curve together with the associated tangent planes can be regarded as forming a strip, the *characteristic strip*, and the equations are often referred to as the characteristic strip equations.

Substituting from equation (2.22) into equation (2.31) shows that the initial curve Γ must be such that

$$F_p y_s - F_q x_s \neq 0.$$

If the curve Γ is such that $F_p y_s - F_q x_s = 0$ then, if u is differentiable, equation (2.23) with t replaced by s must also hold and, as $F \equiv 0$ for the initial data, it follows that equations (2.26) and (2.27) also must hold. Hence all the characteristic strip equations hold on the initial curve and the data in (x, y, u)-space define a characteristic strip. There will be an infinity of solutions satisfying the initial data as prescribing data on a characteristic gives no information about (or constraint on) the solution in a neighbourhood of the characteristic.

If the condition $F_p y_s - F_q x_s = 0$ is satisfied with Γ not being a characteristic, then, as before, u is not differentiable on Γ and Γ is the envelope of projections of characteristics.

The solution of the characteristic strip equations can be somewhat complicated and the process can often be helped by supplementing them with equation (2.19), which is, as has been shown, an integral of the equations. The first step in the calculation in any specific problem is the determination of the values of p and q at $t = 0$.

Example 2.9

Solve

$$pq = u$$

with $u(0, s) = s^2$.

$F = pq - u$ and the strip equations are

$$x_t = q, \qquad y_t = p, \qquad u_t = 2pq, \qquad p_t = p, \qquad q_t = q.$$

Differentiating $u(0, s) = s^2$ with respect to s gives $q(s, 0) = 2s$ and substituting for $q(s, 0)$ in the differential equation gives $p(s, 0) = \frac{1}{2}s$. The equations for p and q can be integrated immediately to give

$$p = \tfrac{1}{2}se^t, \qquad q = 2se^t.$$

Substituting these values of p and q into the other three equations and carrying out the relevant integration then gives

$$x = 2s(e^t - 1), \qquad y = \tfrac{1}{2}s(e^t + 1), \qquad u = s^2 e^{2t}.$$

Finally, eliminating s and t gives

$$16u = (4y + x)^2.$$

Example 2.10

Solve

$$p^2 x + qy - u = 0$$

with $u(s, 1) = -s$.

The strip equations are

$$x_t = 2px, \qquad y_t = y, \qquad u_t = 2p^2x + qy, \qquad p_t = p - p^2, \qquad q_t = 0.$$

Differentiating the identity $u(s, 1) = -s$ with respect to s gives $p(s, 0) = -1$ and substitution in the differential equation shows that $q(s, 0) = -2s$. Hence, as $q_t = 0$, $q \equiv -2s$, the equation for p can be integrated by separation of variables and its solution is

$$p = \frac{1}{1 - 2e^{-t}}.$$

The equation for y can also be integrated to give $y = e^t$, and substituting for p in the equation for x gives, after integration, $x = s(e^t - 2)^2$. The right-hand side of the equation for u_t is now determined in terms of s and t and can be integrated to give

$$u = se^t(e^t - 2).$$

Eliminating s and t gives

$$xy = u(y - 2),$$

showing that u is singular for $y = 2$. The equations of the characteristics are $x = s(y - 2)^2$; they all intersect at $(0, 2)$, suggesting that some kind of singularity might occur for $x = 0$ or $y = 2$.

Exercise 2.3

Determine, in parametric form, the solutions of the following problems and hence, in each case, obtain a relation between u, x, and y.

1. $p^2 + qy - u = 0$, $u = \frac{1}{4}s^2 + 1$ when $x = s$, $y = 1$.
2. $px + qy - p^2q - u = 0$, $u = s + 1$ when $x = s$, $y = 2$.
3. $p^2 = qu$, $u = 1$ when $x = s$, $y = 1 - s$.
4. $(p^2 + q^2)x - pu = 0$, $u = 2s$ when $x = 0$, $y = s^2$.

2.6. Discontinuous and 'weak' solutions and shocks

We have seen that solutions of apparently reasonable problems for partial differential equations (e.g. Examples 2.1, 2.3, 2.5, 2.8) are not differentiable in some regions of the (x, y)-plane and hence, in these regions, cease to be solutions of the original equation. We now examine the possibility of widening the notion of a solution so as to include discontinuous and non-differentiable functions. The general process will be illustrated by considering the solution of

$$u_x + u_y = 0, \qquad (2.32)$$

with $u(x, 0) = f(x)$, where f is a discontinuous function. The formal solution is $u = f(x-y)$, which is discontinuous in some parts of the plane. In order to dispense with differentiability it is necessary to remove, in some sense, some of the derivatives from equation (2.32). One simple way of doing this is to integrate the equation with respect to x from $x = x_1$ to $x = x_2$ giving

$$u(x_2, y) - u(x_1, y) + \frac{\partial}{\partial y} \int_{x_1}^{x_2} u(t, y)\, dt = 0. \qquad (2.33)$$

If u is a differentiable function then setting $x_2 = x$ in equation (2.33) and differentiating with respect to x gives equation (2.32) and hence a differentiable solution of equation (2.33) is a solution of equation (2.32). It is however possible for a non-differentiable function to satisfy equation (2.33), though it clearly is not a solution of equation (2.32). This can be seen by setting $u = (x-y)^{\frac{1}{2}}$, when equation (2.33) becomes

$$(x_2 - y)^{\frac{1}{2}} - (x_1 - y)^{\frac{1}{2}} + \frac{2}{3} \frac{\partial}{\partial y} [(x_2 - y)^{\frac{3}{2}} - (x_1 - y)^{\frac{3}{2}}] = 0,$$

which is an identity, though u is not differentiable when $x = y$.

There are many integral identities which can be derived from equation (2.32) and each identity can be used in some sense to extend the definition of a solution of the equation. We now consider the procedure necessary to define what is normally called a 'weak' solution of equation (2.32) and the process will then be generalized to include more general equations. Multiplying the left-hand side of equation (2.32) by an arbitrary function v gives

$$v(u_x + u_y) = (vu)_x + (vu)_y - uv_x - uv_y,$$

and integrating this identity over some closed bounded region D gives

$$\int_D v(u_x + u_y)\,dx\,dy = \int_D [(vu)_x + (vu)_y]\,dx\,dy$$
$$- \int_D (uv_x + uv_y)\,dx\,dy.$$

The divergence theorem shows that

$$\int_D [(vu)_x + (vu)_y]\,dx\,dy = -\int_C uv(dx - dy),$$

where C is the boundary of D. Hence, if v is assumed to vanish identically on C,

$$\int_D v(u_x + u_y)\,dx\,dy = -\int_D u(v_x + v_y)\,dx\,dy. \qquad (2.34)$$

Therefore, when u satisfies equation (2.32),

$$\int_D u(v_x + v_y)\,dx\,dy = 0, \qquad (2.35)$$

for all differentiable v with $v \equiv 0$ on C. Reversing the above analysis shows that a differentiable function u satisfying equation (2.35) for all functions v (vanishing on C) will satisfy equation (2.32). (It follows immediately that the left-hand side of equation (2.34) vanishes identically and arguments similar to those of §1.3 then give equation (2.32).)

A 'weak' solution of equation (2.32) in a region D is now defined to be a function satisfying equation (2.35) for all differentiable v with $v \equiv 0$ on C. As an example it will now be shown that the unit step function $H(x - y)(H(x) = 1, x > 0, H(x) = 0, x < 0)$ is a weak solution of equation (2.32). The left-hand side of equation (2.35) becomes

$$\int_{D'} (v_x + v_y)\,dx\,dy,$$

where D' is the intersection of D and the region $x > y$ and is the shaded region of Fig. 2.9. The boundary C' of D' will be

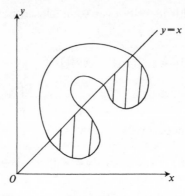

FIG. 2.9.

composed of parts of C and of parts of the line $y = x$ and

$$\int_{D'} (v_x + v_y)\, dx\, dy = -\int_{C'} v(dx - dy).$$

Since $v \equiv 0$ on C and $dx = dy$ on $y = x$, the line integral over C' vanishes, thus showing that $H(x - y)$ is a weak solution of equation (2.32).

The idea of a weak solution can be extended to include the general linear equation

$$au_x + bu_y + cu = 0. \tag{2.36}$$

Multiplying the left-hand side of equation (2.36) by an arbitrary function v (with $v \equiv 0$ on C) gives

$$v(au_x + bu_y + cu) = (avu)_x + (bvu)_y - u(av)_x - u(bv)_y + cuv$$

and

$$\int_D v(au_x + bu_y + cu)\, dx\, dy$$
$$= \int_D [(avu)_x + (bvu)_y]\, dx\, dy + \int_D u[-(va)_x - (vb)_y + cv]\, dx\, dy. \tag{2.37}$$

The first integral on the right-hand side of equation (2.37) may be transformed into a line integral over C and this vanishes as

$v \equiv 0$ on C. Hence, when u satisfies equation (2.36),

$$\int_D u\{-(av)_x - (bv)_y + cv\} \, dx \, dy = 0. \qquad (2.38)$$

It follows again, by reversing all the above arguments, that if equation (2.38) holds for all v, with $v \equiv 0$ on C, when u is a differentiable function, then u is a solution of equation (2.36). A 'weak' solution of equation (2.36) in a region D is thus a function satisfying equation (2.38) for all differentiable v with $v \equiv 0$ on C.

The left-hand side of equation (2.36) and the expression multiplying u in the integrand of equation (2.38) define two differential operators L and L^* such that

$$L(u) = au_x + bu_y + cu,$$
$$L^*(v) = -(av)_x - (bv)_y + cv,$$

and $vL(u) - uL^*(v)$ is a divergence (i.e. is of the form $P_x + Q_y$, for some functions P and Q). The operator L^* is defined to be the operator *adjoint* to L. Also

$$\int_D \{vL(u) - uL^*(v)\} \, dx \, dy = \int_C uv(a \, dy - b \, dx). \qquad (2.39)$$

The next point to consider is whether there are any restrictions on the curve of discontinuity of a sectionally discontinuous 'weak' solution within a region D. We assume the required curve G divides D into two regions D_1 and D_2 (Fig. 2.10) and that u is differentiable in each of the regions D_1 and D_2. Hence u is a solution in the normal sense in D_1 and D_2 so that $L(u) \equiv 0$ and applying equation (2.39) to each of the separate regions gives

$$\int_{D_1} uL^*(v) \, dx \, dy = + \int_G vu(a \, dy - b \, dx), \qquad (2.40)$$

$$\int_{D_2} uL^*(v) \, dx \, dy = - \int_G vu(a \, dy - b \, dx). \qquad (2.41)$$

(The line integral over the boundary of D vanishes as $v \equiv 0$ on C.) The different signs on the right-hand sides of equations (2.40) and (2.41) reflect that the sense of traversing G for applying the divergence theorem to D_2 is opposite to that applicable to D_1.

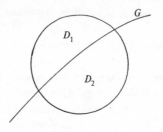

FIG. 2.10.

However, as u is a weak solution,

$$\int_{D} uL^*(v)\,dx\,dy = 0 = \int_{D_1} uL^*(v)\,dx\,dy + \int_{D_2} uL^*(v)\,dx\,dy$$

so that

$$0 = \int_{G} v[u][a\,dy - b\,dx], \qquad (2.42)$$

where $[u]$ denotes the discontinuity in u across G. Therefore, as equation (2.42) has to hold for all v and, by assumption, $[u] \neq 0$,

$$a\,dy = b\,dx, \qquad (2.43)$$

showing that a 'weak' solution of a linear equation can only be discontinuous across the characteristic traces.

A theory of 'weak' solutions can also be developed for a class of non-linear equations of the type

$$a_x + b_y = 0, \qquad (2.44)$$

where a, b, are now twice differentiable functions of x, y, and u. (In fact 'weak' solutions are only really relevant in the practical sense for such non-linear equations.) Equation (2.44) is said to have the form of a conservation law since integrating with respect to x from $x = x_1$ to $x = x_2$ gives

$$\frac{\partial}{\partial y}\int_{x_1}^{x_2} b\,dx = -(a)_{x=x_2} + (a)_{x=x_1}. \qquad (2.45)$$

Hence, on interpreting y as a time variable, equation (2.45) can be regarded as stating that the rate of change of some total quantity b is proportional to the difference of some 'flux' a at the end points and many conservation laws in physics are of this form.

A simple practical example is provided by the continuum model of traffic flow along a straight road. In this model the x-axis is taken along the road, the traffic assumed to flow in positive x-direction, and the density (cars per unit length) at a point x at a time t is denoted by $\rho(x, t)$ whilst $q(x, t)$ denotes the rate (cars per unit time) at which cars go past the point x at time t. The total number of cars in the segment $x_1 < x < x_2$ is $\int_{x_1}^{x_2} \rho \, dx$ and the rate at which the number of cars in this segment is increasing is

$$\frac{d}{dt} \int_{x_1}^{x_2} \rho \, dx = \int_{x_1}^{x_2} \rho_t \, dx.$$

This quantity must be equal, assuming that there are no side roads, to the difference between the number of cars entering at $x = x_1$ and the number leaving at $x = x_2$, i.e. $q(x_1, t) - q(x_2, t)$; thus

$$\int_{x_1}^{x_2} \rho_t \, dt = q(x_1, t) - q(x_2, t) = - \int_{x_1}^{x_2} q_x \, dx.$$

This result has to hold for all x_1 and x_2, giving

$$\rho_t + q_x = 0. \tag{2.46}$$

Multiplying the left-hand side of equation (2.44) by a function v and integrating over D, as for the previous cases, gives

$$\int_D (va_x + vb_y) \, dx \, dy + \int_D (av_x + bv_y) \, dx \, dy = \int_C v[a \, dy - b \, dx].$$

Hence, proceeding by analogy, a 'weak' solution of equation (2.44) is one which satisfies

$$\int_D (av_x + bv_y) \, dx \, dy = 0, \tag{2.47}$$

for all differentiable v vanishing on C. Comparison of equations (2.39) and (2.47) and examination of the argument leading to equation (2.43) shows that, if in a given region D, u is assumed to be a weak solution of equation (2.44) and differentiable everywhere except across a curve G where it is discontinuous, then, on G,

$$[b]\,dx - [a]\,dy = 0, \qquad (2.48)$$

where square brackets denote the discontinuity across G. Since a, b are functions of u the curve defined by equation (2.48) is not a characteristic and is termed a '*shock*'. The magnitude of the discontinuity and the slope of the shock curve are now related and not independent. The term shock is used by analogy with gas dynamics where a curve of discontinuity separating two different states is said to be a shock curve.

The widening of the concept of solution to include weak solutions and particularly sectionally continuous solutions separated by shocks enables solutions to be obtained for problems like that of Example 2.8, where the solution breaks down in certain regions. This form of extension of a solution is not a purely mathematical exercise as the concept of a shock is of considerable significance in areas like gas dynamics where it represents an idealization of a narrow transition region separating two regions where solutions of the flow equations can be found in a reasonably straightforward way. The significance of shocks will be discussed at slightly greater length subsequently, and we first show how admitting weak solutions can lead to a solution of the boundary-value problem posed in Example 2.8 for the equation

$$u u_x + u_y = 0. \qquad (2.49)$$

Inspection of Fig. 2.8 suggests that the most likely weak solution is one discontinuous across some curve emanating from $(1, 1)$ with $u \equiv 1$ to its left and $u \equiv 0$ to its right. The equation of the shock can be found be writing equation (2.49) in conservation form so that the slope can be found from equation (2.48). One possible conservation form of equation (2.49) is

$$(\tfrac{1}{2}u^2)_x + u_y = 0, \qquad (2.50)$$

and we deduce from this equation and equation (2.48) that the slope of the shock is 2; the shock is therefore the line $y = 2x - 1$. The complete weak solution is then as depicted in Fig. 2.11. It should be noted from the solution of Example 2.8 that shocks will only occur in $y > 0$ when $u(x, 0)$ is a decreasing function of x.

In order to see the practical significance of shocks and weak solutions it is necessary to examine the kind of situation which

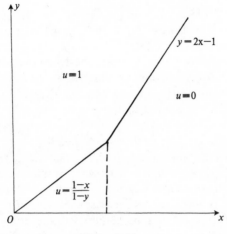

FIG. 2.11.

leads to the need for their introduction. Any model which leads to multi-valued solutions such as that of Example 2.8 is clearly not valid in the breakdown region and the assumptions leading to the model do not hold in this region, though the governing equation may still be valid away from the breakdown region. The shock in this kind of situation represents an idealization of a transition region between two regions where a model is valid. The introduction of a shock means that a model is being used which does not exhibit the fine structure in a transition region (often the analysis of this region proves to be intractable), but which nevertheless shows the main features of the solution. Equation (2.49) can, for example, be regarded as a model representing

$$uu_x + u_y = \varepsilon u_{xx} \qquad (2.51)$$

for very small ε. The breakdown of the solution of equation (2.49) for some problems suggests that, in such cases, there is always a region where εu_{xx} in equation (2.5) is not negligible even for small ε. Away from this region equation (2.49) can be regarded as a good approximation to equation (2.51) for small ε, but the full equation is necessary to exhibit the structure in this transition region.

The significance of a shock is best seen by reference to some practical situation governed by an equation like equation (2.49) and for which some kinds of problems can only be solved by introducing shocks. It turns out that a reasonably plausible theory of traffic flow leads to equation (2.49) and we investigate this problem briefly in order to illustrate the type of situation that can only be resolved by introducing shocks. A reasonable model of traffic flow is one where the velocity $v(x, t)$ of a car passing the point x at time t is related to the density by

$$v = c(1 - \rho/\rho_0),$$

where c is a constant and ρ_0 is the maximum density that the road can take (i.e. bumper-to-bumper traffic). This particular model shows the essential features of the flow in that car velocity decreases as traffic density increases, is zero for bumper-to-bumper traffic, and has a maximum value c for zero density. (There is also some empirical evidence to justify the above relation.) The flow rate q is $v\rho$ and equation (2.46) can be rewritten as

$$d_t + c(1 - 2d)d_x = 0,$$

where d $[= \rho/\rho_0]$ is a relative density; setting $u = 1 - 2d$ and $y = ct$ gives equation (2.49).

In the traffic model a shock is characterized by a sharp discontinuity in density, and hence in car speed. A simple example of this occurs when freely-moving traffic comes upon the rear part of a long queue and has to decelerate rapidly to a virtual standstill (this deceleration region effectively represents the transition region between the queue and the free traffic). y is effectively a time variable and a shock like that of Fig. 2.11 shows the variation with time of the position of the shock (i.e. the back of the queue). dx/dy is the ratio of the shock speed to c and hence $c\,dx/dy$ $(= dx/dt)$ represents the rate at which the back of the queue is moving and is therefore the shock speed. In this case the shock (or disturbance) moves gradually further and further back and this has led to the use of the term shock wave for the disturbance propagated. (We have used the term shock generally to refer to the curve separating two different solutions.) It has previously been noted that shocks only occur in $y > 0$ when $u(x, 0)$ decreases with x and, in the traffic model, this means shocks can only occur for $t > 0$ when the initial density is an increasing function of x and hence car speed is a decreasing function of x. Thus a shock wave is formed when a fast-moving stream catches up with a slowly moving one and this is entirely analogous to the situation which leads to shock waves in gas flow. In a real traffic situation, drivers are likely to alter speed if the density ahead is increasing and therefore a more realistic model would be one where $q = \rho c (1 - \rho/\rho_0) - \varepsilon \rho_x (\varepsilon > 0)$. This in fact gives an equation like equation (2.51).

Extending the concept of solution to include weak solutions has the disadvantage of producing lack of uniqueness. This can be seen most directly by noticing that there is no unique way of writing equation (2.49) in a conservation form and that each different form can give rise to a different shock. Equation (2.49) can, for example, be written as

$$v_y + (e^v)_x = 0 \qquad (2.52)$$

where $v = \log u$ and, if u_L and u_R denote the values of u immediately to the left and to the right of a shock, then equations (2.48) and (2.51) give the shock to be defined by

$$\frac{\mathrm{d}x}{\mathrm{d}y} = \frac{u_L - u_R}{\log u_L/u_R}. \qquad (2.53)$$

On the other hand, using equations (2.48) and (2.50) gives the shock to be defined by

$$\frac{\mathrm{d}x}{\mathrm{d}y} = \tfrac{1}{2}(u_L + u_R). \qquad (2.54)$$

Equations (2.53) and (2.54) demonstrate that different conservation laws can lead to different shocks and hence to non-uniqueness. The appropriate form of conservation law for a given equation has to be determined from the practical context in which the equation occurs.

It is still possible, however, to get more than one weak solution even when the correct conservation law is used. For example both of the functions

$$u = \begin{cases} 0, & x/y < \tfrac{1}{2}, \\ 1, & x/y > \tfrac{1}{2}, \end{cases} \qquad (2.55)$$

$$u = \begin{cases} 0, & x < 0, \\ x/y, & 0 < x/y < 1, \\ 1, & x/y > 1, \end{cases} \qquad (2.56)$$

can be verified to be weak solutions of equation (2.49) and to satisfy the conditions

$$u(x, 0) = 0 \quad \text{for} \quad x < 0, \qquad u(x, 0) = 1 \quad \text{for} \quad x > 0.$$
$$(2.57)$$

The solution given by equation (2.55) is that obtained using equation (2.54).

It is therefore necessary to establish a criterion to enable us to tell, in a practical context, which is the appropriate weak solution to use for an equation in a given conservation form. It is clearly not possible to establish such a criterion without reference to the practical situation which is being modelled. One method of obtaining a criterion in the case of equation (2.49) is to require the solution to be the limiting form as $\varepsilon \to 0$ of the solution of an appropriate boundary-value problem for equation (2.51). It can be shown that this means that, on the shock,

$$u_L \geq \frac{\mathrm{d}x}{\mathrm{d}y} \geq u_R.$$
$$(2.58)$$

Another way of deriving equation (2.58) is to regard y as a time variable (this restricts it to take positive values), and to require that at each point of the shock the solution must be affected by initial conditions on both sides. This is equivalent to requiring all traces from both sides to intersect the shock. The characteristic equations of Example 2.8 show that, on the traces, $\mathrm{d}x/\mathrm{d}y = u$ and hence it follows that equation (2.58) represents the condition that traces from both sides all intersect the shock. In gas flow, applying the condition that entropy must increase across a shock leads to an equation equivalent to equation (2.58) so that the condition defined by this equation is often referred to as the 'entropy' condition. It can be shown that, for any given $u(x, 0)$, there is at most one weak solution of equation (2.49) satisfying equation (2.58) and such that the shock is defined by equation (2.54) (this corresponds to the conservation law of equation (2.50)); this solution is referred to as the permissible solution of the conservation law of equation (2.50). It can also be shown that if such a solution exists then equation (2.49) (or (2.50)) does not also possess a solution in the normal sense of the word.

u defined by equation (2.55) violates equation (2.58) and it is

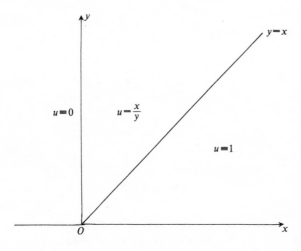

FIG. 2.12.

therefore not a permissible solution of the boundary-value problem it purports to solve. It can be verified by trial and error that it is not possible to construct, in $0 < x/y < 1$, a solution of equations (2.50) and (2.57) which is continuous except across shocks and which satisfies the entropy condition of equation (2.58). The function defined in equation (2.56), however, represents a weak solution and is continuous except at $(0, 0)$ and differentiable except on $x = 0$ and $y = x$. Thus, the appropriate weak solution of equations (2.50) and (2.57) is as depicted in Fig. 2.12. A solution of the form $(x - x_0)/(y - y_0)$, by analogy with gas dynamics, is referred to as a rarefaction wave.

Fig. 2.12 illustrates one kind of additional complication that can arise when attempting to solve boundary-value problems for non-linear equations when the boundary conditions are discontinuous. To provide a further illustration of these complications we consider the problem of solving equation (2.50) with

$$u(x, 0) = 1, \quad x < 0, \quad u(x, 0) = 0, \quad x > 0.$$

Taking $u(x, 0)$ to be unity for $x < 0$ and zero for $x > 0$ gives, as in Example 2.8, $u = 1$ on all lines $y = x + c$ $(c > 0)$ and $u = 0$ on lines $x = $ constant. These lines intersect in $y \geq 0$ and hence a shock must emanate from the origin. Equation (2.54) gives the slope of the shock to be 2 and this satisfies equation (2.58) so that the

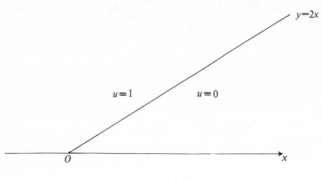

FIG. 2.13.

permissible weak solution of the problem is as shown in Fig. 2.13.

It turns out to be possible, by using an appropriate combination of shocks and rarefaction waves, to determine a unique weak solution of equation (2.50) corresponding to any form of $u(x, 0)$. Effectively shocks are used to link regions of constant state where the value on the left is greater than that on the right whilst rarefaction waves are used when the value on the right is greater than that on the left. The process can be quite complicated and we illustrate it by solving a problem which models the development and eventual dispersion of a traffic jam produced by a complete stoppage at a point (e.g. by an accident) for a period.

It will be assumed that the stoppage takes place at $x = 0$ and that the stream of traffic has uniform relative density a. The hold-up reduces the car velocity to zero and hence the relative density d becomes unity. Therefore in the region $x < 0$ there will be two uniform regions $d = a$ and $d = 1$ and separated by a shock. The corresponding values for u satisfying equation (2.49) are ($u \equiv 1 - 2d$) $1 - 2a$ and -1. The derivation of equation (2.46) shows that the appropriate conservation form is that of equation (2.50) and hence the shock slope is found from equation (2.54). This gives $dx/dy = -a$, showing that the shock wave has velocity $-ac$ ($y = ct$), the form of the solution is shown in Fig. 2.14. If it is assumed that the hold-up lasts for a time τ then we have from Fig. 2.14 that the relative density d in $x < 0$ will be as shown in Fig. 2.15. In a real traffic situation the step discontinuity will be replaced by some kind of smooth transition of the type shown by dotted lines in Fig. 2.15.

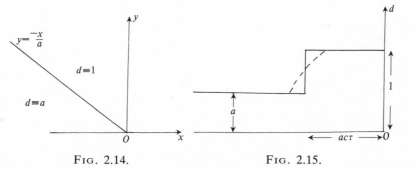

FIG. 2.14.   FIG. 2.15.

In order to see how the traffic jam disperses we solve the problem for $d(x, t)$ (or, equivalently, $u(x, t)$) with $d(x, 0)$ given by Fig. 2.15 for $x < 0$ and $d(x, 0) = 0$ for $x > 0$. (This means effectively neglecting the effect of traffic which has passed through $x = 0$ before the stoppage. This traffic will have no effect in $x < 0$ and hence the results we obtain will only be valid in this region.) The problem for u is thus the solution of equation (2.49) with

$$u(x, 0) = \begin{cases} 1 - 2a, & x < -ac\tau, \\ -1, & -ac\tau < x < 0, \\ 1, & x > 0. \end{cases}$$

The solutions depicted in Figs. 2.12 and 2.13 suggest that, near $y = 0$, the form of the solution will be as shown in Fig. 2.16. This solution breaks down however when the line $x = -y$ intersects the shock $ay = -(x + ac\tau)$, this occurs when $y = ac\tau/(1 - a)$, i.e. after a time $t_0 = a\tau/(1 - a)$. Fig. 2.16 shows that, at any time t, u varies linearly with x from -1 to 1 as x goes from $x = -ct$ to $x = ct$ and hence the effect of removing the hold-up is propagated to the left and right with speed c and the instant t_0 is when the effect of releasing the hold-up catches up with the back of the traffic queue and is effectively the instant at which the jam perceptibly starts to ease all along its length.

In order to complete the solution for $y > ac\tau/(1 - a)$ we attempt to construct an appropriate shock emanating from the point of intersection of the shock and the line $x = -y$ and separating the regions $u = 1 - 2a$ and $u = x/y$. Equation (2.48) shows that the

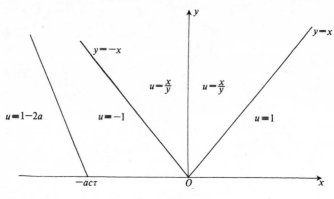

FIG. 2.16.

shock curve is defined by

$$\frac{dx}{dy} = \frac{1}{2}\frac{[(1-2a)y + x]}{y};$$

this equation can be integrated by making the substitution $x = vy$ and the equation of the shock is

$$x = (1-2a)y - 2(ac\tau)^{\frac{1}{2}}(1-a)^{\frac{1}{2}}y^{\frac{1}{2}}.$$

The general shape of the shock curve differs considerably for the two cases $a \geq \frac{1}{2}$ and $a < \frac{1}{2}$, and the curve in the first case is shown by a dotted line in Fig. 2.17 and the solid line in the same figure corresponds to $a < \frac{1}{2}$. Immediately to the right of the shock we have

$$u = x/y = (1-2a) - 2(ac\tau)^{\frac{1}{2}}y^{-\frac{1}{2}}(1-a)^{\frac{1}{2}},$$

and hence, reverting to d and t as variables,

$$d = a + [a(1-a)\tau]^{\frac{1}{2}}t^{-\frac{1}{2}}.$$

It follows from this expression for d (which only holds for $t > t_0$) that the discontinuity across the shock decreases with t (the shock is said to weaken). Fig. 2.17 shows that for $a < \frac{1}{2}$ the shock eventually passes the point $x = 0$ so that the traffic jam clears the section $x = 0$ but that for $a > \frac{1}{2}$ the shock continually recedes back but steadily becomes less perceptible. For $a < \frac{1}{2}$ the shock returns to $x = 0$ after a time $2\tau a(1-a)/(1-2a)^2$ and thus for $a = \frac{3}{8}$ it takes a time 15τ for the jam to clear the road $x < 0$.

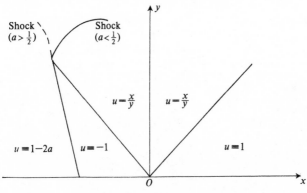

Fig. 2.17.

It is possible to obtain, from Figs. 2.16, and 2.17, diagrams which show, at any given time, the relation between the relative density and x. Such diagrams can be constructed by drawing a line parallel to the x-axis for some value of y and reading off the values of u (and hence d) on each segment of the line. Figs. 2.18–2.23 represent typical situations; in each diagram the relative density at $x = 0$ is $\frac{1}{2}$ and the form of d for $x > 0$ is shown though it will not be completely correct in this region as there will be some interaction with the traffic which had passed before the hold-up. The dotted lines show the more likely behaviour of the solution in a real situation, the slope of the straight line being $-(1/2ct)$ in all cases.

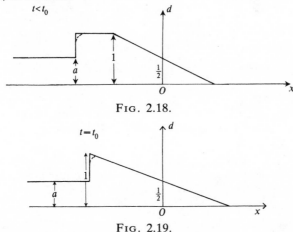

Fig. 2.18.

Fig. 2.19.

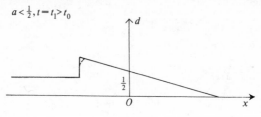

Fig. 2.20.

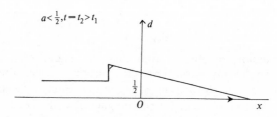

Fig. 2.21.

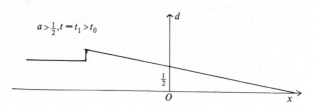

Fig. 2.22.

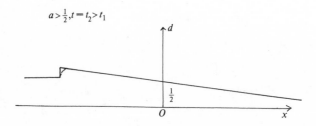

Fig. 2.23.

Exercises 2.4

1. Find in $y > 0$, when $\lambda > 0$, that solution $u(x, y)$ of equation (2.50) such that

$$u(x, 0) = \begin{cases} -\lambda, & x \leq 0, \\ -\lambda(1 - 2x), & 0 \leq x \leq 1, \\ \lambda, & x \geq 0. \end{cases}$$

Show that, for $\lambda < 0$, this problem cannot have a single-valued solution for $y \geq -1/2\lambda$ and construct a suitable weak solution, valid in $y \geq 0$ and satisfying equation (2.58) at any curve of discontinuity.

2. Determine, in the region $y \geq 0$, an appropriate weak solution of equation (2.50) such that

$$u(x, 0) = -\lambda, \qquad x \leq 0, \qquad u(x, 0) = 2\lambda, \qquad x > 0.$$

Consider both positive and negative values of λ.

3. Determine in $y \geq 0$, an appropriate weak solution of equation (2.50) when

$$u(x, 0) = 1, \qquad 0 \leq x \leq 1, \qquad u(x, 0) = 0, \qquad x < 0 \text{ and } x > 1.$$

3 Linear second-order equations in two independent variables

LINEAR partial differential equations of the second order can be classified as one of three types, elliptic, hyperbolic, and parabolic and reduced to an appropriate canonical form. The classification and method of reduction is described in §3.1 and applied in §3.2 to obtain simple properties of the one-dimensional wave equation, which is the archetypal hyperbolic equation. In §3.3 the analysis is extended to obtain the solution of the initial-value–boundary-value problem for the one-dimensional wave equation and energy integral methods are used to prove uniqueness theorems for this problem and for the initial-value problem in the absence of boundaries. In §3.4 the results obtained for the hyperbolic equation are transformed (effectively replacing the wave velocity c by i) to investigate the question of the 'well-posedness' of the Cauchy problem for Laplace's equation, the latter equation being the archetypal elliptic equation. The use in solving equations of factorizing a second-order operator as a product of two first-order operators is sketched in §3.5.

3.1. Classification of linear second-order equations and their reduction to a canonical form

In Chapter 1 the typical problems described for three different second-order partial differential equations differed considerably from each other, thus suggesting that there might be some fundamental difference between the equations. In the cases described the appropriate problems arose naturally from the physical context but it is clearly desirable to be able to determine, directly from the form of the equation, which types of problems have solutions and are well-posed. This is of course particularly important in the practical context when considering a particular partial differential equation which has arisen in some modelling context. It turns out that investigating the solubility of a particular class of boundary-value problems for linear second-order partial differential equations leads naturally to a classification of such equations. This classification, in turn, provides a method of identifying, for a given differential equation, the types of boundary-value problems for which there exist unique solutions which vary continuously with the boundary data.

It has been shown in Chapter 2 that first-order equations have unique solutions when the unknown function is prescribed on a curve. This, coupled with the fact that the solution of second-order ordinary differential equations is determined when the unknown function and its first derivative are given at a point, suggests that an appropriate condition for a second-order equation is prescription of the function and its first derivatives on a curve or, equivalently, of the function and its normal derivative on a curve (cf. §1.2(iv)). The typical problem would therefore appear to be the Cauchy problem.

We now investigate the conditions under which it might be expected to be possible to solve such a problem for the general second-order linear equation

$$au_{xx} + 2bu_{xy} + cu_{yy} + du_x + eu_y + fu = g, \qquad (3.1)$$

or

$$Lu + fu = g,$$

where L is the differential operator

$$a\frac{\partial^2}{\partial x^2} + 2b\frac{\partial^2}{\partial x\,\partial y} + c\frac{\partial^2}{\partial y^2} + d\frac{\partial}{\partial x} + e\frac{\partial}{\partial y},$$

and a to g are real functions of x and y.

The nature of the difficulties that can be expected in the general case can be seen by looking at two particular examples. We consider first the Cauchy problem on the line $y = 0$ for equation (3.1) with $c \equiv 1$; in this case $u(x, 0)$ and $u_y(x, 0)$ are known functions of x and differentiation with respect to x shows that $u_x(x, 0)$, $u_{xx}(x, 0)$, $u_{xxx}(x, 0)$, $u_{xy}(x, 0)$, $u_{xxy}(x, 0)$ are all determined. Furthermore, we have that $u_{yy}(x, 0)$ can then be determined from $u_x(x, 0)$ etc. by using equation (3.1) and repeated differentiation of this equation shows that $u_{yyy}(x, 0)$, etc. are all determined in terms of $u_x(x, 0)$, etc. This suggests that a Taylor expansion of the form

$$u(x, y) = u(x, 0) + yu_y(x, 0) + \tfrac{1}{2}y^2 u_{yy}(x, 0) \cdots$$

can be found and hence that a solution of the Cauchy problem exists in a neighbourhood of $y = 0$. It is possible, in certain circumstances, to make the above arguments rigorous and the resulting existence theorem is a particular case of a general theorem known as the Cauchy–Kowalesky theorem. The most general statement of this theorem is given later but in the present context it asserts the existence of the solution of the Cauchy problem on $y = 0$ for equation (3.1) (with $c \equiv 1$) in a neighbourhood of any point $(x_0, 0)$ at which the functions a to f and $u(x, 0)$ and $u_y(x, 0)$ are analytic (i.e. developable as power series in $x - x_0$ and y).

If we consider the same Cauchy problem for the case $b \equiv 1$ and $a \equiv c \equiv 0$ then it is clearly impossible to determine u_{yy} on $y = 0$ and also equation (3.1) may give a different value to $u_{xy}(x, 0)$ from that obtained by differentiating $u_y(x, 0)$ with respect to x. It therefore appears to be inappropriate to attempt to solve the Cauchy problem in this case.

We now look at the problem of solving equation (3.1) with Cauchy conditions imposed on some curve C defined by $y = y(x)$. Imposition of Cauchy conditions implies that u_x and u_y are known on C, i.e.

$$u_x(x, y(x)) = f(x), \qquad u_y(x, y(x)) = g(x),$$

where f and g are known functions of x.

Differentiating these relations with respect to x gives

$$u_{xx}(x, y(x)) + \frac{\mathrm{d}y}{\mathrm{d}x} u_{xy}(x, y(x)) = f_x, \qquad (3.2)$$

$$u_{xy}(x, y(x)) + \frac{\mathrm{d}y}{\mathrm{d}x} u_{yy}(x, y(x)) = g_x, \qquad (3.3)$$

and we have, from equation (3.1), that

$$a u_{xx}(x, y(x)) + 2b u_{xy}(x, y(x)) + c u_{yy}(x, y(x)) = h, \qquad (3.4).$$

where h is also a known function of x.

Equations (3.2) to (3.4) provide a set of three linear equations to determine u_{xx}, u_{xy}, u_{yy}, and these three quantities can be found unless the determinant of the system vanishes, i.e. unless

$$a\left(\frac{\mathrm{d}y}{\mathrm{d}x}\right)^2 - 2b\frac{\mathrm{d}y}{\mathrm{d}x} + c = 0. \qquad (3.5)$$

Equation (3.5) is a quadratic in dy/dx and hence, when $b^2 > ac$, there exist two families of curves such that no solution can be found when Cauchy conditions are imposed on them. The equation is then said to be *hyperbolic* and, by analogy with the case of first-order equations, the families of curves are known as the characteristics. There is a slight non-uniformity in nomenclature in that the term characteristics is used only for space curves for first-order equations but the term is used to describe plane curves for second-order linear equations.

No characteristics exist when $b^2 < ac$ and the equation is said to be *elliptic* and, when $b^2 = ac$, only one family of characteristics exists and the equation is said to be *parabolic*. It should be noted that, as a, b, and c are functions of x and y, an equation need not be of the same type everywhere in space.

The above arguments only show that solutions of Cauchy problems on characteristics do not exist but do not prove anything about the existence of solutions when conditions are imposed on non-characteristic curves. The basic idea of attempting to form power series solutions can, in many cases, be made rigorous and leads to the general Cauchy–Kowalesky theorem which asserts that, if a given curve is not parallel to a characteristic at a particular point (x_0, y_0) and if the boundary values and the coefficients of the equation are analytic in a neighbourhood of (x_0, y_0), then there exists, in some neighbourhood of (x_0, y_0), an analytic solution of the Cauchy problem. A detailed proof is given in Garabedian (1964). The Cauchy–Kowalesky theorem gives no information about the extent of the domain of existence or about the 'well-posedness' of the problem.

The existence of two families of characteristics for hyperbolic equations suggests that it may be profitable to rewrite equation (3.1) in terms of independent variables ϕ and ψ where the curves $\phi = \text{constant}$ and $\psi = \text{constant}$ are the characteristics. Equation (3.1) can be rewritten, after straightforward application of the chain rule, in terms of two new independent variables ϕ and ψ as

$$A(\phi_x, \phi_y)u_{\phi\phi} + 2B(\phi_x, \phi_y, \psi_x, \psi_y)u_{\phi\psi} + A(\psi_x, \psi_y)u_{\psi\psi}$$

$$+ u_\phi L\phi + u_\psi L\psi + fu = g, \quad (3.6)$$

where

$$A(v, w) = av^2 + 2bvw + cw^2, \quad (3.7)$$

$$B(v_1, w_1, v_2, w_2) = av_1v_2 + b(v_1w_2 + v_2w_1) + cw_1w_2. \quad (3.8)$$

The identity

$$A(\phi_x, \phi_y)A(\psi_x, \psi_y) - B^2(\phi_x, \phi_y, \psi_x, \psi_y) = -(b^2 - ac)(\phi_x\psi_y - \psi_x\phi_y)^2$$
(3.9)

can, by direct expansion of both sides, be shown to be true. If $\phi = $ constant and $\psi = $ constant are the characteristics then, on these curves,

$$\phi_x \, dx + \phi_y \, dy = 0 = \psi_x \, dx + \psi_y \, dy,$$

i.e.

$$\frac{dy}{dx} = -\frac{\phi_x}{\phi_y}, \qquad \frac{dy}{dx} = -\frac{\psi_x}{\psi_y}.$$

Equation (3.5) shows that the values of dy/dx on the characteristics are the roots λ_1 and λ_2 of

$$a\lambda^2 - 2b\lambda + c = 0,$$
(3.10)

so that

$$\frac{\phi_x}{\phi_y} = -\lambda_1,$$
(3.11)

$$\frac{\psi_x}{\psi_y} = -\lambda_2.$$
(3.12)

Substituting the values of λ defined by equations (3.11) and (3.12) into equation (3.10) gives

$$a\phi_x^2 + 2b\phi_x\phi_y + c\phi_y^2 = a\psi_x^2 + 2b\psi_x\psi_y + c\psi_y^2 = 0.$$

Hence the coefficients of $u_{\phi\phi}$ and $u_{\psi\psi}$ in equation (3.6) are identically zero. Furthermore, as $\lambda_1 \neq \lambda_2$, the Jacobian $\phi_x\psi_y - \psi_x\phi_y$ is non-zero and it follows from equation (3.9) that $B \neq 0$. Thus, with the above choice of ϕ and ψ, equation (3.1) has the form

$$a'u_{\phi\psi} + b'u_{\phi} + c'u_{\psi} + f'u = g'.$$
(3.13)

Equation (3.13) is the canonical form for a hyperbolic equation and is characterized by the absence of terms involving $u_{\phi\phi}$ and $u_{\psi\psi}$.

Example 3.1

Reduce the equation

$$yu_{xx} + (x+y)u_{xy} + xu_{yy} = 0$$

to a canonical form.

In this case $b^2 - ac = \frac{1}{4}(x-y)^2$ so the equation is hyperbolic and the characteristics are defined by

$$y\left(\frac{dy}{dx}\right)^2 - (x+y)\frac{dy}{dx} + x = 0.$$

Hence

$$\frac{dy}{dx} = \frac{x}{y} \quad \text{or} \quad \frac{dy}{dx} = 1$$

whose solutions are

$$\phi = y^2 - x^2 = \text{constant} \quad \text{and} \quad \psi = y - x = \text{constant}.$$

Substituting into equation (3.8) gives

$$B(\phi_x, \phi_y, \psi_x, \psi_y) = -(x-y)^2 = -\psi^2,$$
$$L\phi = -2(y-x) = -2\psi, \qquad L\psi = 0,$$

and the required canonical form is

$$\psi u_{\phi\psi} + u_\phi = 0.$$

This equation is fairly simple and can be factorized into a pair of first-order differential equations by writing it as

$$\frac{\partial}{\partial\phi}(\psi u_\psi + u) = 0.$$

A first integration gives $\psi u_\psi + u$ to be a function of ψ and a further integration gives ψu to be the sum of an arbitrary function of ϕ and an arbitrary function of ψ.

Equations (3.8) and (3.9) are far too complicated to remember and in any particular example the canonical form can be obtained by direct substitution and carrying out the algebra inherent in deriving equation (3.6).

Equations (3.11) and (3.12) will still be valid when equation (3.1) is elliptic though λ_1 and λ_2, and hence ϕ and ψ, will now be complex. Thus equation (3.1) will still have the canonical form of equation (3.13) with ϕ and ψ complex. λ_1 and λ_2 will, from the theory of quadratic equations, be a complex conjugate pair and therefore ϕ and ψ will be a pair of complex conjugate functions and can be written as $\lambda + i\mu$ and $\lambda - i\mu$ respectively, where λ and μ are real functions. In this case

$$2u_\phi = u_\lambda - iu_\mu, \qquad 2u_\psi = u_\lambda + iu_\mu, \qquad 4u_{\phi\psi} = u_{\lambda\lambda} + u_{\mu\mu},$$
(3.14)

and equation (3.1) becomes, in terms of the real variables μ and λ,

$$a''(u_{\lambda\lambda} + u_{\mu\mu}) + b''u_\lambda + c''u_\mu + f''u = g''. \qquad (3.15)$$

Equation (3.15) is the canonical form for an elliptic equation.

It should be noted that Laplace's equation in two independent variables is already in canonical form and is, in fact, the archetypal elliptic equation.

Example 3.2

Reduce to canonical form

$$u_{xx} + yu_{yy} = 0, \qquad y > 0.$$

In this case $b^2 - ac = -y$ so that the equation is elliptic when $y > 0$. The characteristics are defined by

$$\left(\frac{dy}{dx}\right)^2 + y = 0, \quad \text{i.e.} \quad \frac{dy}{dx} = \pm iy^{\frac{1}{2}}$$

so that $\phi = x + 2iy^{\frac{1}{2}}$, $\psi = x - 2iy^{\frac{1}{2}}$.

In order to reduce manipulation it is convenient one again to use equation (3.8) and we have that

$$B(\phi_x, \phi_y, \psi_x, \psi_y) = 2, \qquad L\phi = -\tfrac{1}{2}iy^{-\frac{1}{2}}, \qquad L\psi = \tfrac{1}{2}iy^{-\frac{1}{2}},$$

giving

$$4u_{\phi\psi} - \frac{i}{2y^{\frac{1}{2}}}(u_\phi - u_\psi) = 0.$$

Setting $u = 2y^{\frac{1}{2}}$ and $\lambda = x$ and using equations (3.14) gives the canonical form to be

$$\mu(u_{\lambda\lambda} + u_{\mu\mu}) - u_\mu = 0.$$

For a parabolic equation there is only one (repeated) root of equation (3.10), namely b/a, and hence only ϕ, say, can be found by integrating

$$\frac{dy}{dx} = \frac{b}{a},$$

so that only the coefficient of $u_{\phi\phi}$ in equation (3.6) is zero.

Once ϕ has been found then ψ can be chosen to be any other variable independent of ϕ (i.e. such that $\phi_x\psi_y - \psi_x\phi_y \not\equiv 0$). Equation (3.11) gives

$$a\phi_x + b\phi_y = 0;$$

multiplying this equation by b and using the condition $b^2 = ac$ gives

$$b\phi_x + c\phi_y = 0.$$

Hence

$$B(\phi_x, \phi_y, \psi_x, \psi_y) = \psi_x(a\phi_x + b\phi_y) + \psi_y(b\phi_x + c\phi_y) = 0,$$

so that the term in $u_{\phi\psi}$ is also absent. The canonical form for the parabolic equation is thus

$$\alpha u_{\psi\psi} + \beta u_\phi + \gamma u_\psi + \delta u = \eta, \tag{3.16}$$

and we see that the heat-conduction equation is already in canonical form.

Example 3.3

Reduce to canonical form

$$x^2 u_{xx} + 2xy u_{xy} + y^2 u_{yy} = 0.$$

The equation is parabolic and the characteristic is defined by

$$\frac{dy}{dx} = \frac{y}{x},$$

giving $\phi = y/x$. For simplicity the second variable ψ is taken to be x and therefore

$$A(\psi_x, \psi_y) = x^2, \qquad L\psi = L\phi = 0,$$

giving

$$u_{\phi\phi} = u_{xx} = 0.$$

It should be remembered that, in this equation, x and ϕ are the independent variables so that its general solution is $xf(\phi) + g(\phi)$, where f and g are arbitrary functions.

The above method of reducing a second-order equation to canonical form is valid also for half-linear and quasi-linear equations. In the former case the coefficients a, b, and c are still functions of x and y only and the characteristics are defined explicitly. In the quasi-linear cases a, b, c will be functions of u, u_x, u_y and hence the characteristics are not defined explicitly, or independently of u.

Exercises 3.1

1. Reduce to canonical form

   (i) $u_{xx} + 5u_{xy} + 6u_{yy} = 0$,

  (ii) $u_{xx} + 2u_{xy} + 5u_{yy} = 0$,

 (iii) $u_{xx} + 2u_{xy} + u_{yy} = 0$,

 (iv) $x^2 u_{xx} - y^2 u_{yy} = 0$,

  (v) $y^2 u_{xx} + x^2 u_{yy} = 0$,

 (vi) $u_{xx} \sin^2 x - 2y u_{xy} \sin x + y^2 u_{yy} = 0$,

(vii) $u_{xx} + y u_{yy} = 0$.

2. Find the general solutions of

 (i) $u_{xx} - 2u_{xy} \sin x - u_{yy} \cos^2 x - u_y \cos x = 0$,

(ii) $y^2 u_{xx} - 2y u_{xy} + u_{yy} = u_x + 6y$.

3. Show that, on the characteristics, the equation

$$au_{xx} + 2bu_{xy} + cu_{yy} = 0,$$

reduces to

$$a\frac{dy}{dx}\frac{dp}{dx} + c\frac{dq}{dx} = 0,$$

where $p = u_x$, $q = u_y$.

3.2. The one-dimensional wave equation

The one-dimensional wave equation is (§1.2(iv))

$$c^2 u_{xx} - u_{tt} = 0, \tag{3.17}$$

where c is a constant. It is obviously hyperbolic with the characteristics in the (x, t)-plane being defined by

$$c^2\left(\frac{\mathrm{d}t}{\mathrm{d}x}\right)^2 = 1.$$

The characteristics are therefore the lines $x \pm ct = \text{constant}$ so that appropriate choices for ϕ and ψ are $\phi = x - ct$ and $\psi = x + ct$. Substituting for ϕ and ψ in equation (3.17) gives

$$u_{\phi\psi} = 0. \tag{3.18}$$

The general solution of equation (3.18) is

$$u = F(\phi) + G(\psi), \tag{3.19}$$

or

$$u = F(x - ct) + G(x + ct), \tag{3.20}$$

where F and G are arbitrary functions. The nature of this solution is possibly best seen by comparing the form of $u \equiv F(x - ct)$ at $t = 0$ (i.e. $F(x)$) with that at a subsequent time $t = T$ [i.e. $F(x - cT)$; Fig. 3.1]. We have that u at time T at any point $x = x_0$ is equal to $F(x_0 - cT)$, which is the value of u at $t = 0$ at the point $x_0 - cT$, i.e. at a distance cT to the left of $x = x_0$. The curve representing u at $t = T$ is therefore obtained by displacing that representing u at $t = 0$ a distance cT to the right. Hence

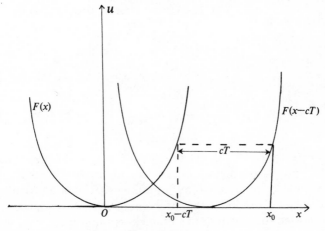

FIG. 3.1.

$u \equiv F(x - ct)$ represents a disturbance (or wave) moving to the right with uniform speed c. Similarly $u \equiv G(x + ct)$ represents a disturbance moving to the left with speed c and the general solution of equation (3.17) is a linear superposition of two such disturbances, hence the nomenclature 'wave equation'.

The typical problem (the initial-value problem) for equation (3.17) is the Cauchy problem defined by

$$u(x, 0) = f(x), \qquad u_t(x, 0) = g(x), \tag{3.21}$$

where f and g are given functions. Equation (3.21) imply that

$$F(x) + G(x) = f(x), \tag{3.22}$$

$$-cF'(x) + cG'(x) = g(x), \tag{3.23}$$

where the prime denotes the derivative with respect to the argument. It follows on integrating equation (3.23) that,

$$-cF(x) + cG(x) = \int_{x_0}^{x} g(w) \, dw, \tag{3.24}$$

where x_0 is some arbitrary constant. Equations (3.22) and (3.24) now give

$$G(x) = \tfrac{1}{2}f(x) + \frac{1}{2c} \int_{x_0}^{x} g(w) \, dw,$$

$$F(x) = \tfrac{1}{2}f(x) - \frac{1}{2c} \int_{x_0}^{x} g(w) \, dw,$$

and

$$u = \tfrac{1}{2}[f(x - ct) + f(x + ct)] + \frac{1}{2c} \int_{x-ct}^{x+ct} g(w) \, dw; \tag{3.25}$$

it can be verified by direct substitution that u satisfies equation (3.17) provided that f is twice differentiable and g is differentiable. u defined by equation (3.25) is D'Alembert's solution for the initial-value problem for the wave equation and can be used to show that this problem is well-posed. If u_1 and u_2 are taken to represent the solutions of equation (3.17) with $f = f_i$

and $g = g_i$ $(i = 1, 2)$ with $|f_1 - f_2| < \varepsilon$ and $|g_1 - g_2| < \varepsilon$, then equation (3.25) gives

$$u_1 - u_2 = \tfrac{1}{2}[f_1(x - ct) - f_2(x - ct) + f_1(x + ct) - f_2(x + ct)]$$

$$+ \frac{1}{2c} \int_{x-ct}^{x+ct} [g_1(w) - g_2(w)] \, dw$$

and

$$|u_1 - u_2| < \tfrac{1}{2}(\varepsilon + \varepsilon) + \frac{1}{2c} \int_{x-ct}^{x+ct} \varepsilon \, dw = \varepsilon(1 + t).$$

Hence, for any finite time interval, a small change in the boundary conditions only produces a proportionately small change in the solution showing that the problem is well-posed. The solution defined by equation (3.25) is valid for positive and negative values of t but physical sense decrees that, when t is a time variable, u can only be found for $t > 0$.

Fig. 3.2 shows an arbitrary point $P(x_0, t_0)$ with the characteristics $x - x_0 = \pm c(t - t_0)$ through P intersecting the line $t = 0$ at A and B, respectively, where A and B are the points $(x_0 - ct_0, 0)$ and $(x_0 + ct_0, 0)$, respectively. Equation (3.25) shows that the value of u at P depends only on f at A and B and on the values of g between A and B. u is therefore dependent only on the initial values on the segment AB of the line $t = 0$ and hence this segment is defined to be the *domain of dependence* of P on the initial curve. The initial values outside AB have no effect at P;

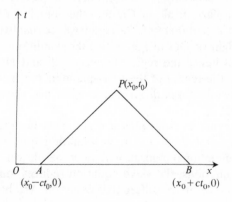

FIG. 3.2.

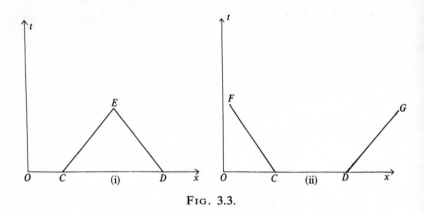

FIG. 3.3.

this corresponds to the fact that solutions of equation (3.17) represent disturbances propagated with a finite speed so that regions outside AB on $t = 0$ are too far away from P for any disturbance emanating from them to have reached P by $t = t_0$. The concept of a finite domain of dependence is common to all hyperbolic equations.

Fig. 3.3(i) shows two points C and D on $t = 0$ with C to the left of D and the characteristics $x - ct = \text{constant}$ through C and $x + ct = \text{constant}$ through D are shown. Fig. 3.3(ii) shows the same points together with the characteristics $x + ct = \text{constant}$ through C and $x - ct = \text{constant}$ through D. The above arguments concerning the domain of dependence shows that, within the triangle ECD, the solution is determined completely by the initial conditions on part, or all, of CD and the domain CDE is termed the *domain of determinacy of CD*. Again we see that to the left of CF and to the right of DG in Fig. 3.3(ii) the conditions on CD have no effect and hence the region between CF and DG represents the whole of the region of the (x, t)-plane in which conditions on CD have any effect *and this region is called the region of influence of CD.*

Thus the characteristics play a major part in determining the form of the solution of the wave equation (and in fact of all hyperbolic equations) and it will now be shown that, as for first-order equations, the wave equation reduces, on the characteristics, to an ordinary differential equation. We have that

$$d(cu_x \pm u_t) = (cu_{xx} \pm u_{xt}) \, dx + (cu_{xt} \pm u_{tt}) \, dt$$

and hence, when $x + ct$ is constant,

$$d(cu_x + u_t) = dt(-c^2 u_{xx} + u_{tt}) = 0,$$

which is an ordinary differential equation along the characteristics $x + ct = $ constant, and integrating gives $cu_x + u_t$ to be constant on these characteristics. Similarly $cu_x - u_t$ is constant on the characteristics $x - ct = $ constant. These results can be applied to give an alternative derivation of D'Alembert's solution. $cu_x - u_t$ is constant along the characteristic AP of Fig. 3.2 so that

$$cu_x(x_0, t_0) - u_t(x_0, t_0) = cu_x(x_0 - ct_0, 0) - u_t(x_0 - ct_0, 0)$$

$$= c\frac{\partial}{\partial x_0} f(x_0 - ct_0) - g(x_0 - ct_0).$$

Similarly $cu_x + u_t$ is constant along BP giving

$$cu_x(x_0, t_0) + u_t(x_0, t_0) = c\frac{\partial}{\partial x_0} f(x_0 + ct_0) + g(x_0 + ct_0).$$

Therefore

$$2cu_x(x_0, t_0) = c\frac{\partial}{\partial x_0}[f(x_0 - ct_0) + f(x_0 + ct_0)] + g(x_0 + ct_0) - g(x_0 - ct_0),$$

$$2u_t(x_0, t_0) = c\frac{\partial}{\partial x_0}[f(x_0 + ct_0) - f(x_0 - ct_0)] + g(x_0 + ct_0) + g(x_0 - ct_0),$$

and the common solution of these equations is

$$u = \tfrac{1}{2}[f(x_0 + ct_0) + f(x_0 - ct_0)] + \frac{1}{2c}\int_{x_0 - ct_0}^{x_0 + ct_0} g(w)\,\mathrm{d}w,$$

which is equation (3.25).

The above alternative derivation of equation (3.25) shows the characteristics to be the curves along which the initial values are propagated. This particular property of the characteristics forms the basis of the numerical method referred to as the 'method of characteristics' (§10.2).

It is also possible to generalize D'Alembert's solution to find, when F is a given function of x and t, that solution of

$$c^2 u_{xx} - u_{tt} = F(x, t), \tag{3.26}$$

which satisfies equations (3.21). Equation (3.26) is the inhomogeneous one-dimensional wave equation and, in the context of the vibrating string problem, F is proportional to the force applied to a point x of the string at time t. Equation (3.26) can be rewritten in terms of ϕ and ψ as

$$4c^2u_{\phi\psi} = F[x(\phi, \psi), t(\phi, \psi)] \equiv G(\phi, \psi). \qquad (3.27)$$

The line $t = 0$ is the line $\phi = \psi$ in the (ϕ, ψ)-plane and in this plane Fig. 3.2 transforms into Fig. 3.4 with $\phi_0 = x_0 - ct_0$ and $\psi_0 = x_0 + ct_0$.

In view of the linearity of the left-hand side of equation (3.26) it is possible to write u as the sum of two functions, the first satisfying equations (3.17) and (3.21) (i.e. the solution of the original initial-value problem) and the second being the solution of equations (3.26) and (3.21) with f and g set equal to zero. The first solution is given by equation (3.25) and we therefore need to solve equation (3.27) with $u = u_\phi = u_\psi = 0$ on $\phi = \psi$.

Integrating equation (3.27) with respect to ψ from ϕ to ψ_0 (i.e. along CD) gives

$$4c^2u_\phi(\phi, \psi_0) - 4c^2u_\phi(\phi, \phi) = 4c^2u_\phi(\phi, \psi_0) = \int_\phi^{\psi_0} G(\phi, \psi)\,\mathrm{d}\psi,$$

and integrating this equation with respect to ϕ from ψ_0 to ϕ_0

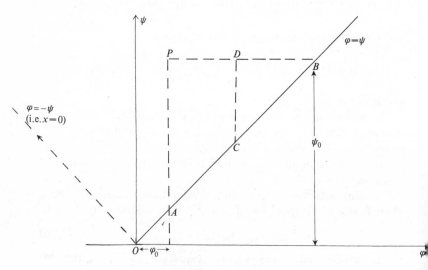

FIG. 3.4.

gives

$$4c^2 u(\phi_0, \psi_0) = \int_{\psi_0}^{\phi_0} \int_{\phi}^{\psi_0} G(\phi, \psi) \, d\psi \, d\phi. \qquad (3.28)$$

Reverting to x and t and combining equations (3.25) and (3.28) gives

$$\begin{aligned} u(x_0, t_0) = \tfrac{1}{2}[f(x_0 - ct_0) + f(x_0 + ct_0)] \\ + \frac{1}{2c} \int_{x_0 - ct_0}^{x_0 + ct_0} g(w) \, dw - \frac{1}{2c} \int_{APB} F(x, t) \, dx \, dt. \end{aligned} \qquad (3.29)$$

(In deriving equation (3.29) the result $d\phi \, d\psi = 2c \, dx \, dt$ has been used. The change of sign in the last term is due to the fact that, for the configuration of Fig. 3.4, in the integration with respect to ϕ in equation (3.28) the upper limit is smaller than the lower one.) Equation (3.29) shows that $u(x_0, t_0)$ involves only the values of F within the triangle APB and hence the domain of dependence is, in this case, the whole of the triangle.

In the following section we obtain the solution of the other typical problem associated with equation (3.17), namely the mixed initial-value–boundary-value problem described in Chapter 1. Before doing this, however, we consider briefly the possibility of there being other general classes of problems, not suggested by a physical situation, for which solutions might be expected to exist. In particular there is no *ab initio* reason why Dirichlet problems, which arise naturally for elliptic equations like Laplace's equation, should not be soluble for equation (3.17) or, equivalently, equation (3.18). It transpires that it is not possible to establish existence theorems for such problems and some of the difficulties encountered can be seen by looking at two particular examples.

We consider first the Dirichlet problem for equation (3.18) in $0 \le \phi \le a$, $0 \le \psi \le a$. The general solution is $u = F(\phi) + G(\psi)$ and

$$u(0, 0) = F(0) + G(0),$$
$$u(0, a) = F(0) + G(a),$$
$$u(a, 0) = F(a) + G(0),$$
$$u(a, a) = F(a) + G(a).$$

These equations show that the Dirichlet problem cannot be

solved when $u(0, 0) + u(a, a) \neq u(a, 0) + u(0, a)$ and hence that general existence theorems are unlikely to be available.

Another possible difficulty that can arise for Dirichlet problems can be seen by considering

$$u_{xx} - u_{yy} = 0$$

in $0 < x < 1$, $0 < y < 1$, with Dirichlet conditions imposed on the boundary. The difference of any two solutions will satisfy the equation and be zero on the boundary and a possible form for this difference is $\sin \pi x \sin \pi y$. Hence the difference of two possible solutions is not zero and hence the problem does not have a unique solution.

It would also not make physical sense to attempt to solve the wave equation in a rectangular region $0 \leq x \leq a$, $0 \leq t \leq T$, with conditions imposed on all four sides of the rectangle. The existence of a solution for such a problem would imply that present events (i.e. in $0 \leq t < T$) could be affected by future ones (i.e. at $t = T$).

3.3. The mixed initial-value–boundary-value problem for the one-dimensional wave equation

The problem to be solved is the determination of a solution of equation (3.17) subject to the initial conditions prescribed by equation (3.21) and to the additional boundary conditions $u(0, t) = u(a, t) = 0$. This problem is equivalent to finding, when the initial displacement and velocity is known at all points of its length, the shape of a taut string whose ends are fixed. It is shown in §5.4 that discontinuities in u or its first derivatives will be propagated along the characteristics through the point of discontinuity and hence in order to avoid discontinuities (u could not be a solution in the normal sense) we require that $f(0) = g(0) = f(a) = g(a) = 0$.

It will be shown that, in principle, the solution can still be found by integrating along the characteristics though it is not possible using this method to obtain easily an explicit closed form of solution. In order to obtain such a solution it is more direct to use the method of separation of variables and this leads to a closed-form solution in terms of a Fourier sine series.

In this case u has to be determined within the U-shaped region shown in Fig. 3.5. The characteristics from the point $P(x_0, t_0)$ are drawn in the direction $t < t_0$ and intersect $x = 0$ and $x = a$ at P_1 and P_2. The characteristics from these points into the region $0 \le x \le a$ are drawn to their points of intersection with $x = 0$, $x = a$, or $t = 0$. The process is then continued until two of the characteristics intersect $t = 0$, one possible configuration is shown in Fig. 3.5. We let p and q denote u_x and u_t respectively and use a suffix to denote values calculated at a correspondingly numbered point. We have already seen that $cp - q$ is constant along characteristics of positive slope whilst $cp + q$ is constant on those of negative slope, so that, on $P_5 P_3$,

$$cp_3 - q_3 = cp_5 - q_5 = cp_3 (q_3 = 0 \text{ from } u(a, t) \equiv 0).$$

Hence u_x at P_3 is determined from the conditions on $t = 0$; if a Cauchy condition had been imposed on $x = a$ we would have had a contradiction. Continuing along $P_3 P_1$ and $P_1 P$ gives

$$cp_1 = cp_3 \quad \text{and} \quad cp - q = cp_1,$$

and combining all the above equations gives

$$cp - q = cu_x(x_0, t_0) - u_t(x_0, t_0) = cp_5 - q_5.$$

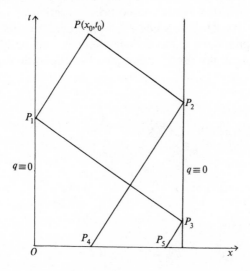

Fɪɢ. 3.5.

Similarly following the path P_4P_2P gives

$$cu_x(x_0, t_0) + u_t(x_0, t_0) = cp_4 - q_4.$$

Hence the first two derivatives of u at any point (x_0, t_0) can be found from the given values of p_4, p_5, q_4, q_5; the identity

$$u(x_0, t_0) = \int_0^{x_0} u_x(w, t_0) \, dw$$

then enables u to be found, at least in principle.

For small values of t_0 (more precisely if P lies in the triangle bounded by the characteristics through $(0, 0)$ and $(a, 0)$) the form of u will be simple, and in fact will be given by equation (3.25), as it is only necessary to integrate along two characteristics. Increasing t_0 means an increase in the number of characteristics along which it is necessary to integrate and physically this corresponds to initial disturbances reaching the ends and being reflected. The path $P_5P_3P_1P$ for example, corresponds to a disturbance moving to the right starting from P_5, being reflected at $x = a$ and this latter disturbance being further reflected at $x = 0$. As t_0 increases, the paths from $t = 0$ to a given point P become extremely complicated and it is better to try and find a different method which avoids some of the complications; such an approach is provided by the method of separation of variables.

In the particular case when $a = \infty$, however, the above method of integrating along characteristics remains practicable and, before describing the use of the method of separation of variables

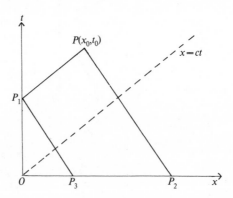

FIG. 3.6.

for the case of finite a, we look briefly at the special case of infinite a. If $x_0 > ct_0$, then both of the characteristics, drawn backwards in time from P, intersect the positive x-axis and D'Alembert's solution is again obtained. When $x_0 < ct_0$, however, one of the characteristics drawn backwards in time from P intersects the line $x = 0$ at $P_1(0, t_0 - x_0/c)$ (Fig. 3.6) whilst the other characteristic drawn backwards from P intersects the line $t = 0$, as before, at $P_2(x_0 + ct_0, 0)$. Fig. (3.6) shows that one of the characteristics drawn backwards from P_1 intersects the positive x-axis at the point $P_3(ct_0 - x_0, 0)$. Integrating along $P_2 P$ and $P_3 P_1 P$ gives, since $u \equiv 0$ on $x = 0$,

$$cu_x(x_0, t_0) + u_t(x_0, t_0) = c \frac{\partial}{\partial x_0} f(x_0 + ct_0) + g(x_0 + ct_0),$$

$$cu_x(x_0, t_0) - u_t(x_0, t_0) = -c \frac{\partial}{\partial x_0} f(ct_0 - x_0) + g(ct_0 - x_0).$$

The solution of these equations is

$$u = \tfrac{1}{2}\{f(x_0 + ct_0) - f(ct_0 - x_0)\} + \frac{1}{2c} \int_{ct_0 - x_0}^{ct_0 + x_0} g(z) \, dz.$$

Physically, the above problem corresponds to determining the motion of a semi-infinite string, one end of which is kept fixed. The relative simplicity of the solution is due to the fact that there is only one reflection (at $x = 0$). The problem can also be solved directly by using the general solution quoted in equation (3.20).

The first step in the method of separation of variables is to look for solutions of the form $X(x)T(t)$ and substituting this form into equation (3.17) gives

$$TX'' = \frac{1}{c^2} XT'',$$

where the primes denote derivatives with respect to the arguments.

This equation can be rearranged in the form

$$\frac{X''}{X} = \frac{1}{c^2} \frac{T''}{T}. \tag{3.30}$$

The left-hand side of equation (3.30) is a function of x whilst the right-hand side is a function of t and, as x and t are independent variables, therefore both sides must be equal to a constant $-\lambda^2$, say. Hence,

$$X'' + \lambda^2 X = 0, \tag{3.31}$$

$$T'' + c^2\lambda^2 T = 0. \tag{3.32}$$

The general solution of equation (3.31) is $X = A\cos\lambda x + B\sin\lambda x$, where A and B are constants; the condition $u(a, t) = 0$ implies $X(0) = 0$ so that $A = 0$. Similarly the condition $u(a, t) = 0$ shows that $X(a) = 0$ and hence, unless $B = 0$ (this would give $u \equiv 0$),

$$\sin\lambda a = 0.$$

Thus the only possible values of λ are $n\pi/a$, where n is an integer, and, as we have no method of rejecting any values of n, we have to let n range over all positive values. The equation is linear and hence the sum of constant multiples of solutions is also a solution (this is the principle of superposition of §1.1). Therefore a possible form of u is given by

$$u = \sum_{n=1}^{\infty}\left(C_n\cos\frac{n\pi ct}{a} + D_n\sin\frac{n\pi ct}{a}\right)\sin\frac{n\pi x}{a}, \tag{3.33}$$

where C_n and D_n are constants, the term in brackets in equation (3.33) being the general solution of equation (3.32) with $\lambda = n\pi/a$. The conditions on $x = 0$ and $x = a$ have been satisfied by construction and it now only remains to satisfy the conditions of equation (3.21). Substituting $t = 0$ into the series on the right-hand side of equation (3.33) and into the corresponding series obtained by differentiating with respect to t shows that C_n and D_n have to satisfy

$$f(x) = \sum_{n=1}^{\infty}C_n\sin\frac{n\pi x}{a}, \qquad 0 \le x \le a, \tag{3.34}$$

$$g(x) = \sum_{n=1}^{\infty}\frac{n\pi}{a}D_n\sin\frac{n\pi x}{a}, \qquad 0 \le x \le a. \tag{3.35}$$

The right-hand sides of equations (3.34) and (3.35) are the Fourier sine series of f and g respectively, and it now follows immediately from equation (1.27) that

$$C_n = \frac{2}{a} \int_0^a f(x) \sin \frac{n\pi x}{a} \, \mathrm{d}x, \qquad (3.36)$$

$$D_n = \frac{2}{n\pi} \int_0^a g(x) \sin \frac{n\pi x}{a} \, \mathrm{d}x. \qquad (3.37)$$

The integrations in equations (3.36) and (3.37) can be carried out (analytically or numerically) for any given f and g and hence a series solution to any specific problem can be obtained.

It should be noted that implicit in the above analysis is the assumption that the series in equation (3.33) is 'term-by-term' twice differentiable with respect to both x and t and it should be verified that C_n and D_n are such that term-by-term differentiation is justified. It can be shown in general that term-by-term differentiation is justified and that the twice-differentiated series is convergent in $0 < x < a$ provided that f is four times differentiable, and g is three times differentiable, in $0 < x < a$, and $f^{(p)}(0) = f^{(p)}(a) = 0 (0 \le p \le 3) g^{(p)}(0) = g^{(p)}(a) = 0$, $0 \le p \le 2$, where the affix indicates the pth derivative. These conditions will hold in most practical contexts for the vibrating string problem. It is also possible to define 'weak' solutions for second-order equations, and by extending to notion of a solution to include weak solutions it is then possible to relax some of the above conditions on f and g. The existence of weak solutions of hyperbolic equations is discussed in §5.4.

The derivation of the above Fourier series solution tacitly assumes that the solution has a certain form, and though the physical circumstances indicate that the solution is unique, there is no reason why there might not be a second solution which could not be expressed in the form on the right-hand side of equation (3.33). In such situations, when one assumes a certain form of solution, it is necessary to have available an appropriate uniqueness theorem and such a theorem can be established fairly easily in this case by considering the function $E(t)$ defined by

$$E(t) = \int_0^a \left[u_t^2 + c^2 u_x^2 \right] \mathrm{d}x.$$

Then

$$\frac{\mathrm{d}E}{\mathrm{d}t} = 2 \int_0^a \left[u_t u_{tt} + c^2 u_x u_{xt} \right] \mathrm{d}x$$

and hence, using

$$u_x u_{xt} = (u_t u_x)_x - u_t u_{xx},$$

$$\frac{dE}{dt} = 2\int_0^a u_t[u_{tt} - c^2 u_{xx}]\,dx + 2c^2\{(u_t u_x)_{x=a} - (u_t u_x)_{x=0}\}.$$

Thus dE/dt will be zero for any solution of equation (3.17) which vanishes at $x = 0$ or $x = a$ (or whose derivative vanishes at these points). For a vibrating string E is proportional to the total energy and the vanishing of its time derivative is a consequence of energy conservation. If there are two solutions u_1 and u_2 of the initial-value–boundary-value problem then their difference u will also satisfy equation (3.17) and, at $t = 0$, $u_x \equiv u_t \equiv u \equiv 0$. For this particular choice of u, $E(0) = 0$ and hence $E \equiv 0$, showing that $u_x \equiv u_t \equiv 0$ and that u must be a constant. The condition on u at $t = 0$ then gives $u \equiv 0$ thus proving uniqueness.

The above method of proof (which is often referred to as an energy integral method) may be adapted to prove the uniqueness of the solution of the Cauchy problem in the absence of boundaries and to give, at the same time, an independent demonstration of the extent of the domain of dependence on $t = 0$ of the solution at P. In this case we consider, for some fixed (x_0, t_0), $E(t)$

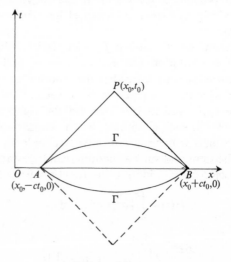

FIG. 3.7.

defined by

$$E(t) = \int_{x_0+c(t-t_0)}^{x_0-c(t-t_0)} [u_t^2 + c^2 u_x^2]\, dx, \qquad t < t_0,$$

the integral is therefore along a line parallel to AB in the triangle PAB of Fig. 3.7 (i.e. the region formed by the line $t = 0$ and the characteristics drawn backwards from $P(x_0, t_0)$ to $t = 0$). We have, on using the above results, that, when u satisfies equation (3.17),

$$\frac{dE}{dt} = -c[u_t^2 + c^2 u_x^2 - 2cu_t u_x]_{x=x_0-c(t-t_0)}$$
$$\qquad\qquad -c[u_t^2 + c^2 u_x^2 + 2cu_t u_x]_{x=x_0+c(t-t_0)}$$
$$= -c[u_t + cu_x]_{x=x_0-c(t-t_0)}^2 - c[u_t - cu_x]_{x=x_0+i(t-t_0)}^2$$

and therefore

$$\frac{dE}{dt} \le 0.$$

Hence, if $u = u_t = 0$ at $t = 0$, (or just on the segment AB) we see that $E = 0$ for $t > 0$ and hence that $u_t = u_x = 0$ on every line in APB parallel to AB and therefore u is constant in the triangle APB. The condition $u = 0$ at $t = 0$ then gives $u \equiv 0$. Therefore Cauchy conditions on $t = 0$ determine u uniquely (the difference between any two solutions would have to satisfy $u = u_t = 0$ on $t = 0$) and in fact Cauchy conditions on the segment AB determine u uniquely within the triangle APB, thus showing that the segment AB is the domain of dependence on $t = 0$ of the solution at P. It should also be noted that, by considering the solution in the region formed by reflecting the triangle APB in the line $t = 0$, the above analysis can be used to prove that Cauchy conditions on AB determine u uniquely in this region also.

It is also possible to modify the above proof so as to prove uniqueness of the solution of the Cauchy problem posed on any non-characteristic curve Γ joining A and B and which is completely contained within the parallelogram formed by APB and its reflection in the line $t = 0$. It will also be shown that the solution is determined uniquely in the region between Γ and PA and PB and in fact in the whole of the above parallelogram.

We consider first the case when Γ only intersects AB at A and B and has, therefore, one of the two forms shown in Fig. 3.7. The proof can be modified in a fairly obvious was to cover the case of multiple intersections with AB. The identity

$$u_t(u_{tt} - c^2 u_{xx}) = (c^2 u_x^2 + u_t^2)_t - 2c^2(u_x u_t)_x,$$

becomes, when u satisfies equation (3.17),

$$(c^2 u_x^2 + u_t^2)_t - 2c^2(u_x u_t)_x = 0. \tag{3.38}$$

Integrating this identity over the region bounded by Γ and AB and applying the divergence theorem gives

$$\int_\Gamma \{(c^2 u_x^2 + u_t^2)\, \mathrm{d}x + 2c^2(u_x u_t)\, \mathrm{d}t\} = \int_{x_0-ct_0}^{x_0+ct_0} (c^2 u_x^2 + u_t^2)\, \mathrm{d}x = E(0).$$

Homogeneous Cauchy conditions on Γ therefore give $E(0) = 0$ and therefore it follows from our previous analysis that u is constant in the parallelogram referred to above. The condition that $u = 0$ on Γ then gives $u \equiv 0$ in the region stated. The uniqueness of the solution of the Cauchy problem then follows immediately. The only difficulty associated with multiple intersections of Γ with AB is that equation (3.38) has to be integrated separately over the regions bounded by arcs of Γ and segments of AB joining two successive points of intersection of Γ with AB. Adding the integrals over the successive segments gives the relation obtained above for $E(0)$.

If Γ were characteristic at some point (i.e. parallel to lines $x \pm ct = \text{constant}$) then it would be possible to join some points of it by segments of a characteristic. $cu_x + u_t$ or $cu_x - u_t$ would be constant on such segments and this implies some relationship between u_x and u_t at the end points of the segments. Therefore u_x and u_t cannot be prescribed arbitrarily on Γ.

Exercises 3.2

1. Obtain, in each of the following cases, the function u which satisfies equation (3.17) and the conditions stated. Find, in the first two cases, the regions in which u is determined by the given conditions.

(i) $u(ct, t) = \phi(t)$, $0 \le t \le a$; $u(-ct, t) = \psi(t)$, $0 \le t \le b$, where $\phi(0) = \psi(0)$.

(ii) $u(ct, t) = \phi(t)$, $0 \le t \le a$, $u(2ct, t) = \psi(t)$, $t \ge 0$, where $\phi(0) = \psi(0)$.

(iii) $u(x, 0) = \phi_0(x)$, $u_t(x, 0) = \phi_1(x)$, $x \geq 0$, $u(x, 2x/c) = \phi_2(x)$, $x \geq 0$, when $\phi_0(0) = \phi_2(0)$. Determine the relationship that has to be satisfied by ϕ_0, ϕ_1, and ϕ_2 in order that the solution obtained has continuous first derivatives across the line $x = ct$.

2. Find, in $0 < x < a$, $t > 0$, that solution of equation (3.17) that satisfies the conditions

$$u(0, t) = u_x(a, t) = 0; \qquad u(x, 0) = x, \qquad u_t(x, 0) = 0.$$

3. Determine a relationship between the functions f and g which is sufficient to ensure that the equation

$$u_{xx} + 2u_{xy} - 3u_{yy} + 2u_x + 6u_y = 0$$

possesses a solution satisfying the conditions $u(t, 3t) = f(t)$, $u_x(t, 3t) = g(t)$.

3.4. The Cauchy problem for Laplace's equation

It has already been pointed out in §1.4 that Cauchy problems are not necessarily well-posed and an appropriate form of equation (3.25) will now be used to examine this situation in slightly more detail. Replacing t by y and c by i in equation (3.17) gives Laplace's equation

$$u_{xx} + u_{yy} = 0, \tag{3.39}$$

and carrying out the same substitution in equation (3.25) gives the solution of equation (3.39) with

$$u(x, 0) = f(x), \qquad u_y(x, 0) = g(x) \tag{3.40}$$

to be

$$u = \tfrac{1}{2}[f(x + iy) + f(x - iy)] - \tfrac{1}{2}i \int_{x-iy}^{x+iy} g(w)\, \mathrm{d}w.$$

Hence, assuming f and g to be real functions,

$$u = \operatorname{Re} f(x + iy) - \tfrac{1}{2}i\left[\int_x^{x+iy} g(w)\, \mathrm{d}w - \int_x^{x-iy} g(w)\, \mathrm{d}w\right]$$

$$= \operatorname{Re} f(x + iy) - \tfrac{1}{2}i\left[\int_0^{iy} g(x + v)\, \mathrm{d}v - \int_0^{-iy} g(x - v)\, \mathrm{d}v\right]$$

$$= \operatorname{Re} f(x + iy) + \tfrac{1}{2}\int_0^y [g(x + it) + g(x - it)]\, \mathrm{d}t$$

$$= \operatorname{Re} f(x + iy) + \operatorname{Re} \int_0^y g(x + it)\, \mathrm{d}t. \tag{3.41}$$

The arguments of f and g in equations (3.25) are real and hence boundedness, etc. of these functions on the real line (which is where they are defined) is sufficient to ensure that u is bounded. Inspection of equation (3.41) shows that the arguments of f and g are now complex and thus for u to be non-singular it is necessary that f and g, regarded as functions of a complex variable, are regular in an appropriate neighbourhood. This is a very strong condition to impose and thus the solutions of Cauchy problems for Laplace's equation can, at best, only exist for a restricted range of conditions. This can be illustrated by setting $f = 1/(1+x^2)$, $g = 0$, giving

$$u = \frac{1+x^2-y^2}{(1+x^2-y^2)^2+4x^2y^2}.$$

This function is unbounded at $(0, \pm 1)$ and this reflects the fact that, as a function of a *complex variable*, f is singular at $x = \pm i$.

The difficulties associated with 'well-posedness' can be illustrated by considering the difference $v = u' - u$, where u satisfies equation (3.40) and

$$u'(x, 0) = f(x) + \delta(x), \qquad u_y'(x, 0) = g(x).$$

Equation (3.41) then gives

$$v = \text{Re } \delta(x + iy).$$

δ, as a function of a real argument, can be extremely small for all values of its argument and yet $\delta(x+iy)$ can be unbounded arbitrarily close to $y = 0$. This can be seen by setting $\delta = \varepsilon/(x^2+a^2)$; over any part of the real axis δ can be made as small as desired by taking ε sufficiently small but $\varepsilon/[(x+iy)^2+a^2]$ is unbounded at $(0, \pm a)$. Thus the problem is certainly not well-posed, suggesting, at least for Laplace's equation, that the Cauchy problem is not well-posed unless the initial data is analytic in a region of the complex plane containing the initial curve.

3.5. Factorization of second-order linear partial differential equations

No general methods are available for obtaining the general solutions of second-order partial differential equations, but it seems worthwhile to point out that progress can often be made with those particular differential equations which can be factorized into two first-order equations. Equation (3.18) is a trivial example of such an equation, a more complicated example is that of Example 3.1.

It is possible to give criteria which the coefficients would have to satisfy in order that the differential equation be factorable but direct inspection of the equation should achieve the same end. It is often much easier to factorize an equation when in its canonical form, but for equations with constant coefficients the factorization can often be seen directly. As a general rule it is only for hyperbolic and parabolic equations that factorization can be a useful method of solution.

The equation

$$u_{xx} - u_{yy} + 2u_x + u = 0$$

can be written as

$$\left(\frac{\partial}{\partial x} + \frac{\partial}{\partial y} + 1\right)\left(\frac{\partial}{\partial x} - \frac{\partial}{\partial y} + 1\right)u = 0,$$

and it is therefore equivalent to the pair of first-order equations

$$u_x - u_y + u = v,$$

and

$$v_x + v_y + v = 0.$$

The hyperbolic equation

$$acu_{xy} + adu_x + (bc + ac_x)u_y + u(bd + ad_x) = 0$$

can be written as

$$\left(a\frac{\partial}{\partial x} + b\right)\left(c\frac{\partial}{\partial y} + d\right)u = 0.$$

It is therefore equivalent to

$$cu_y + du = v,$$

$$av_x + bv = 0.$$

It should be noted that, unlike the case when the coefficients are constant, the differential operators need not commute, and that the order of the factors can be important.

Exercises 3.3

1. By factorizing the differential operator, or otherwise, obtain the general solution of the following equations:

(i) $6u_{xx} + 11u_{xy} + 3u_{yy} = 0,$

(ii) $u_{xx} + u_{xy} - 2u_{yy} + 3u_x + 2u_y + 2u = 0,$

(iii) $x^2 u_{xx} - y^2 u_{yy} - yu_y = 0.$

2. Express

$$u_{xx} - u_{yy} + 6xu_x - 2yu_y + (9x^2 - y^2 + 2)u = 0$$

in canonical form. Solve the resulting equation by factorizing the differential operator, given that

$$u = 2, \qquad u_x = u_y = 0, \quad \text{when} \quad x = 0, \qquad y > 0.$$

In what region is the solution valid?

Bibliography

Garabedian, P. R. (1964). *Partial differential equations*. Wiley, New York.

4 Elliptic equations

THIS chapter considers various aspects of the solution of boundary-value problems for second-order linear elliptic equations in two independent variables. Some aspects of existence and uniqueness are discussed briefly in §4.1 and various properties of Laplace's equation (which is the archetypal equation) are established in §4.2, together with proofs of uniqueness theorems, and some elementary methods of solving boundary-value problems are illustrated. One useful method of solving such boundary-value problems for Laplace's equation is by means of suitable integral representations of solutions and these representations are obtained most directly in terms of particular singular solutions, termed Green's functions; the properties and applications of these functions are described in §4.3. Laplace's equation is a particular member of a general class of equations which are said to be self-adjoint and in §4.4 the results of the previous sections are generalized to include a more general class of self-adjoint equations. In §4.5 some eigenvalue problems associated with these latter equations are discussed and the use of eigenfunctions to solve boundary-value problems is illustrated. The notion of weak solutions of elliptic equations is again introduced in §4.6, where it is shown that, for the class of equations considered, weak solutions which are discontinuous across a curve cannot exist. The chapter concludes in §4.7 with a brief sketch of the way the analysis of §4.4 has to be modified when the 'self-adjoint' condition is removed.

4.1. Preliminary remarks on existence and uniqueness

Boundary-value problems for elliptic equations are, in general, only well-posed when boundary conditions are imposed on closed curves as in the problems described for Laplace's equation in §1.2, and are either of the Dirichlet or Neumann type or are such that a linear combination of the dependent variable and its normal derivative is known on the boundary. The proof of existence theorems for such problems is technically extremely difficult and no attempt will be made to provide such proofs. In fact, other than for Dirichlet problems, the exact conditions of validity of existence theorems are difficult to state for the general second-order linear elliptic equation.

In the subsequent sections we shall confine ourselves almost entirely to an examination of the properties of solutions of Laplace's and Poisson's equation and of a slight generalization of these equations, viz.

$$Eu = [a(x, y)u_x]_x + [c(x, y)u_y]_y + f(x, y)u = g(x, y). \quad (4.1)$$

The differential operator in equation (4.1) has the particular property that, for all twice differentiable functions u and v, $vEu - uEv$ is a divergence and the operator E is therefore referred to as a self-adjoint operator (cf. §2.6). This particular property of the differential operator considerably eases the detailed analysis necessary to obtain properties and representations of solutions of equation (4.1) and simplifies the statements and proofs of the various uniqueness theorems. The extra complications arising when the differential operator is not self-adjoint are sketched in §4.7. Unless it is stated to the contrary it may be assumed that solutions of all differential equations are being sought in a simple closed region D bounded by a smooth curve C.

It should be noted that the fact that a solution can be found by only prescribing u on a boundary does not contradict the Cauchy–Kowalesky theorem, as this theorem only ensures the existence of the solution in a neighbourhood of the boundary and not everywhere in the interior of a region.

It is possible to state a fairly concise existence and uniqueness theorem for Dirichlet problems for elliptic equations and we have

Theorem 4.1

The equation

$$a(x, y)u_{xx} + 2b(x, y)u_{xy} + c(x, y)u_{yy} + d(x, y)u_x$$
$$+ e(x, y)u_y + f(x, y)u = g(x, y),$$

possesses one and only one solution in the given region D with

$u = h(x, y)$ on C provided that, in D and on C,

  (i) a, b, c are continuously differentiable,
 (ii) d, e, f, g, h are continuous,
(iii) $f \leq 0$.

Also, under these conditions, the boundary-value problem is well-posed. Theorem 4.1 is also valid in the unbounded region external to a smooth closed curve provided that the coefficients are bounded at infinity and that u and g tend to zero at infinity. It is also valid in bounded, multiply-connected regions, provided the boundary curves are smooth.

Existence theorems can be proved for particular differential equations under much weaker conditions, for example it can be proved that Theorem 4.1 holds for Laplace's equation for any curve C such that, at every point P of C, it is possible to draw a line segment which, apart from P, lies completely outside D. Therefore the theorem is valid, in particular, for Laplace's equation in regions with polygonal boundaries. It is also possible to relax the condition that h is continuous and yet obtain a twice differentiable solution everywhere in D; this is illustrated by a specific example in §4.2.

Detailed statements and proof of existence theorems for different types of boundary-value problems may be found in Miranda (1970) and Gilbarg and Trudinger (1977).

4.2. Some properties and solutions of Laplace's equation

Laplace's equation, namely,

$$u_{xx} + u_{yy} = 0,$$

is the archetypal elliptic equation and occurs in steady-state heat conduction, irrotational incompresible flow, and electrostatics. It differs from the archetypal hyperbolic equation (i.e. the one-dimensional wave equation) in that it is not possible, using elementary methods, to solve simple boundary-value problems of the type occurring in practical situations. It is, however, helpful to investigate some of the properties of its solutions before looking at the more general equation, particularly as the techniques necessary for dealing with the latter are often generalizations of those originally developed for Laplace's equation.

There is one particular feature, however, of Laplace's equation which is not common to all elliptic equations and which enables solutions to be obtained for many boundary-value problems. This special feature is its relation to complex variable theory, particularly the fact that as a direct consequence of the Cauchy–Riemann equations, the real and imaginary parts of any analytic function of the complex variable $z(=x+\mathrm{i}y)$ are both solutions of Laplace's equation (solutions of Laplace's equation are often referred to as harmonic functions). It is, therefore, possible to use some of the powerful tools of complex variable theory to solve boundary-value problems for Laplace's equation. As an example of this, we now use Cauchy's integral theorem to obtain an explicit solution for the Dirichlet problem for Laplace's equation in a circular domain. This solution will also be used to illustrate some of the features which are common to the solutions of Dirichlet problems for all elliptic equations.

If G is a closed contour bounding a region R of the complex z-plane then it follows from Cauchy's integral theorem that a function $f(z)$, which is analytic in R and on G, is given at any point of R by

$$f(z) = \frac{1}{2\pi\mathrm{i}} \int_G \frac{f(\zeta)}{\zeta - z} \, \mathrm{d}\zeta. \tag{4.2}$$

Equation (4.2) demonstrates that an analytic function within a region can be expressed in terms of its values on the boundary of the region. For the particular case when G is the circle of radius a with centre at the origin we have that, when z is within G, the point $a^2/\bar{z}$, where the bar denotes the complex conjugate, is outside G. The function $f(\zeta)/(\zeta - a^2/\bar{z})$ is therefore analytic for $|\zeta| \leq a$ and, by Cauchy's theorem,

$$\int_G \frac{f(\zeta) \, \mathrm{d}\zeta}{\zeta - a^2/\bar{z}} = 0. \tag{4.3}$$

Combining equations (4.2) and (4.3) gives

$$f(z) = \frac{1}{2\pi\mathrm{i}} \int_G f(\zeta) \left[\frac{1}{\zeta - z} - \frac{1}{\zeta - a^2/\bar{z}} \right] \mathrm{d}\zeta, \tag{4.4}$$

and equation (4.4) can be simplified and written in the alternative

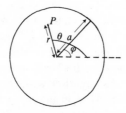

Fig. 4.1

form

$$f(re^{i\theta}) = \frac{1}{2\pi} \int_0^{2\pi} \frac{f(ae^{i\phi})(a^2 - r^2)\,d\phi}{a^2 + r^2 - 2\,ar\cos(\phi - \theta)}, \qquad (4.5)$$

where ϕ, θ, and r are defined by (see Fig. 4.1) $\zeta = ae^{i\phi}$, $z = re^{i\theta}$, and P in Fig. 4.1 represents the point z.

$f(ae^{i\phi})$ can be written as $f_1(\phi) + if_2(\phi)$, where f_1 and f_2 are real, the real part of the right-hand side of equation (4.5) is the real part of an analytic function whose real part on G is equal to $f_1(\phi)$. The real part of an analytic function is a harmonic function, and therefore u, defined by

$$u = \frac{1}{2\pi} \int_0^{2\pi} \frac{f_1(\phi)(a^2 - r^2)\,d\phi}{a^2 + r^2 - 2\,ar\cos(\theta - \phi)}, \qquad (4.6)$$

is harmonic within the above circle and equal to $f_1(\phi)$ on it.

It is also possible to show, by direct calculation, that u defined by equation (4.6) satisfies Laplace's equation for $0 \le r < a$ and that, for continuous functions f_1, $\lim_{r \to a} u = f_1$. Equation (4.6) therefore demonstrates the existence of a solution of the Dirichlet problem for Laplace's equation within a circular region. Furthermore its form shows that a slight change in f_1 only produces a change of the same order in u, thereby showing that the Dirichlet problem is well-posed. This is a particular example of the general result that Dirichlet problems for elliptic equations are well-posed. It is also possible to show that equation (4.6) defines a continuous solution of Laplace's equation in $0 \le r < a$ even when f_1 is discontinuous, in this case $\lim_{r \to a} u$ at a point of discontinuity varies with the direction of approach to the discontinuity. Thus for Laplace's equation, and indeed for all elliptic equations, discontinuities in boundary data are smoothed out in the interior of the domain, in marked contrast to the case of linear hyperbolic equations where any discontinuity in data is propagated along the characteristics (§5.4).

Another simple solution of a Dirichlet problem for Laplace's equation can be obtained by taking R to be the upper half-plane and G to be the real axis together with the semi-circle at infinity in the upper half-plane. If z is taken to be in the upper half-plane, then $f(\zeta)/(\zeta - \bar{z})$ is analytic for ζ in the upper half-plane and equation (4.3) has to be replaced by

$$\int_G \frac{f(\zeta)}{\zeta - \bar{z}} \, d\zeta = 0,$$

so that

$$f(z) = \frac{1}{2\pi i} \int_G f(\zeta) \left(\frac{1}{\zeta - z} - \frac{1}{\zeta - \bar{z}} \right) d\zeta. \qquad (4.7)$$

If $f(\zeta)$ is assumed to be bounded as $|\zeta| \to \infty$, then the integral over the semi-circle at infinity in equation (4.7) can be shown to be zero so that the only non-zero contribution is that from integration along the real axis. Writing $\zeta = x'$ and $z = x + iy$ in equation (4.7) gives

$$f(z) = \frac{y}{\pi} \int_{-\infty}^{\infty} \frac{f(x') \, dx'}{(x' - x)^2 + y^2}$$

and, by arguments identical to those used in deriving equation (4.6) from equation (4.5), we deduce that

$$u = \frac{y}{\pi} \int_{-\infty}^{\infty} \frac{f_1(x') \, dx'}{(x' - x)^2 + y^2} \qquad (4.8)$$

is the solution of the Dirichlet problem for Laplace's equation in the upper half-plane with $u = f_1(x)$ on $y = 0$ and u bounded as $x^2 + y^2 \to \infty$.

We can, by letting $f_1(x)$ be equal to unity for $0 \le x \le 1$ and zero otherwise, use equation (4.8) to obtain more explicit information about the behaviour of a solution in the neighbourhood of a discontinuity in boundary data. The integration in equation (4.8) is now elementary and we obtain

$$u = \frac{1}{\pi} \left[\tan^{-1} \left(\frac{1-x}{y} \right) + \tan^{-1} \frac{x}{y} \right]; \qquad (4.9)$$

the first term in equation (4.9) tends to $\frac{1}{2}$ as y and x approach zero, whilst the second tends to $(\frac{1}{2}\pi - \theta)/\pi$, where θ is the usual polar angle, and hence tends to any value between $-\frac{1}{2}$ and $\frac{1}{2}$ as the origin is approached. The first term displays a similar behaviour near $(1, 0)$; u defined by equation (4.9), however, is continuous and differentiable in $y > 0$, thus demonstrating explicitly that the discontinuity in the boundary data is not propagated into the domain of interest, the sole effect of such a discontinuity being to make the limiting value of u at a point of discontinuity vary with the direction of approach to that point.

It is possible by means of conformal transformations to use equations (4.6) and (4.8) to derive solutions for Dirichlet problems within more general regions, but we shall not pursue this point as conformal mapping is only applicable to Laplace's equation and cannot be extended to more general equations.

A more generally applicable method is that of the separation of variables combined with the use of Fourier series, and the following example illustrates the use of this technique for solving certain boundary-value problems for Laplace's equation within rectangular regions.

Example 4.1

Determine u such that

$$u_{xx} + u_{yy} = 0, \qquad 0 < x < a, \qquad 0 < y < b,$$

and

$$u(x, 0) = 0, \qquad u(x, b) = f(x), \qquad u(0, y) = u(a, y) = 0.$$

Seeking, as in §3.3, a product solution of the form $X(x)Y(y)$ gives

$$X_{xx} + \lambda^2 X = 0, \qquad Y_{yy} - \lambda^2 Y = 0,$$

where λ is a constant. The conditions on u when $x = 0$ and $x = a$ give $X(0) = X(a) = 0$ and hence, exactly as in §3.3, λ is equal to $n\pi/a \, (n = 1, \ldots)$ and X is proportional to $\sin n\pi x/a$. Solving the equation for Y (with $\lambda = n\pi/a$) gives

$$u = \sum_{n=1}^{\infty} \sin \frac{n\pi x}{a} \left(A_n \sinh \frac{n\pi y}{a} + B_n \cosh \frac{n\pi y}{a} \right);$$

the condition $u(x, 0) = 0$ gives $B_n = 0$. The condition at $y = b$ is satisfied provided that

$$f(x) = \sum_{n=1}^{\infty} A_n \sinh \frac{n\pi b}{a} \sin \frac{n\pi x}{a},$$

and equation (1.27) then gives

$$A_n \sinh \frac{n\pi b}{a} = \frac{2}{a} \int_0^a f(x) \sin \frac{n\pi x}{a} \, dx.$$

Thus a formal series solution can be obtained and it is possible to show that, for most 'reasonable' functions $f(x)$, the series obtained does satisfy Laplace's equation and takes the correct values on the boundary. This example, therefore, provides an explicit solution of a Dirichlet problem in a polygonal region.

The representation of u in the above problem as a sine series arose naturally as a result of seeking product solutions satisfying the boundary conditions at $x = 0$ and $x = a$. An alternative method of dealing with the problem would have been to use the fact that a function $u(x, y)$ vanishing at $x = 0$ and $x = a$ can be represented by a sine series in x (the coefficients being functions of y) so that an appropriate representation for u is

$$u = \sum_{n=1}^{\infty} Y_n(y) \sin \frac{n\pi x}{a}.$$

Substituting this into Laplace's equation gives

$$\sum_{n=1}^{\infty} \left(\frac{d^2 Y_n}{dy^2} - \frac{n^2 \pi^2}{a^2} Y_n \right) \sin \frac{n\pi x}{a} = 0,$$

and the equation for Y_n follows on equating all the coefficients of the series to zero. It might appear that this method is unduly arbitrary as there are an infinite number of ways of obtaining a representation for u which satisfies the conditions at $x = 0$ and $x = a$. The important additional feature of the Fourier series representation is that it gives a simple equation for the Y_n, and the form of the Laplacian operator shows that the Fourier series representation is the best one to use for Laplace's equation in a rectangular region.

This alternative method of deriving a suitable representation for u shows how problems can be solved with other types of

conditions imposed at $x = 0$ and $x = a$; for example, requiring
that $u_x(0, y) = u_x(a, y) = 0$, means that $\sin n\pi x/a$ has to be replaced by $\cos n\pi x/a$ and equation (1.20) has then to be used to
determine the arbitrary constants occurring. Similarly the boundary conditions $u(0, y) = u_x(a, y) = 0$ give a series involving
$\sin(2n - 1)\pi x/2a$, whilst the conditions $u_x(0, y) = u(a, y) = 0$ produce a series involving $\cos(2n - 1)\pi x/2a$.

Since Laplace's equation is linear the more complicated problem where $u(0, y) = 0$ is replaced by $u(0, y) = g(y)$ can be solved
by writing u as $u_1 + u_2$, where u_1 denotes the solution just found
above and u_2 is a solution of Laplace's equation such that

$$u_2(x, 0) = 0, \qquad u_2(x, b) = 0, \qquad u_2(0, y) = g(y), \qquad u_2(a, y) = 0.$$

u_2 can be found in a similar way to u_1, the main difference being
that u_2 is a sine series involving $\sin n\pi y/b$. This process of adding
solutions with u or u_x vanishing at $x = 0$ and $x = a$ to solutions with
u or u_y vanishing at $y = 0$ and $y = b$ can be used to obtain formal
series solutions to problems where u or its normal derivative is
prescribed on the sides of the rectangle.

Integral representations similar to those of equations (4.6) and
(4.8) play a significant part in the theory of solutions of more
general elliptic equations. In the latter case general theorems and
representations cannot be deduced using complex variable theory
and a different method has to be used. Therefore, as a prelude to
studying equation (4.1), we prove, without recourse to complex
variable theory, uniqueness and other theorems for Laplace's
equation. The derivation of suitable integral representations requires the use of particular functions known as Green's functions
and these are defined, and their role in deriving integral representations described, in the subsequent section. The only tools
necessary for obtaining all the relevant results are the well-known
Green's identities

$$\int_D (v\nabla^2 u - u\nabla^2 v)\,dS = \int_C \left(v\frac{\partial u}{\partial n} - u\frac{\partial v}{\partial n}\right) ds, \qquad (4.10)$$

$$\int_D (u\nabla^2 u + \text{grad}^2 u)\,dS = \int_C u\frac{\partial u}{\partial n}\,ds. \qquad (4.11)$$

In equations (4.10) and (4.11) ∇^2 denotes the two-dimensional
Laplacian operator, $\partial/\partial n$ is the derivative on C in the direction of

the outward normal and the relationship between the direction of the outward normal and the sense of integration round C is that appropriate for the validity of the divergence theorem (roughly the domain is on the left as C is traversed in the positive sense). Equations (4.10) and (4.11) also hold for multiply-connected regions; in this case the right-hand side consists of a sum of line integrals over the bounding curves.

The following uniqueness theorem can be proved almost immediately from equation (4.11).

Theorem 4.2

There is at most one solution of Laplace's equation in D and taking prescribed values on C.

Proof

If there are two solutions u_1 and u_2 then $u = u_1 - u_2$ will also satisfy Laplace's equation and is zero on C. In this case the line integral in equation (4.11) vanishes and therefore, as $\nabla^2 u = 0$,

$$\int_D \text{grad}^2 u \, dS = 0,$$

Therefore grad $u \equiv \mathbf{0}$, giving u to be a constant and, as $u = 0$ on C, it follows that u is identically zero, thus proving uniqueness.

The difference between two solutions of Poisson's equation will be a solution of Laplace's equation and hence uniqueness also holds for Poisson's equation.

The fact that there can only be the trivial solution $u \equiv 0$ of the homogeneous Dirichlet problem for Laplace's equation in a closed region provides the theoretical basis of the practice of 'screening' electrical instruments by enclosing them in an earthed container. In these circumstances it can be shown that the electric intensity within the container is the gradient of a function u which satisfies Laplace's equation and vanishes on the container. Thus $u \equiv 0$ and hence there is no electric intensity within the container.

Theorem 4.2 also holds if $\partial u/\partial n + \mu u$, $\mu > 0$, is prescribed on C. In this case the difference of two solutions will satisfy $\partial u/\partial n +$

$\mu u = 0$ on C and equation (4.11) gives

$$\int_D \mathrm{grad}^2 u \, \mathrm{d}S + \mu \int_C u^2 \, \mathrm{d}s = 0.$$

Both integrals have to be separately zero, giving $u = 0$ on C and grad $u = \mathbf{0}$ within D and therefore, as before, $u \equiv 0$. The case $\mu = 0$, which is the Neumann problem, is not as straightforward since the most that can be deduced is that u is constant. In order to have uniqueness in this case it is necessary to specify u at one point of D, which then determines the constant.

An additional difficulty with the Neumann problem for Poisson's equation is that $\partial u / \partial n$ on C cannot be prescribed arbitrarily but has to satisfy a consistency condition. The need for such a condition can be seen by setting $v = 1$ in equation (4.10) giving

$$\int_D \nabla^2 u \, \mathrm{d}S = \int_C \frac{\partial u}{\partial n} \, \mathrm{d}s. \tag{4.12}$$

Hence, when

$$\nabla^2 u = g,$$

$\partial u / \partial n$ must be such that

$$\int_D g \, \mathrm{d}S = \int_C \frac{\partial u}{\partial n} \, \mathrm{d}s. \tag{4.13}$$

The necessity for equation (4.13) in a practical context can be seen by assuming u to be the steady-state temperature in D in the absence of heat sources. In this case the rate of flow of heat is proportional to $\int_C (\partial u / \partial n) \, \mathrm{d}s$ and u satisfies Laplace's equation. If equation (4.13) (with $g \equiv 0$) were not true this would mean that heat was flowing into D at a non-zero rate, thus contradicting the steady-state assumption.

Theorem 4.2 can be modified to give a result valid for unbounded regions. Equation (4.11) holds for all closed regions and, in particular, is valid for the region D' between C and a circle C_r of large radius r. In this case the boundary is $C + C_r$ and we have that

$$\int_{D'} (u \nabla^2 u + \mathrm{grad}^2 u) \, \mathrm{d}S = \int_C u \frac{\partial u}{\partial n} \, \mathrm{d}s + \int_{C_r} u \frac{\partial u}{\partial n} \, \mathrm{d}s,$$

If it is assumed that u is bounded at infinity then it is possible (cf. end of Example 7.2) to show that, as $r \to \infty$, $|\partial u/\partial r| < A/r^2$ for some constant A, so that

$$\left| \int_{C_r} u \frac{\partial u}{\partial n} \, ds \right| < \int_{C_r} \frac{B}{r^2} \, ds < \frac{2\pi B}{r}, \quad \text{for some constant } B.$$

The integral over C_r tends to zero as $r \to \infty$ and therefore, with the additional condition that u is bounded, Theorem 4.2 and its various extensions hold for unbounded regions.

Existence theorems can be proved for the Dirichlet and Neumann problems (provided equation (4.13) holds) and also when $\partial u/\partial n + \mu u$ is prescribed on C.

We conclude this section with a direct proof that the Dirichlet problem for Laplace's equation is well-posed and, in order to do this, need the following theorem (often referred to as the mean-value theorem):

Theorem 4.3

The value of a harmonic function at a point P is equal to the mean of its values on any circle with centre at P.

Proof

There is no loss of generality in taking P to be the origin and the mean value of u on a circle C of radius r and centre at the origin is

$$w(r) = \frac{1}{2\pi} \int_0^{2\pi} u(r \cos \theta, r \sin \theta) \, d\theta.$$

Differentiating this expression with respect to r gives

$$w_r = \frac{1}{2\pi} \int_0^{2\pi} u_r \, d\theta.$$

On C, $\partial/\partial r = \partial/\partial n$ and hence

$$w_r = \frac{1}{2\pi r} \int_0^{2\pi} u_r r \, d\theta = \frac{1}{2\pi r} \int_C \frac{\partial u}{\partial n} \, ds,$$

and it now follows from equation (4.12) that, when u is harmonic, $w_r = 0$. Thus w is independent of r and, in particular, is equal to $w(0)$ which is just $u(0, 0)$, thus proving the mean-value theorem.

The mean-value theorem can be used to prove a very powerful result (which also holds for a large class of elliptic equations) known as the *maximum–minimum principle* which asserts that a harmonic function cannot have a maximum or a minimum within a region D. If it is assumed that a harmonic function has a maximum M at a point P of D then u must be equal to M on all circles with centre P and lying entirely in D since $u < M$ at points of such a circle would lead to a contradiction of the mean-value theorem. If u is not identically constant then there will be a point Q, not on the above circles, where $u < M$. If P and Q are joined by some path then in traversing this path from Q to P, there will be some point R at which u just becomes equal to M and therefore $u < M$ at some points on a sufficiently small circle centred at R and this contradicts the mean-value theorem. Therefore, unless u is constant, it cannot attain its maximum within D; repeating the argument with u replaced by $-u$ shows that a harmonic function also cannot attain its minimum value within D. (A more general maximum principle associated with the Laplacian operator is proved in Exercises 4.1, 4.)

We are now in a position, having established the maximum–minimum principle, to prove fairly simply that the Dirichlet problem for Poisson's equation is well-posed. If we have two solutions u_1 and u_2 of Poisson's equation and equal to f_1 and f_2 respectively on C, then $u = u_1 - u_2$ is a solution of Laplace's equation and is equal to $f = f_1 - f_2$ on C. A non-constant harmonic function attains its maximum value on the boundary so that

$$\max_{\text{in } D} |u| \le \max_{\text{on } C} |f|,$$

$$\max_{\text{in } D} |u_1 - u_2| \le \max_{\text{on } C} |f_1 - f_2|.$$

If two sets of boundary values only differ slightly then the deviation between the corresponding solutions will be of the same order so we deduce that Dirichlet problems for Laplace's equation are well-posed.

Exercises 4.1

1. Solve Laplace's equation in $0 < x < a$, $0 < y < b$ when

  (i) $u(0, y) = 0 = u_x(a, y)$, $u(x, 0) = 1$, $u(x, b) = 0$,

  (ii) $u(0, y) = u(x, 0) = 1$, $u(a, y) = u(x, b) = 0$.

2. Show that, in a region bounded by a closed curve C, the only solution of

$$\nabla^2 u = u^{2n+1}, \quad n \text{ a positive integer},$$

which vanishes on C is $u \equiv 0$.

3. Prove that if u and v are both harmonic functions within the region D bounded by a closed curve C and $u \geq (\leq)v$ on C then $u \geq (\leq)v$ in D. Show also that, when $a \leq u - v \leq b$ on C, where a, b are constant, then $a \leq u - v \leq b$ in D.

4. If $\nabla^2 u \geq 0$ for all points of a bounded region D, show that the mean value of u over any circle completely contained in D cannot be less than the value of u at the centre of the circle. Hence deduce that such a function cannot attain its maximum at an interior point of D.

5. If u is harmonic in $|x| < 1$, $|y| < 1$, and $u = x^2 + y^2$ on $|x| = 1$, $|y| = 1$, find lower and upper bounds for $u(0, 0)$.

6. Verify that v defined by

$$v = \tfrac{49}{40} - \tfrac{1}{5}(x^4 - 6x^2y^2 + y^4)$$

is harmonic and that, when $|x| < 1$ and $|y| = 1$, $-0.025 \leq v - 1 - x^2 \leq 0.025$.

Hence deduce that $u(0, 0)$ of the previous exercise satisfies $1.15 < u(0, 0) < 1.2$.

7. u is harmonic in $a^2 \leq x^2 + y^2 \leq b^2$ and $m(r)$ denotes its maximum value on the circle $x^2 + y^2 = r^2$. Show that

$$m(r)\log\frac{b}{a} \leq m(b)\log\frac{r}{a} - m(a)\log\frac{r}{b}.$$

4.3. Integral representations and Green's functions

In deriving integral representations similar to those of equations (4.6) and (4.8) it is important to use a notation which distinguishes clearly between the point at which a function is being determined and a variable point of integration. We let (x, y) denote the coordinates of the point at which a function is to be found and use (x', y') to refer to integration points within a domain or on a curve; (x', y') are therefore dummy variables. It is also necessary to be able to distinguish between the Laplacian operator acting on the variables x and y and that acting on the variables x' and y'. The former will be denoted by ∇^2 and the latter by ∇'^2.

The essential feature of the derivation, for general elliptic equations, of integral representations akin to equations (4.6) and (4.8) is the use of singular solutions of Laplace's equation. It turns out that one suitable singular solution is $\log R$, where $R^2 = (x - x')^2 + (y - y')^2$. It can be verified by direct differentiation that

$$\nabla^2 \log R = \nabla'^2 \log R = 0, \qquad R \neq 0,$$

$\nabla^2 \log R$ is clearly not defined when $R = 0$ though it is its behaviour near $R = 0$ which makes it useful in deriving integral representations. The special role of $\log R$ is best seen by carrying out the necessary calculation and the first step is to use equation (4.10) with $v = \log R$ and D being replaced by the region $D - D_\delta$ between C and a small circle C_δ of radius δ surrounding the point $P(x, y)$. We thus obtain

$$\int_{D-D_\delta} (\log R \nabla'^2 u - u \nabla'^2 \log R) \, dx' \, dy'$$

$$= \int_C \left(\log R \frac{\partial u}{\partial n'} - u \frac{\partial}{\partial n'} \log R \right) ds'$$

$$+ \int_{C_\delta} \left(\log R \frac{\partial u}{\partial n'} - u \frac{\partial}{\partial n'} \log R \right) ds', \qquad (4.14)$$

where the normals on C and C_δ are drawn out of the region $D - D_\delta$ and their directions, together with the appropriate senses of integration round the curves, are shown in Fig. 4.2.

On C_δ,

$$R = \delta \quad \text{and} \quad \partial/\partial n' = -\partial/\partial R$$

so that

$$\int_{C_\delta} \left(\log R \frac{\partial u}{\partial n'} - u \frac{\partial}{\partial n'} \log R \right) ds' = \int_{C_\delta} \left(\log \delta \frac{\partial u}{\partial n'} + \frac{u}{\delta} \right) ds'.$$

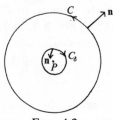

FIG. 4.2

On C_δ, $\partial u/\partial n$ will be bounded and hence less in modulus than some constant A so that

$$\left| \int_{C_\delta} \log \delta \frac{\partial u}{\partial n'} ds' \right| \le \int_{C_\delta} A \left| \log \delta \right| ds'$$

$$= A \int_0^{2\pi} \delta \left| \log \delta \right| d\theta = 2\pi A \delta \left| \log \delta \right|. \qquad (4.15)$$

If u_p denotes the value of u at P then

$$\int_{C_\delta} \frac{u}{\delta} ds' = u_p \int_{C_\delta} \frac{ds'}{\delta} + \int_{C_\delta} \frac{(u - u_p)}{\delta} ds', \qquad (4.16)$$

and also, by continuity, there exists a constant B such that, on C_δ, $|u - u_p| < B\delta$, giving

$$\left| \int_{C_\delta} \frac{|u - u_p|}{\delta} ds' \right| < B \int_{C_\delta} ds' = 2\pi B \delta. \qquad (4.17)$$

Equations (4.15) to (4.17) show that

$$\lim_{\delta \to 0} \int_{C_\delta} \left(\log R \frac{\partial u}{\partial n'} - u \frac{\partial}{\partial n'} \log R \right) ds' = 2\pi u(x, y);$$

the point $R = 0$ is excluded from $D - D_\delta$ and therefore $\log R$ is harmonic in this region and the second term in the double integral in equation (4.14) is identically zero. Thus, letting $\delta \to 0$ in equation (4.14) gives

$$u(x, y) = -\frac{1}{2\pi} \int_C \left(\log R \frac{\partial u}{\partial n'} - u \frac{\partial \log R}{\partial n'} \right) ds'$$

$$+ \frac{1}{2\pi} \int_D \log R \nabla'^2 u \, dx' \, dy'. \qquad (4.18)$$

If u satisfies Laplace's equation then the last term in equation (4.18) vanishes and, when

$$\nabla^2 u = g(x, y), \tag{4.19}$$

equation (4.18) becomes

$$u(x, y) = -\frac{1}{2\pi} \int_C \left(\log R \frac{\partial u}{\partial n'} - u \frac{\partial \log R}{\partial n'} \right) ds'$$

$$+ \frac{1}{2\pi} \int_D \log R g(x', y') \, dx' \, dy'. \tag{4.20}$$

Equations (4.18) and (4.20) are not equivalent to equations (4.6) and (4.8) as, unlike these latter equations, they involve the values of both u and $\partial u/\partial n$ on C. Equations (4.18) and (4.20) therefore cannot be used immediately to solve Dirichlet problems for Laplace's equation as $\partial u/\partial n$ will not usually be known on C.

It is, however, possible to modify the above analysis so as to obtain representations in terms of either u, or $\partial u/\partial n$, on C, but before doing this we need to look more carefully at the above analysis in order to establish the precise role played by $\log R$.

Comparison of equations (4.18) and (4.10) shows that the former could have been obtained immediately from the latter by setting $v \equiv (2\pi)^{-1}\log R$ provided that

$$\frac{1}{2\pi} \int_D u \nabla'^2 \log R \, dx' \, dy' = u(x, y). \tag{4.21}$$

$\nabla'^2 \log R$ vanishes except when $x = x'$ and $y = y'$; therefore, as it is impossible for the integral of a function which vanishes at all but one point to be non-zero, the left-hand side of equation (4.21) cannot represent an integral in the normal sense. The above analysis however shows that treating $\nabla'^2 \log R$ to be such that equation (4.21) holds, does produce a correct result. This kind of situation, where 'legitimate' operations of the kind leading to equation (4.18) produce the same effect as improper processes of the type shown in equation (4.21), often occurs and $\nabla'^2 \log R$ is an example of an entity known as a generalized function.

Such entities are not functions in the normal sense of the word but a functional notation is employed in handling them and they can be formally treated as functions provided that manipulations involving them follow certain precise rules. In the nomenclature of generalized functions, and using a vector notation, we have that

$$\frac{1}{2\pi}\nabla'^2\log R = \delta_2(\mathbf{r}'-\mathbf{r}),\qquad(4.22)$$

where $\delta_2(\mathbf{r}'-\mathbf{r})$ is the two-dimensional Dirac delta function which has the property that

$$\int_D \delta_2(\mathbf{r}'-\mathbf{r})g(x',y')\,\mathrm{d}x'\,\mathrm{d}y' = \begin{cases} g(x,y), & (x,y) \text{ in } D, \\ 0 & \text{otherwise.} \end{cases}\qquad(4.23)$$

For the present purposes the only property required of the delta function is that it satisfies equation (4.23) and it only has a meaningful interpretation in the context of this equation. For practical purposes it may be treated as a function vanishing except at one point and it may also be assumed that $\delta_2(\mathbf{r}'-\mathbf{r})=\delta_2(\mathbf{r}-\mathbf{r}')$. $\delta_2(\mathbf{r}'-\mathbf{r})$ can also be taken to be equal to the product of two one-dimensional delta functions $\delta(x'-x)\delta(y'-y)$ with the property that

$$\int_a^b f(x')\delta(x'-x)\,\mathrm{d}x' = \begin{cases} f(x), & a < x < b, \\ 0 & \text{otherwise.} \end{cases}\qquad(4.24)$$

The delta function will be used from now on without any attempt to justify its use, as the justification is provided by the now well developed theory of generalized functions.

It may be helpful in handling the delta function to remember that the familiar point mass represents a delta function mass density distribution. There are ways other than the above for rigorously deriving equation (4.18); one way is to set v in equation (4.10) equal to $(2\pi)^{-1}\log(R+\varepsilon)$ and take the limit as $\varepsilon\to 0$ after carrying out the necessary differentiations. The main object of introducing the delta function is to avoid having to carry out such complicated limiting processes.

It should also be noted that we have already encountered the one-dimensional delta function in equation (4.8) where, on taking the limit as $y\to 0$, we have that

$$f_1(x) = \lim_{y\to 0}\frac{y}{\pi}\int_{-\infty}^{\infty}\frac{f_1(x')\,\mathrm{d}x'}{(x-x')^2+y^2},$$

and hence $\lim_{y \to 0} y\pi^{-1}\{(x'-x)^2+y^2\}^{-1}$ is equivalent to $\delta(x'-x)$. It can be proved rigorously that the limiting value of the above integral is $f_1(x')$ even though $\lim_{y \to 0} y\pi^{-1}\{(x'-x)^2+y^2\}^{-1}$ is not a normal function as it vanishes except when $x'=x$ where it is unbounded.

We thus have that $(2\pi)^{-1}\log R$ is a solution of

$$\nabla'^2 F = \delta_2(\mathbf{r}'-\mathbf{r}) \qquad (4.25)$$

and any solution $F(x, y, x', y')$ of equation (4.25) is said to be a *fundamental solution* of Laplace's equation, the convention being adopted that the last two named variables are those on which the differential operator acts. Equation (4.25) does not define a fundamental solution uniquely as it is clearly possible to add a harmonic function to any fundamental solution and still have a solution of equation (4.25). An equivalent definition of a fundamental solution is that $F - (2\pi)^{-1}\log R$ is a regular harmonic function, i.e. F behaves like $(2\pi)^{-1}\log R$ near $R = 0$.

Taking u in equation (4.10) to be a solution of equation (4.19) and $v \equiv F$ gives, on using equations (4.23) and (4.25),

$$u(x, y) = -\int_C \left(F\frac{\partial u}{\partial n'} - u\frac{\partial F}{\partial n'}\right)ds' + \int_D Fg\,dx'\,dy', \quad (4.26)$$

where F is any fundamental solution. Equation (4.26) suffers from the same deficiency as equation (4.20) in that it requires knowledge of both u and $\partial u/\partial n$ on C. There is, however, an element of arbitrariness in the choice of F and it is possible, by judicious choice of F, to obtain a more suitable form of solution. Taking, for example, $F \equiv G_1(x, y, x', y')$, where $G_1 = 0$ when (x', y') is on C, gives

$$u = \int_C u\frac{\partial G_1}{\partial n'}\,ds' + \int_D gG_1\,dx'\,dy'. \qquad (4.27)$$

Equation (4.27) shows that, if a solution of equation (4.25) can be found which vanishes on C, it is possible to express u in terms of its values on C. G_1 can, by definition, be written as $u + (2\pi)^{-1}\log R$ where u is harmonic, so the problem of determining G_1 is equivalent to finding a regular harmonic function equal to

$-(2\pi)^{-1}\log R$ for (x', y') on C. Our previously quoted existence theorem shows therefore that, in principle, G_1 can be found.

A fundamental solution which is required to satisfy a boundary condition on a curve is referred to as a Green's function and, as equation (4.27) shows, the use of such functions provides a powerful method of obtaining solutions of boundary-value problems. A Green's function is not restricted to be a function vanishing on a boundary and defining Green's functions G_2 and G_3 such that $\partial G_2/\partial n = 0$ and $\partial G_3/\partial n + \mu G_3 = 0$ on C, gives

$$u = -\int_C G_2 \frac{\partial u}{\partial n'} \, ds' + \int_D g G_2 \, dx' \, dy' + \text{constant}, \quad (4.28)$$

(For G_2 to exist and be unique $-1/(\text{area of } D)$ has to be included in equation (4.25) and the condition $\int_D G_2 \, dx' \, dy' = 0$ imposed; for unbounded regions a condition at infinity is necessary.)

$$u = -\int_C G_3 \left(\frac{\partial u}{\partial n'} + \mu u \right) ds' + \int_D g G_3 \, dx' \, dy'. \quad (4.29)$$

Equations (4.28) and (4.29) show that, once G_2 and G_3 have been found, the problems of finding harmonic functions with $\partial u/\partial n + \mu u$, $\mu \geq 0$, prescribed on C, are reduced to evaluating integrals. It is, however, important to realize that the determination of Green's functions is often extremely difficult, and it can sometimes be quicker to solve specific problems directly using an *ad hoc* method rather than by first finding a Green's function and then carrying out the various integrations. The great advantage of Green's functions is that, once found for a particular boundary and particular type of boundary condition, they enable solutions to be found, in principle, for all boundary values. They also provide a useful method of representing solutions of problems and can prove particularly useful in the reduction of boundary-value problems to integral equations.

An important useful additional property of the functions G_1 to G_3 is that they all satisfy the symmetry relation

$$G(x, y, x', y') = G(x', y', x, y). \quad (4.30)$$

Equation (4.30) can be proved by setting $v = G_i(x_1, y_1, x', y')$, $u = G_i(x_0, y_0, x', y')$ $(i = 1, 2, 3)$ in equation (4.10) in the dashed variables, giving

$$\int_D \{G_i(x_1, y_1, x', y')\delta(x' - x_0)\delta(y' - y_0)$$

$$- G_i(x_0, y_0, x', y')\delta(x' - x_1)\delta(y' - y_1)\} \, dx' \, dy'$$

$$= \int_C \left\{ G_i(x_1, y_1, x', y') \frac{\partial G_i}{\partial n'}(x_0, y_0, x', y') \right.$$

$$\left. - G_i(x_0, y_0, x', y') \frac{\partial G_i}{\partial n'}(x_1, y_1, x', y') \right\} ds'.$$

The line integral in this equation vanishes in view of the boundary conditions satisfied by G_1 to G_3 when (x', y') is on C, and evaluating the left-hand side using the basic property of the delta function gives

$$G_i(x_1, y_1, x_0, y_0) = G_i(x_0, y_0, x_1, y_1)$$

which is equivalent to equation (4.30).

In those practical situations which give rise to Poisson's equation (i.e. equation (4.19)) the right-hand side of the equation represents some kind of source term (e.g. in electrostatics it corresponds to a charge density). Hence, in a general sense, Green's functions can be interpreted as being produced by some kind of singular source distribution (G_1 for example can be interpreted physically as the potential produced by a line charge at (x, y) when C is earthed). The notation adopted shows that $G_i(x, y, x', y')$ is the effect at (x', y') of some source at (x, y) and equation (4.30) is effectively a reciprocity condition in that it states that the effect at (x', y') of a source at (x, y) is the same as that at (x, y) due to a source at (x', y'). Simple reciprocity relations of this type are often found associated with self-adjoint differential operators.

It is possible with only a very slight modification of the above analysis to extend the concept of fundamental solution to solve boundary-value problems for

$$\nabla^2 u + au = g, \tag{4.31}$$

where a may depend on x and y. The appropriate fundamental solution in this case is defined to be a solution F of

$$\nabla'^2 F + aF = \delta_2(\mathbf{r}' - \mathbf{r}). \tag{4.32}$$

When u satisfies equation (4.31) and F satisfies equation (4.32) we have that

$$(F\nabla'^2 u - u\nabla'^2 F) = Fg - u\delta_2(\mathbf{r}' - \mathbf{r}).$$

Hence setting $F \equiv v$ in equation (4.10) shows that u will have the same general form as in equation (4.26) (F will be different from that in the latter equation). In any region excluding $x = x'$, $y = y'$, F will satisfy the homogeneous equation but it will have some kind of singular behaviour at $x = x'$, $y = y'$. The most singular part of the left-hand side of equation (4.32) near $x = x'$, $y = y'$ will be the first term (differentiation generally 'increases' a singularity) so that, near $\mathbf{r}' = \mathbf{r}$,

$$\nabla'^2 F = \delta_2(\mathbf{r}' - \mathbf{r}) = \nabla'^2 \left(\frac{1}{2\pi} \log R \right),$$

showing that, when $\mathbf{r}' \approx \mathbf{r}$, $F \approx (2\pi)^{-1}\log R$. Hence an alternative way of defining F is that, except when $x' = x$ and $y' = y$, it satisfies equation (4.31) (with $g = 0$) and near $x' = x$, $y' = y$

$$F \approx \frac{1}{2\pi} \log R. \tag{4.33}$$

The above derivation of the singular behaviour is a trifle heuristic but it can be shown quite rigorously that the alternative definition is entirely equivalent to that of equation (4.32).

In the following examples we adopt the convention used in deriving equations (4.27) to (4.29) that the differential equation is solved in terms of the variables (x', y'), and that (x, y) refer to the coordinates of the point at which equations (4.27) to (4.29) define u. Equation (4.30) shows that (x, y) and (x', y') are interchangeable but it seems preferable at this stage to avoid confusion by adhering rigidly to the above convention.

There are no general methods available for finding Green's functions, but for certain geometries the methods of images and inversion are useful and Fourier series are appropriate for determining Green's functions in rectangular regions. These methods are illustrated in the following examples.

Example 4.2

Find the solution of

$$\nabla'^2 G = \delta(x' - x)\delta(y' - y)$$

in $y' > 0$, with $G(x', 0) = 0$ and $G \to 0$ as $x'^2 + y'^2 \to \infty$. Hence find the solution u of Laplace's equation in $y > 0$, equal to $f(x)$ on $y = 0$ and tending to zero at infinity.

The Green's function $G(x, y, x', y')$ is, by definition, of the form

$$\frac{1}{2\pi} \log[(x - x')^2 + (y - y')^2]^{\frac{1}{2}} + v(x', y'),$$

where v satisfies Laplace's equation in $y' > 0$ (and hence is non-singular) and

$$v(x', 0) = -\frac{1}{2\pi} \log[(x - x')^2 + y^2]^{\frac{1}{2}}.$$

Direct differentiation shows that, for any (x_0, y_0), $\log[(x' - x_0)^2 + (y' - y_0)^2]^{\frac{1}{2}}$ satisfies Laplace's equation except at $x' = x_0$, $y' = y_0$. Choosing $x_0 = x$, $y_0 = -y$ gives a harmonic function which is non-singular in $y' \geq 0$ and satisfies the boundary condition on $y' = 0$. Thus the appropriate choice for v is

$$-\frac{1}{2\pi} \log[(x - x')^2 + (y + y')^2]^{\frac{1}{2}},$$

so that

$$G = \frac{1}{2\pi} \log \frac{[(x - x')^2 + (y - y')^2]^{\frac{1}{2}}}{[(x - x')^2 + (y + y')^2]^{\frac{1}{2}}}.$$

Clearly $G \to 0$ at infinity and, as it can be proved that G is uniquely determined by its values on $y' = 0$ and at infinity, is the required Green's function. u can now be calculated from equation (4.27), $\partial/\partial n'$ being in this case equal to $-\partial/\partial y'$. There will be a contribution to the right-hand side of equation (4.27) from a circle C_∞ of large radius. As $r = (x'^2 + y'^2)^{\frac{1}{2}}$ tends to infinity G is $O(1/r)$ and ds on C_∞ is $r \, d\theta$; hence the contribution from C_∞ tends to zero. Hence,

$$u(x, y) = -\int_{-\infty}^{\infty} \left(\frac{\partial G}{\partial y'}\right)_{y'=0} f(x') \, dx',$$

and carrying out the differentiation gives

$$u(x, y) = \frac{y}{\pi} \int_{-\infty}^{\infty} \frac{f(x') \, dx'}{(x - x')^2 + y^2},$$

which is just equation (4.8).

The above method of finding the Green's function is the method of images used in electrostatics (v is singular at the image, in $y'=0$, of the point (x, y)) and is applicable to problems in wedge-shaped regions of angle π/n, where n is an integer.

Example 4.3

Find, for Laplace's equation within a circle of radius a, the Green's function which vanishes on the boundary of the circle.

This problem is most easily solved by taking polar coordinates with their origin at the centre of the circle and (r, θ) and (ρ, ϕ) are defined by

$$x = r \cos \theta, \qquad y = r \sin \theta, \qquad x' = \rho \cos \phi, \qquad y' = \rho \sin \phi.$$

In Fig. 4.3, P and Q denote the points corresponding to the coordinates (x, y) and (x', y'), respectively, and r_1 denotes the distance PQ, i.e.

$$r_1^2 = r^2 - 2r\rho \cos(\theta - \phi) + \rho^2.$$

The Green's function $G(x, y, x', y')$ will be of the form

$$\frac{1}{2\pi} \log r_1 + v$$

where v is a solution of Laplace's equation (in the variables x' and y') and we try to find v by using the fact that the logarithm of the distance from any fixed point is a solution of Laplace's equation. The boundary condition on G implies, that whenever Q is on the circle,

$$v = -\frac{1}{2\pi} \log r_1.$$

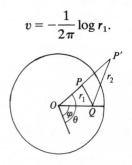

Fig. 4.3

The distance r_2 of Q from a point P' on OP, and at a distance R from O, is

$$r_2{}^2 = \rho^2 - 2\rho R \cos(\theta - \phi) + R^2,$$

and clearly it is impossible to find R such that, on $\rho = a$, $r_2 = r_1$. We therefore try to find R such that, on $\rho = a$, $r_1 = \lambda r_2$ for some λ, i.e.

$$a^2 - 2ar \cos(\theta - \phi) + r^2 = \lambda^2(a^2 - 2aR \cos(\theta - \phi) + R^2);$$

equating the coefficients of $\cos(\theta - \phi)$ gives $r = \lambda^2 R$ so that either $r = R$ or. $Rr = a^2$. The first possibility gives $r_2 \equiv r_1$ which is of no use but the choice $R = a^2/r$ and $\lambda = r/a$ gives

$$r_2 = (a/r)r_1 \quad \text{on} \quad \rho = a.$$

Therefore

$$v = -\frac{1}{2\pi} \log \frac{rr_2}{a}$$

is a regular harmonic function (the point P', corresponding to $R = a^2/r$, is outside the circle and hence r_2 is not zero for $\rho \le a$) and takes the correct value for $\rho = a$. The point P' is the inverse point of P with respect to the circle and the above results can be obtained more directly using the geometric properties of inverse points.

The Green's function is therefore

$$\frac{1}{2\pi} \log \frac{ar_1}{rr_2}.$$

If a solution u of Laplace's equation in $r \le a$ is equal to a given function $f(\theta)$ on $r = a$ then the Green's function of Example 4.3 can be combined with the integral representation of equation (4.27) to give a formal representation of u within the circle. We have therefore that

$$u(x, y) = \int_C f(\theta) \left(\frac{\partial G}{\partial n'}\right)_{\rho=a} ds'.$$

On the circle, $\partial/\partial n' = \partial/\partial \rho$, $ds' = a\,d\phi$, and therefore

$$\left(\frac{\partial G}{\partial n'}\right)_{\rho=a} = \frac{1}{2\pi} \left\{\frac{\partial}{\partial \rho} \log \frac{ar_1}{rr_2}\right\}_{\rho=a} = \frac{1}{4\pi} \left\{\frac{1}{r_1{}^2} \frac{\partial r_1{}^2}{\partial \rho} - \frac{1}{r_2{}^2} \frac{\partial r_2{}^2}{\partial \rho}\right\}_{\rho=a}.$$

Carrying out the differentiations and using the result that, on $\rho = a$, $r_2^2 = a^2 r_1^2 / r^2$, gives

$$u(r \cos \theta, r \sin \theta) = \frac{1}{2\pi} \int_0^{2\pi} \frac{(a^2 - r^2) f(\phi) \, d\phi}{a^2 + r^2 - 2ar \cos(\theta - \phi)} \, .$$

This is the result previously given in equation (4.6) and obtained by complex variable methods.

Example 4.4

Determine the Green's function suitable for solving the Dirichlet problem for Laplace's equation in the rectangle $0 \le x \le a$, $0 \le y \le b$.

Equation (4.27) shows that the required function $G(x, y, x', y')$ satisfies

$$G_{x'x'} + G_{y'y'} = \delta(x' - x)\delta(y' - y) \tag{4.34}$$

and vanishes on $x' = 0$, $x' = a$, $y' = 0$, $y' = b$.

The most direct method of solving problems of this type is to adopt the approach outlined after Example 4.1, of using a Fourier series representation which automatically satisfies some of the boundary conditions. In this case G satisfies homogeneous conditions on all four parts of the boundary and hence it is entirely arbitrary as to whether the representation chosen is such as to satisfy the conditions at $x' = 0$ and $x' = a$ or those at $y' = 0$ and $y' = b$. If we choose to satisfy the former pair of conditions then the appropriate representation for G is

$$G = \sum_{n=1}^{\infty} Y_n(y') \sin \frac{n\pi x'}{a} \, .$$

The technique of substituting this representation into equation (4.34) and equating coefficients of $\sin n\pi x'/a$ on both sides of the equation only works if the right-hand side of the equation can be expressed as a sine series. We therefore need to find c_n such that

$$\delta(x' - x) = \sum_{n=1}^{\infty} c_n \sin \frac{n\pi x'}{a} \, ,$$

and treating the delta function as an ordinary function gives

$$c_n = \frac{2}{a} \int_0^a \delta(x' - x) \sin \frac{n\pi x'}{a} \, dx' = \frac{2}{a} \sin \frac{n\pi x}{a} \, .$$

Hence,

$$\sum_{n=1}^{\infty} \sin\frac{n\pi x'}{a}\left\{\frac{\mathrm{d}^2 Y_n}{\mathrm{d}y'^2} - \frac{n^2\pi^2}{a^2}Y_n\right\} = \frac{2}{a}\,\delta(y'-y)\sum_{n=1}^{\infty}\sin\frac{n\pi x}{a}\sin\frac{n\pi x'}{a}.$$

(The series on the right-hand side does not define a function in the normal sense and is just a representation of the delta function.) Equating the coefficients of $\sin n\pi x'/a$ gives

$$\frac{\mathrm{d}^2 Y_n}{\mathrm{d}y'^2} - \frac{n^2\pi^2}{a^2}Y_n = \frac{2}{a}\sin\frac{n\pi x}{a}\,\delta(y'-y),$$

a novel feature of this equation being the delta-function term on the right-hand side. The integral with respect to y' of the delta function over any interval is non-zero when the interval contains $y' = y$ and is zero otherwise. Therefore, at any point,

$$\frac{\mathrm{d}Y_n}{\mathrm{d}y'} = \frac{n^2\pi^2}{a^2}\int_0^{y'} Y_n(w)\,\mathrm{d}w + \text{constant} + \frac{2}{a}\sin\frac{n\pi x}{a}H(y'-y),$$

where $H(z)$ is the Heaviside step function defined by $H(z) = 1$, $z \geq 0$, $H(z) = 0$, $z < 0$. It follows from this equation that

$$\left(\frac{\mathrm{d}Y_n}{\mathrm{d}y'}\right)_{y'=y+0} - \left(\frac{\mathrm{d}Y_n}{\mathrm{d}y'}\right)_{y'=y-0} = \frac{2}{a}\sin\frac{n\pi x}{a}, \qquad (4.35)$$

so that $\mathrm{d}Y_n/\mathrm{d}y'$ is discontinuous at $y' = y$. A further integration of the expression for $\mathrm{d}Y_n/\mathrm{d}y'$ shows that Y_n is continuous everywhere. Y_n satisfies the homogeneous equation

$$\frac{\mathrm{d}^2 Y_n}{\mathrm{d}y'^2} - \frac{n^2\pi^2}{a^2}Y_n = 0,$$

except at $y' = y$, and the boundary conditions at $y' = 0$ and $y' = b$ imply that $Y_n(0) = Y_n(b) = 0$. The continuous function Y_n satisfying these boundary conditions will thus be of the form

$$Y_n = \begin{cases} \dfrac{A\sinh\dfrac{n\pi}{a}(y'-b)}{\sinh\dfrac{n\pi}{a}(y-b)}, & y' \geq y, \\[4ex] \dfrac{A\sinh\dfrac{n\pi y'}{a}}{\sinh\dfrac{n\pi y}{a}}, & y' \leq y, \end{cases}$$

where A is a constant. Equation (4.35) will be satisfied provided that

$$A \frac{n\pi}{a} \left(\coth \frac{n\pi}{a}(y-b) - \coth \frac{n\pi y}{a} \right) = \frac{2}{a} \sin \frac{n\pi x}{a},$$

which gives

$$A = \frac{2}{n\pi} \frac{\sinh \dfrac{n\pi y}{a} \sinh \dfrac{n\pi}{a}(y-b)}{\sinh \dfrac{n\pi b}{a}} \sin \frac{n\pi x}{a}.$$

Hence

$$G(x, y, x', y') = \begin{cases} \dfrac{2}{\pi} \displaystyle\sum_{n=1}^{\infty} \dfrac{1}{n} \operatorname{cosech} \dfrac{n\pi b}{a} \sinh \dfrac{n\pi}{a}(y'-b) \sinh \dfrac{n\pi y}{a} \\ \qquad\qquad \times \sin \dfrac{n\pi x}{a} \sin \dfrac{n\pi x'}{a}, \qquad y' \geq y, \\[2mm] \dfrac{2}{\pi} \displaystyle\sum_{n=1}^{\infty} \dfrac{1}{n} \operatorname{cosech} \dfrac{n\pi b}{a} \sinh \dfrac{n\pi}{a}(y-b) \sinh \dfrac{n\pi y'}{a} \\ \qquad\qquad \times \sin \dfrac{n\pi x}{a} \sin \dfrac{n\pi x'}{a}, \qquad y' \leq y. \end{cases}$$

Substitution of this expression into equation (4.27) will give the solution for the Dirichlet problem for the rectangle in terms of integrals of the boundary values. The form obtained will be identical with that found by expressing the solution as the sum of solutions of the type found in Example 4.1.

The above derivation of the solution for Y_n is blatantly non-rigorous in that the delta function is represented as a divergent Fourier series. It is, however, the most direct method of finding the equation for Y_n, and a slightly less direct method, not involving obviously divergent series, is to multiply equation (4.34) by $(2/a)\sin n\pi x'/a$ and integrate with respect to x' from $x' = 0$ to $x' = a$. The term involving $G_{x'x'}$ is then integrated by parts twice and, in view of the conditions satisfied by G at $x' = 0$ and $x' = a$, this gives

$$\frac{2}{a} \frac{\mathrm{d}^2}{\mathrm{d}y'^2} \int_0^a G \sin \frac{n\pi x'}{a} \, \mathrm{d}x' - 2n^2 \frac{\pi^2}{a^3} \int_0^a G \sin \frac{n\pi x'}{a} \, \mathrm{d}x'$$
$$= \frac{2}{a} \sin \frac{n\pi x}{a} \delta(y' - y).$$

It follows from the definition of Y_n that

$$Y_n = \frac{2}{a} \int_0^a G \sin \frac{n\pi x'}{a} \, \mathrm{d}x',$$

so that the above equation reduces to that previously found for Y_n.

The above analysis can be very easily modified to give the Green's function appropriate for solving the Dirichlet problem in the strip $0 \le x \le a$, $-\infty < y < \infty$, when the harmonic function vanishes as $y \to \pm\infty$. In this case, in order to make sure that the integrals in equation (4.27) vanish over the segments $y' \to \pm\infty$, $0 \le x < a$, G has to vanish as $|y'| \to \infty$. The appropriate representation is the same as that of Example 4.4, the equation for Y_n and the conditions at $y' = y$ remain unchanged, and the only real difference is that we now require $\lim_{|y'| \to \infty} Y_n = 0$. The solution for Y_n will thus be of the form $Be^{-n\pi|y'-y|/a}$, where B is constant, and equation (4.35) shows that B will be given by

$$B = -\frac{1}{n\pi} \sin \frac{n\pi x}{a}$$

so that, in this case,

$$G = -\frac{1}{\pi} \sum_{n=1}^{\infty} \frac{1}{n} e^{-n\pi|y'-y|/a} \sin \frac{n\pi x}{a} \sin \frac{n\pi x'}{a}. \qquad (4.36)$$

Example 4.5

Find the solution of

$$(\nabla'^2 - k^2)G = \delta(\mathbf{r}' - \mathbf{r}), \quad k \text{ real}, \qquad (4.37)$$

which vanishes as $x'^2 + y'^2 \to \infty$.

We thus have to find a function G which satisfies

$$(\nabla'^2 - k^2)G = 0, \qquad \mathbf{r} \ne \mathbf{r}', \qquad (4.38)$$

and is such that $G - (2\pi)^{-1}\log R$, where $R^2 = (x - x')^2 + (y - y')^2$, is bounded near $R = 0$. The singular part of the solution depends only on R and this suggests using polar coordinates with origin at (x, y) and, assuming radial symmetry, equation (4.38) becomes

$$R^2 \frac{\mathrm{d}^2 G}{\mathrm{d}R^2} + R \frac{\mathrm{d}G}{\mathrm{d}R} - k^2 R^2 G = 0.$$

Independent solutions of this equation are $I_0(kR)$ and $K_0(kR)$, where I_0 and K_0 are the modified Bessel functions of the first and second kind and of order zero. The condition as $R \to \infty$ implies that G has to be a multiple of $K_0(kR)$, and it is known that, near $z = 0$, $K_0(z) \sim -\log z$, so that

$$G = -\frac{1}{2\pi} K_0(kR).$$

This Green's function can be substituted into equation (4.26) to give a representation, in the region exterior to C, of the solution of equation (4.38) which vanishes at infinity. (The conditions on u and G at infinity ensure that there is no contribution from the circle at infinity.) The method of images can also be used, as described in Example 4.2, to define Green's functions suitable for solving Dirichlet and Neumann problems for equation (4.38) in a sector of angle π/n when the unknown function vanishes at infinity.

Exercises 4.2

1. Find, for Laplace's equation in the region $x' > 0$, $y' > 0$, the Green's function which vanishes when $x' = 0$ and when $y' = 0$ and which tends to zero as $(x'^2 + y'^2) \to \infty$. Hence obtain, in the form of an infinite integral, that solution of Laplace's equation in $x > 0$, $y > 0$ which vanishes on $x = 0$ and is equal to a given function $f(x)$ on $y = 0$.

2. Determine the Green's function appropriate for solving the Dirichlet problem for Laplace's equation in the semi-circular region $x^2 + y^2 < a^2$, $y > 0$.

3. Find the Green's function $G(x, y, x', y')$ for Laplace's equation in

   (i) $0 < x' < a$, $0 < y' < b$; with $G = 0$, $x' = 0$; $G_{x'} = 0$, $x' = a$; $G_{y'} = 0$, $y' = 0$; $G = 0$, $y' = b$
   (ii) $0 < y' < b$, $x' > 0$, with $G = 0$ on $x' = 0$, $y' = 0$, $y' = b$ and $G \to 0$ as $x' \to \infty$.

4. Obtain, in the form of an infinite integral, the solution of $u_{xx} + u_{yy} - k^2 u = 0$, k real, such that $u(x, 0) = f(x)$ and $u \to 0$ as $x^2 + y^2 \to \infty$.

4.4. Basic theorems for a class of self-adjoint equations

In this section we study some of the properties of the more general equation

$$Eu = (au_x)_x + (cu_y)_y + fu = g \qquad (4.1)$$

and, in particular, generalize to equation (4.1), the basic results obtained for Laplace s equation. Most of the latter results were obtained by using the Green's identities displayed in equations (4.10) and (4.11) and, in order to extend these results to equation (4.1), it is necessary first to obtain suitable generalizations of the Green's identities.

The adjoint M^* of an operator M is defined to be such that, for all u and v, $vMu - uM^*v$ is a divergence (this is the same definition as that given in §2.6 for a first-order operator), and the identity

$$vEu - uEv = v\{(au_x)_x + (cu_y)_y + fu\} - u\{(av_x)_x + (cv_y)_y + fv\}$$
$$= (vau_x - uav_x)_x + (vcu_y - ucv_y)_y \qquad (4.39)$$

therefore shows that the operator E is self-adjoint (i.e. $E^* = E$). The divergence theorem states that

$$\int_D (P_x + Q_y)\, dx\, dy = \int_C (lP + mQ)\, ds,$$

where l, m are the direction cosines of the outward normal to C, and P and Q are differentiable functions. Choosing $P = vau_x - uav_x$ and $Q = vcu_y - ucv_y$ gives, on using equation (4.39),

$$\int_D (vEu - uEv)\, dx\, dy = \int_C \{al(vu_x - uv_x) + cm(vu_y - uv_y)\}\, ds. \qquad (4.40)$$

When $a \equiv c$, equation (4.40) simplifies to

$$\int_D (vEu - uEv)\, dx\, dy = \int_C a\left(v\frac{\partial u}{\partial n} - u\frac{\partial v}{\partial n}\right) ds. \qquad (4.41)$$

Equations (4.40) and (4.41) are the required generalizations of equation (4.10) and writing $a \equiv 1$ in equation (4.41) yields equation (4.10).

Setting $u \equiv u^2$ and $v \equiv 1$ in equation (4.40) gives

$$\int_D (Eu^2 - fu^2)\, dx\, dy = \int_C 2u(alu_x + cmu_y)\, ds, \qquad (4.42)$$

and, from this equation and the identity

$$Eu^2 = 2(auu_x)_x + 2(cuu_y)_y + fu^2$$
$$= 2u(au_x)_x + 2u(cu_y)_y + fu^2 + 2(au_x^2 + cu_y^2)$$
$$= 2uEu - fu^2 + 2(au_x^2 + cu_y^2),$$

we see that

$$\int_D uEu \, dx \, dy + \int_D (au_x^2 + cu_y^2 - fu^2) \, dx \, dy$$
$$= \int_C u(alu_x + cmu_y) \, ds. \quad (4.43)$$

Equation (4.43) simplifies, when $a \equiv c$, to

$$\int_D uEu \, dx \, dy + \int_D \{(u_x^2 + u_y^2) - fu^2\} \, dx \, dy = \int_C au\frac{\partial u}{\partial n} \, ds \quad (4.44)$$

and, when $a = 1$, this equation reduces to equation (4.11).

We are now in a position to prove some basic results for solutions of the homogeneous form of equation (4.1) and we assume that the equation is always written so that $a > 0$, $c > 0$.

Theorem 4.4

When $f \leq 0$ the equation

$$Eu = 0, \quad (4.45)$$

with $u = 0$ on C, possesses only the trivial solution $u \equiv 0$.

Proof

Equation (4.43) simplifies, when u satisfies equation (4.45) and vanishes on C, to

$$\int_D (au_x^2 + cu_y^2 - fu^2) \, dx \, dy = 0. \quad (4.46)$$

Our assumptions regarding a, c, and f mean that the integrand in equation (4.46) is never negative, the integral of a non-negative function can only be zero if the integrand is identically zero and therefore

$$au_x^2 + cu_y^2 - fu^2 = 0. \quad (4.47)$$

Equation (4.47) shows that, when f is not identically zero, $u \equiv 0$, but when f is identically zero the most that can be deduced from equation (4.47) is that u is constant. The fact that $u = 0$ on C then shows that u is identically zero.

It is possible, by considering a somewhat less general form of equation (4.45), to generalize Theorem 4.4 to other boundary value problems and we have

Theorem 4.5

When $a \equiv c$, the only solution of equation (4.45) satisfying one of

(i) $\dfrac{\partial u}{\partial n} = 0$ on C and $f < 0$,

or

(ii) $\dfrac{\partial u}{\partial n} + \mu u = 0$ on C, $\mu > 0$, $f \le 0$,

is the trivial solution.

If $\partial u / \partial n = 0$ on C and $f \equiv 0$ then, when $a \equiv c$, u is a constant.

Proof

When $\partial u / \partial n = 0$ the line integral in equation (4.44) vanishes so that, if $Eu = 0$, equation (4.46) (with $a \equiv c$) holds and therefore, by arguments immediately following equation (4.47), $u \equiv 0$ provided that $f < 0$. If $f \equiv 0$ then the most that can be inferred is that u is constant.

Equation (4.44) reduces, when $\partial u / \partial n + \mu u = 0$ on C, to

$$\int_D u Eu \, \mathrm{d}x \, \mathrm{d}y + \int_D \{a(u_x^2 + u_y^2) - f u^2\} \, \mathrm{d}x \, \mathrm{d}y + \int_C \mu u^2 \, \mathrm{d}s = 0.$$

Hence, when $Eu = 0$, the line integral and the remaining double integral in the above equation must vanish separately. The vanishing of the line integral gives $u = 0$ on C; the vanishing of the double integral yields equation (4.47) (with $a \equiv c$); and repeating the previous arguments then gives $u \equiv 0$.

Various uniqueness theorems now follow directly from Theorems 4.4 and 4.5 and we have

Theorem 4.6

There is at most one solution of equation (4.1) with u taking prescribed values on C.

Proof

If we assume that there are two solutions u_1 and u_2 taking the same values on C then the difference $u = u_1 - u_2$ vanishes on C and satisfies equation (4.45). Theorem 4.4 then gives $u \equiv 0$ so that $u_1 \equiv u_2$, hence proving uniqueness.

Theorem 4.7

Equation (4.1) with $a \equiv c$ possesses at most one solution such that

(i) $\partial u / \partial n$ is prescribed on C, $f < 0$,

or

(ii) $\partial u / \partial n$ is prescribed on C, $f \equiv 0$, u is given at one point of D,

or

(iii) $\partial u / \partial n + \mu u$ is prescribed on C, $\mu > 0$, $f \leq 0$.

If there are two solutions u_1 and u_2 with the same normal derivative on C then the difference $u_1 - u_2$ has zero normal derivative on C and satisfies equation (4.45). It then follows from Theorem 4.5 that, when $f < 0$, this difference is zero, thereby proving uniqueness. For $f \equiv 0$ it is only possible to deduce that $u_1 - u_2$ is constant, prescribing u_1 and u_2 at one point gives this constant to be zero. Similarly, when u_1 and u_2 are such that, on C,

$$\frac{\partial u_1}{\partial n} + \mu u_1 = \frac{\partial u_2}{\partial n} + \mu u_2$$

then their difference u satisfies $\partial u / \partial n + \mu u = 0$ on C. Uniqueness then follows from Theorem 4.5.

Theorems 4.6 and 4.7 give no information about existence but it can be proved that a solution exists for each type of problem enumerated, provided that a and c are differentiable and that f and the prescribed values on C are continuous. It is, however, not possible to prescribe $\partial u / \partial n$ completely arbitrarily when $a \equiv c$ and $f \equiv 0$; setting $v \equiv 1$ in equation (4.41) gives

$$\int_D Eu \, dx \, dy = \int_C a \frac{\partial u}{\partial n} \, ds,$$

so that the prescribed values of $\partial u/\partial n$ on C must be such that

$$\int_D g \, dx \, dy = \int_C a \frac{\partial u}{\partial n} \, ds. \qquad (4.48)$$

Theorems 4.6 and 4.7 are the generalizations for equation (4.1) of Theorem 1 and equation (4.48) represents a generalization of the consistency condition of equation (4.13).

Theorems 4.4 and 4.5 have only been proved when $f \le 0$ and for positive f there may be non-trivial solutions of the various homogeneous problems. That this is possible can be seen by considering

$$u_{xx} + u_{yy} + 2\pi^2 u = 0, \qquad 0 < x < 1, \qquad 0 < y < 1,$$

with $u = 0$ on the boundary. Direct substitution shows that a non-trivial solution of this problem is provided by $\sin \pi x \sin \pi y$. Such non-trivial solutions of homogeneous boundary-value problems for homogeneous equations are known as eigenfunctions or eigensolutions and some of the properties of such functions are investigated in §4.5.

It is not possible to generalize the mean-value theorem for Laplace's equation so as to apply to equation (4.1) but there does exist a maximum–minimum principle very similar to that proved in §4.2. This principle, which will not be proved as the proof for $f \le 0$ is somewhat complicated, asserts that solutions of equation (4.45), when $f \le 0$, attain their maximum and minimum values on the boundary (the case $f < 0$ is discussed in Exercises 4.3). It can therefore be deduced, exactly as for Laplace's equation, that the Dirichlet problem for equation (4.1) is well-posed when $f \le 0$.

The notion of fundamental solution can be generalized very simply and $F(x, y, x', y')$ is said to be a fundamental solution of equation (4.1) if

$$E'F = \delta_2(\mathbf{r}' - \mathbf{r}), \qquad (4.49)$$

where the dash denotes that the differential operator acts on the variables x' and y'. Replacing the variables x and y in the integrals of equation (4.40) by x' and y' and setting $v \equiv F$ gives

$$u(x, y) = \int_D Fg \, dx' \, dy' - \int_C \{al(Fu_{x'} - uF_{x'}) + cm(Fu_{y'} - uF_{y'})\} \, ds'.$$

$$(4.50)$$

Equation (4.50) simplifies, when $a \equiv c$, to

$$u = \int_D Fg \, dx' \, dy' - \int_C a\left(F\frac{\partial u}{\partial n'} - u\frac{\partial F}{\partial n'}\right) ds'. \qquad (4.51)$$

The next point to consider is whether the concept of Green's functions (i.e. fundamental solutions which reduce a given problem to evaluating an integral) is still applicable in this case. The form of equation (4.50) shows that, at least for the Dirichlet problem, the appropriate Green's function is that which vanishes when (x', y') is on C and u is then expressed as an integral involving only its boundary values on C. (It can be shown that this Green's function exists for $f \le 0$.) It is not in general possible to state in a simple fashion the boundary condition satisfied by the Green's function appropriate for solving the Neumann problem and we limit our discussion of this problem to the case $a \equiv c$. In these circumstances we see from equation (4.51) that the required Green's function has to have zero normal derivative on C.

The nature of the singularity near $\mathbf{r} = \mathbf{r}'$ of solutions of equation (4.49) is slightly more complicated than for Laplace's equation, and in order to determine the nature of this singularity a non-rigorous approach will be adopted. The singular behaviour is determined by the behaviour near $\mathbf{r} = \mathbf{r}'$ of the terms involving the highest order derivatives on the left-hand side of equation (4.49). Thus the singular behaviour will be the same as that of a solution of

$$a_0 u_{x'x'} + c_0 u_{y'y'} = \delta_2(\mathbf{r}' - \mathbf{r}),$$

where $a_0 = a(x, y)$, $c_0 = c(x, y)$, this equation can be rewritten as

$$u_{X'X'} + u_{Y'Y'} = \delta(x' - x)\delta(y' - y),$$

where $X' = x'/a_0^{\frac{1}{2}}$, $Y = y'/c_0^{\frac{1}{2}}$. The variables occurring in the arguments of the delta functions are not those with respect to which the differentiations are carried out and it is therefore necessary to express the delta functions in terms of X' and Y'. The simple idea of the delta function as a function which vanishes except when its argument is zero, gives

$$\delta(x' - x)\delta(y' - y) = A\delta(X' - X)\delta(Y' - Y),$$

where $X = x/a_0^{\frac{1}{2}}$, $Y = y/c_0^{\frac{1}{2}}$, and A is a constant. A can now be determined from

$$\int_D \delta(x'-x)\delta(y'-y)\,dx'\,dy' = 1 = A\int \delta(X'-X)\delta(Y'-Y)\,dx'\,dy'$$

where D is any region including the point (x, y). Hence

$$1 = A(a_0 c_0)^{\frac{1}{2}} \int_{D'} \delta(X'-X)\delta(Y'-Y)\,dX'\,dY'$$

where D' is any region including the point (X, Y), and therefore $A = 1/(a_0 c_0)^{\frac{1}{2}}$. Thus the singular behaviour of F is the same as that of the singular solution of

$$u_{X'X'} + u_{Y'Y'} = \frac{1}{(a_0 c_0)^{\frac{1}{2}}} \delta(X'-X)\delta(Y'-Y).$$

Comparison of this equation with equation (4.25) shows that, near $\mathbf{r} = \mathbf{r}'$,

$$F \approx \frac{1}{2\pi(a_0 c_0)^{\frac{1}{2}}} \log\left\{\frac{(x-x')^2}{a_0} + \frac{(y-y')^2}{c_0}\right\}^{\frac{1}{2}}.$$

Exercises 4.3

1. Show that when $a > 0$, $b > 0$, and $f < 0$, $Eu < 0$ at any interior point at which u attains a positive maximum. Determine the sign of Eu at any interior point at which u attains a negative minimum.

2. Use the result of the previous exercise to deduce that the only solution of $Eu = 0$ in a bounded region, with $u = 0$ on the boundary, is $u \equiv 0$.

4.5. Properties of eigenfunctions

For particular values of the coefficients, a partial differential equation which admits the trivial solution $u \equiv 0$ can also have non-trivial solutions which satisfy homogeneous boundary conditions. Such solutions are, as mentioned earlier, referred to as eigenfunctions; they can be of use in solving boundary-value problems and we now therefore investigate the properties of eigenfunctions associated with the operator E.

The general problem considered is the solution of

$$Eu + \lambda wu = 0, \qquad B(u) = 0 \text{ on } C, \qquad (4.52)$$

where λ is a constant, w is a given positive continuous function, and E is the operator of equation (4.1), and $B(u)$ denotes any of the quantities u, $\partial u/\partial n$ or $\partial u/\partial n + \mu u$. When B has one of the two latter forms only the special case $a \equiv c$ will be considered. It is also assumed that $f \leq 0$.

Theorems 4.4 and 4.5 show that there are no non-trivial solutions for $\lambda < 0$ but give no information about the case $\lambda > 0$. It is however possible to prove that there exist non-trivial solutions of equation (4.52) with $B(u) = 0$ on C for an infinitely denumerable set of λ's $(\lambda_1 \leq \lambda_2 \cdots)$, provided that a and c are continuously differentiable and f is continuous. The quantities $\lambda_1 \cdots \lambda_n, \ldots$ are known as the eigenvalues and the corresponding solutions are referred to as eigenfunctions.

Theorem 4.8

The eigenvalues of equation (4.52) are real and positive when either

  (i) $u = 0$ on C,
  (ii) $a \equiv c$, $f \not\equiv 0$, $\partial u/\partial n = 0$ on C,
  (iii) $a \equiv c$, $\partial u/\partial n + \mu u = 0 (\mu > 0)$ on C.

For $a \equiv c$ and $f \equiv 0$ the smallest eigenvalue for the case $\partial u/\partial n = 0$ on C is zero and the corresponding eigenfunction is a constant. Also

$$\int_D wu_r u_s \, dx \, dy = 0,$$

where u_r, u_s are eigenfunctions corresponding to different eigenvalues λ_r, λ_s.

Proof

The line integrals in equations (4.40) and (4.41) vanish when u and v are independent solutions of the eigenvalue problems defined by any one of conditions (i) to (iii) and hence, for all such u and v,

$$\int_D vEu \, dx \, dy = \int_D uEv \, dx \, dy. \qquad (4.53)$$

All the coefficients occurring in equation (4.52) and in the forms for $B(u)$, are real so that, for any eigenfunction u,

$$E\bar{u} + \bar{\lambda}w\bar{u} = 0, \tag{4.54}$$

with $B(\bar{u}) = 0$ on C, where the bar denotes the complex conjugate. Equation (4.54) shows that $\bar{u}$ is the eigenfunction corresponding to the eigenvalue $\bar{\lambda}$. Equations (4.52) and (4.54) give $\bar{u}Eu - uE\bar{u} = (\bar{\lambda} - \lambda)w |u|^2$; hence,

$$\int_D (\bar{u}Eu - uE\bar{u}) \, dx \, dy = (\bar{\lambda} - \lambda) \int_D w |u|^2 \, dx \, dy. \tag{4.55}$$

u and $\bar{u}$ are both eigenfunctions of equation (4.52) so that, from equation (4.53), the left-hand side of equation (4.55) is zero and hence $\bar{\lambda} = \lambda$, i.e. λ is real. The fact that the eigenvalues are real also implies that the real and imaginary parts of any eigenfunction are also eigenfunctions and hence all eigenfunctions may be assumed to be real.

When u satisfies equation (4.52), equation (4.43) gives

$$\lambda = \frac{\displaystyle\int_D \{au_x^2 + cu_y^2 - fu^2\} \, dx \, dy - \int_C u(alu_x + cmu_y) \, ds}{\displaystyle\int_D wu^2 \, dx \, dy}, \tag{4.56}$$

and for $a = c$ this reduces to

$$\lambda = \frac{\displaystyle\int_D \{a(u_x^2 + u_y^2) - fu^2\} \, dx \, dy - \int_C au\frac{\partial u}{\partial n} \, ds}{\displaystyle\int_D wu^2 \, dx \, dy}. \tag{4.57}$$

The line integral in equation (4.56) vanishes when $u = 0$ on C, whilst that in equation (4.57) vanishes when either u or $\partial u/\partial n$ is zero on C and it is negative when $\partial u/\partial n + \mu u = 0$ on C. Thus λ is non-negative; $\lambda = 0$ would imply the existence of a non-trivial solution of equation (4.45) with homogenous conditions on the boundary. Theorem 4.5 shows that this can only occur for $a \equiv c$, $f \equiv 0$, and $\partial u/\partial n = 0$ on C and that the eigenfunction is then a constant.

The eigenfunctions u_r and u_s satisfy

$$Eu_r + \lambda_r w u_r = 0,$$

$$Eu_s + \lambda_s w u_s = 0,$$

so that

$$u_s E u_r - u_r E u_s = (\lambda_s - \lambda_r) w u_r u_s.$$

Hence, after using equation (4.53),

$$(\lambda_s - \lambda_r) \int_D w u_s u_r \, dx \, dy = 0.$$

λ_s and λ_r are not equal, and therefore

$$\int_D w u_r u_s \, dx \, dy = 0. \tag{4.58}$$

Functions $u_1, u_2, \ldots$ satisfying a relation of the form of equation (4.58) are said to form an orthogonal set and the relation is referred to as an orthogonality relation.

Determining the eigenvalues and eigenfunctions for a particular equation is not usually easy and no general methods are available for their exact determination though variational methods can be useful in giving approximate values for the smaller eigenvalues (cf. §9.3). Separation of variables can be helpful in finding some eigenfunctions but it is not always easy to show that any method gives all the eigenvalues and this kind of demonstration often requires use of fairly powerful theorems in analysis. The general processes involved can be seen by considering the eigenvalue problem for

$$u_{xx} + u_{yy} + \lambda u = 0$$

in the square $0 < x < 1$, $0 < y < 1$, with $u = 0$ on the boundary of the square.

Direct substitution shows that some of the eigenfunctions u_{mn} and corresponding eigenvalues λ_{mn} are

$$u_{mn} = \sin m\pi x \sin n\pi y,$$

$$\lambda_{mn} = \pi^2(m^2 + n^2),$$

where m and n are integers.

It is not obvious that all the eigenvalues or all the eigenfunctions are of the above form and we first assume the existence of an eigenfunction v not corresponding to an eigenvalue of the form λ_{mn}. The orthogonality condition of equation (4.58) shows that v must satisfy

$$\int_0^1 v \sin m\pi x \sin n\pi y \, \mathrm{d}x \, \mathrm{d}y = 0, \quad \text{for all } m, n.$$

This can be written as

$$\int_0^1 w_n(x)\sin m\pi x \, \mathrm{d}x = 0, \quad \text{for all } m, n \qquad (4.59)$$

where

$$w_n(x) = \int_0^1 v \sin n\pi y \, \mathrm{d}y. \qquad (4.60)$$

It follows from the theory of Fourier series that equation (4.59) implies $w_n \equiv 0$ (strictly the theory gives $w_n \equiv 0$ almost everywhere, but v has to be twice differentiable and hence w_n has at least to be continuous and thus it must be identically zero). Repeating the same argument for equation (4.60) gives $v \equiv 0$ and hence all the eigenvalues are of the form λ_{mn}.

The other possibility is that there is an eigenfunction v corresponding to a particular λ, e.g. λ_{pq}, and which is not of the form u_{pq}. In this case v cannot be orthogonal to u_{pq}; otherwise by the above argument it would be identically zero; hence a constant c can be found so that $v - cu_{pq}$ is orthogonal to all the eigenfunctions. Applying the previous arguments to $v - cu_{pq}$ gives $v \equiv cu_{pq}$. Thus all the eigenfunctions are of the form u_{mn} with the corresponding eigenvalues being of the form λ_{mn}.

The eigenfunctions of equation (4.52) may be used to obtain a formal series solution for u satisfying equation (4.1), i.e.

$$Eu = g, \qquad (4.1)$$

with $B(u) = h$ on C, where $B(u) = \mu u + \partial u/\partial n$.

For boundary-value problems with $B(u) \not\equiv u$ only the special case $a \equiv c$ is considered. For simplicity it will be assumed that $h \equiv 0$; this involves no loss of generality as, in principle, it is possible to find a twice-differentiable function u_1 such that, on C,

$B(u_1) = h$, and writing $u = u_1 + u_2$ shows that the equation for u_2 is of the same form as equation (4.1) (g is replaced by $g - Lu_1$) and, on C, $B(u_2) = 0$.

It is necessary to quote one further general result relating to the eigenfunctions of equation (4.52), namely that, for any continuously differentiable function g satisfying the same condition as the eigenfunctions u_r on C, it is possible to find constants α_r such that

$$g = \sum_{r=1}^{\infty} \alpha_r u_r \qquad (4.61)$$

holds uniformly in D and on C. The coefficients α_r can be found by multiplying equation (4.61) by wu_s and integrating over D; this gives

$$\int_D gwu_s \, \mathrm{d}x \, \mathrm{d}y = \sum_{r=1}^{\infty} \alpha_r \int_D wu_r u_s \, \mathrm{d}x \, \mathrm{d}y,$$

assuming that summation and integration may be interchanged. The orthogonality relation of equation (4.58) now gives

$$\alpha_r \int_D wu_r^2 \, \mathrm{d}x \, \mathrm{d}y = \int_D gwu_r \, \mathrm{d}x \, \mathrm{d}y.$$

If g does not satisfy the same condition as the u_r on C then equation (4.61) will not be valid on C; a similar situation occurs in the theory of Fourier series where, for example, the sine series of a function which is non-zero at the end points of the interval only converges to the given function in the open interval.

The solution of equation (4.1) with $B(u) = 0$ will satisfy the same conditions as the u_r on the boundary and hence an expansion of the form

$$u = \sum_{r=1}^{\infty} \beta_r u_r \qquad (4.62)$$

will be valid in D and on C. Effectively, in writing u in the form of the right-hand side of equation (4.62), we are choosing a form of solution which automatically satisfies the boundary condition and it then only remains to satisfy the equation. The β's are given by

$$\beta_r \int_D wu_r^2 \, \mathrm{d}x \, \mathrm{d}y = \int_D uwu_r \, \mathrm{d}x \, \mathrm{d}y. \qquad (4.63)$$

The right-hand side of equation (4.63) can, from equation (4.52), be written in the alternative form

$$-\frac{1}{\lambda_r}\int_D uEu_r\,dx\,dy;$$

u and u_r both satisfy the same condition on C and therefore equation (4.53) gives

$$\int_D uEu_r\,dx\,dy=\int_D u_rEu\,dx\,dy=\int_D gu_r\,dx\,dy.$$

Hence

$$\beta_r\int_D wu_r{}^2\,dx\,dy=-\frac{1}{\lambda_r}\int_D gu_r\,dx\,dy. \qquad (4.64)$$

An alternative method of deriving equation (4.64) is to substitute the expansion of equation (4.62) directly into equation (4.64) and use the orthogonality relation to determine the β_r. Equations (4.62) and (4.64) show that a formal series solution of the problem is provided by

$$u=-\sum_{r=1}^{\infty}\frac{1}{\lambda_r}u_r(x,y)\int_D g(x',y')u_r(x',y')\,dx'\,dy'\Big/\int_D wu_r{}^2\,dx'\,dy'$$

$$=\int_D g(x',y')\left\{-\sum_{r=1}^{\infty}\frac{u_r(x,y)u_r(x',y')}{\lambda_r\int wu_r{}^2\,dx\,dy}\right\}dx'\,dy'. \qquad (4.65)$$

The solution is a formal one in the sense that integration and summation were assumed interchangeable in the above derivation. In any particular example it is necessary to verify that the series obtained is convergent and twice differentiable, and that the series obtained for Lu by term-by-term differentiation converges to g in D. The necessary detailed calculation is shown for one particular case in the following example.

Example 4.6

Solve, in the square $0<x<1$, $0<y<1$,

$$u_{xx}+u_{yy}=g,$$

with $u=0$ on the boundary of the square.

The eigenfunctions have been shown to be of the form $\sin m\pi x \sin n\pi y$ with corresponding eigenvalues $\lambda_{mn} = \pi^2(m^2 + n^2)$. In this case $w = 1$ and equation (4.64) shows that the formal solution is given by

$$u = \sum_{m=1}^{\infty} \sum_{n=1}^{\infty} A_{mn} \sin m\pi x \sin n\pi y, \qquad (4.66)$$

where

$$A_{mn} = -\frac{4}{\lambda_{mn}} \int_0^1 \int_0^1 g \sin m\pi x \sin n\pi y \, dx \, dy. \qquad (4.67)$$

It is now necessary to show that the series for u, and that for $u_{xx} + u_{yy}$, both converge; this requires estimating A_{mn}. On assuming that g is such that $|g| < M$ it follows from equation (4.67) that $|A_{mn}| < 4M/\lambda_{mn}$ so that the series for u converges. However, that for $u_{xx} + u_{yy}$ has coefficients of order $(m^2 + n^2)A_{mn}$ and hence the above bound for A_{mn} is not sufficient. Establishing a finer bound requires more information about g and we now assume that it is twice differentiable with $|g_{xx} + g_{yy}| < M'$ and that $g = 0$ on the boundary of the square. As $\sin m\pi x \sin n\pi y$ and g vanish on the boundary it follows from equation (4.53) that

$$\int_0^1 \int_0^1 g \left(\frac{\partial^2}{\partial x^2} + \frac{\partial^2}{\partial y^2} \right) \sin m\pi x \sin n\pi y \, dx \, dy$$

$$= \int_0^1 \int_0^1 \sin m\pi x \sin n\pi y (g_{xx} + g_{yy}) \, dx \, dy$$

thus giving

$$\int_0^1 \int_0^1 g \sin m\pi x \sin n\pi y \, dx \, dy$$

$$= \frac{-1}{\pi^2(m^2 + n^2)} \int_0^1 \int_0^1 \sin m\pi x \sin n\pi y (g_{xx} + g_{yy}) \, dx \, dy.$$

Therefore

$$\int_0^1 \int_0^1 g \sin m\pi x \sin n\pi y \, dx \, dy < \frac{M'}{\lambda_{mn}}$$

and $|A_{mn}| < 4M'/\lambda_{mn}^2$. The series for $u_{xx} + u_{yy}$ can now also be

proved to be uniformly convergent and hence, with the above
assumptions about g, the formal solution has been shown to be
the correct solution.

It is possible in specific cases to relax the conditions on g, this
can be seen by taking $g \equiv xy$ thus giving

$$u = \frac{4}{\pi^4} \sum_{m=1}^{\infty} \sum_{n=1}^{\infty} (-1)^{m+n+1} \frac{\sin m\pi x \sin n\pi y}{mn(m^2+n^2)}.$$

This series is uniformly convergent in $0 \le x \le 1$, $0 \le y \le 1$, but that
for $u_{xx} + u_{yy}$ only converges for $0 < x < 1$, $0 < y < 1$; however this
is still sufficient to ensure that the equation is satisfied within D.

The particular method used in Example 4.6 is of more formal
than practical interest as the solution is obtained in the form of a
double series which is only slowly convergent. In this particular
problem a more direct approach is that described following
Example 7.4.

Equation (4.63) can also be used to give formal eigenfunction
expansions of the Green's functions previously defined for equation (4.1). The solution of equation (4.1) with $B(u) = 0$ on C can
be obtained by replacing F in equation (4.51) by the appropriate
Green's function G. The line integrals will vanish as the Green's
function and u both satisfy homogeneous conditions on C, and
hence

$$u(x, y) = \int G(x, y, x', y')g(x', y')\, dx'\, dy'. \qquad (4.68)$$

Comparison of equations (4.65) and (4.68) gives

$$G(x, y, x', y') = -\sum_{r=1}^{\infty} \frac{u_r(x, y)u_r(x', y')}{\lambda_r \displaystyle\int_D wu_r^2\, dx\, dy}. \qquad (4.69)$$

It would be possible to use equation (4.69) with the eigenfunctions of Example (4.6) to obtain an alternative expression for the
Green's function of Example 4.4, but the resulting double series
is not as useful as the form of solution obtained by the method
described in Example 4.4.

Exercise 4.4

1. Determine, for Laplace's equation, the complete set of eigenfunctions which vanish on $r = (x^2 + y^2)^{\frac{1}{2}} = 1$.
[It may be assumed that, in the notation of §1.5,

$$\int_0^1 x J_m(j_{nm}x)f(x) = 0, \, n = 0, \ldots, \text{ implies } f \equiv 0.]$$

2. Use the result of the previous exercise to obtain the complete eigenfunction expansion for the solution of

$$\nabla^2 u = x^2 - y^2, \quad \text{in} \quad 0 \le r < 1,$$

with $u = 0$ when $r = 1$.

3. Show that all the values of λ such that, in a region D bounded by a closed curve C,

$$\nabla^4 u + \lambda u = 0, \qquad u = \nabla^2 u = 0 \quad \text{on} \quad C,$$

are real and negative and that the eigenfunctions u_n, u_m, corresponding to different eigenvalues λ_n, λ_m, satisfy

$$\int_D u_n u_m \, ds = 0.$$

4. If the eigenfunctions w_n satisfying

$$\nabla^2 w_n + \mu_n w_n = 0$$

in D and vanishing on C are such that

$$\int_D \psi w_n \, ds = 0, \quad \text{all} \quad n,$$

implies $\psi \equiv 0$, then show that all the eigenvalues in the previous exercise are given by $\lambda_n = -\mu_n^2$.

4.6. Discontinuous and weak solutions

For the self-adjoint operator E defined in equation (4.1) we have that, for all functions u and v,

$$\int_D (vEu - uEv) \, dx \, dy = \int_C [al(vu_x - uv_x) + mc(vu_y - uv_y)] \, ds,$$

(i.e. equation (4.40)) and this can be written in a slightly simpler form as

$$\int_D (vEu - uEv)\, dx\, dy = \int_C a\left(v\frac{\partial u}{\partial \nu} - u\frac{\partial v}{\partial \nu}\right) ds, \qquad (4.70)$$

where $\partial/\partial\nu$ denotes the derivative in the direction of the vector ν inclined at an angle θ to the x-axis where $\tan\theta = mc/al$. Taking u in equation (4.70) to be a solution of

$$Eu = g$$

and v to be any function which, together with its derivatives, vanishes on C gives

$$\int_D vg\, dx\, dy = \int_D uEv\, dx\, dy. \qquad (4.71)$$

Reversing the argument shows that if u is a twice-differentiable function satisfying equation (4.71) for all v satisfying the stated conditions, then u is a solution equation (4.1) (i.e. $Eu = g$). Therefore, exactly as for a first-order equation, a weak solution of equation (4.1) is defined to be a solution of equation (4.71).

It was shown in §2.6 that a weak solution of a first-order equation can only be discontinuous across characteristics and, as an elliptic equation has no characteristics associated with it, this suggests that an elliptic equation cannot have weak solutions. It can in fact be proved that, if the functions a, c, f, and g satisfy the conditions of Theorem 4.1, any weak solution is a solution in the normal sense (a strong solution) and that it is twice differentiable. We now prove that the elliptic equation (4.1) cannot have a weak solution discontinuous across a curve Γ. Assuming the existence of such a curve would give, using arguments similar to those following equation (2.39),

$$\int_\Gamma a\left\{v\left[\frac{\partial u}{\partial \nu}\right] - [u]\frac{\partial v}{\partial \nu}\right\} ds = 0, \qquad (4.72)$$

where the square brackets denote jumps in the variables. It follows that, unless ν is tangential to Γ, $\partial v/\partial\nu$ and v are independent on Γ and equation (4.72) can only hold for all v if the jumps in both u and $\partial u/\partial\nu$ are zero. The direction cosines of the tangent to Γ are $-m$ and l and hence ν being tangent to Γ means that

$$\frac{mc}{al} = -\frac{l}{m}.$$

This is impossible for an elliptic equation since $c/a > 0$. Equation (4.1) is hyperbolic when a/c is negative and hence Γ is a possible curve of discontinuity if $l^2/m^2 = -c/a$. Further $(l/m)^2$ is the square of the slope of the tangent to Γ, i.e.

$$\left(\frac{dy}{dx}\right)^2 = -\frac{c}{a}, \quad \text{where } y = y(x) \text{ is the equation of } \Gamma.$$

Comparison with equation (3.10) now shows that for a hyperbolic equation of the general form of equation (4.1), the possible curves of discontinuity are the characteristics.

A further interesting point relating to Dirichlet problems for elliptic equations is that there exist differentiable solutions even when the boundary data is discontinuous. This property was illustrated for Laplace's equation by using equation (4.8) to derive a solution for the case when the boundary values were discontinuous.

4.7. Extension to differential operators that are not self-adjoint

The discussion hitherto has been confined to equations where the differential operator is self-adjoint and we now briefly look at the consequences of removing this restriction. In order to simplify the analysis we only consider the general elliptic equation in its canonical form, i.e.

$$E_2 u = u_{xx} + u_{yy} + d u_x + e u_y + f u = g. \tag{4.73}$$

The adjoint operator E_3 is defined to be such that, for all differentiable u and v, $v E_2 u - u E_3 v$ is a divergence and E_3 is therefore defined by

$$E_3 v = v_{xx} + v_{yy} - (dv)_x - (ev)_y + fv$$

and

$$v E_2 u - u E_3 v = (v u_x - u v_x + d u v)_x + (v u_y - u v_y + e u v)_y. \tag{4.74}$$

Applying the divergence theorem to equation (4.74) gives

$$\int_D (vE_2u - uE_3v)\, \mathrm{d}x\, \mathrm{d}y = \int_C [e(vu_x - uv_x + duv)$$
$$+ m(vu_y - uv_y + euv)]\, \mathrm{d}s.$$
$$= \int_C \left[v\frac{\partial u}{\partial n} - u\frac{\partial v}{\partial n} + uv(ld + me) \right] \mathrm{d}s.$$

$$(4.75)$$

Equation (4.75) is a generalization of the Green's identity of equation (4.10) and, like the latter, can be used to express u, at any point, in terms of integrals involving u and $\partial u/\partial n$ on C. We see from equation (4.75) that, in order to obtain such representations, v must be a solution of

$$E_3v = \delta_2(\mathbf{r}' - \mathbf{r}), \qquad (4.76)$$

and therefore the Green's functions for solving equation (4.73) are those associated with the adjoint operator. This, of course, represents a major change from the situation for self-adjoint operators. We also note from equation (4.75) that the Green's function appropriate for solving the Dirichlet problem will vanish on C, whilst that appropriate to solving the Neumann problem will have, on C, to satisfy the more complicated condition

$$\frac{\partial v}{\partial n} = v(ld + me).$$

We shall not pursue any further the use of Green's functions for solving non-self-adjoint equations as the situation can get rather complicated.

Setting $u \equiv u^2$ and $v \equiv 1$ in equation (4.75) yields a generalization of equation (4.44) and this can be used to deduce some uniqueness theorems for equation (4.73). Direct differentiation gives

$$E_2u^2 = 2uE_2u - fu^2 + 2(u_x^2 + u_y^2), \qquad E_31 = f - d_x - e_y,$$

and therefore

$$\int_D [uE_2u - u^2\{f - \tfrac{1}{2}d_x - \tfrac{1}{2}e_y\}]\, \mathrm{d}x\, \mathrm{d}y + \int_D (u_x^2 + u_y^2)\, \mathrm{d}x\, \mathrm{d}y$$
$$= \int u\left[\frac{\partial u}{\partial n} + \frac{u}{2}(ld + me) \right] \mathrm{d}s.$$

Comparison of this equation with equation (4.44) and repeating the arguments of Theorems 4.4 and 4.6 shows that the Dirichlet problem for equation (4.73) will have a unique solution provided that

$$f - \tfrac{1}{2}(d_x + e_y) \leq 0.$$

Bibliography

Gilbarg, D. and Trudinger, H. S. (1977). *Elliptic partial differential equations of the second order.* Springer, New York.
Miranda, C. (1970). *Partial differential equations of elliptic type.* Springer, New York.

5 Hyperbolic equations

THE theoretical analysis of the solutions of problems for linear hyperbolic equations in two independent variables is technically somewhat easier than for elliptic equations, primarily because most of the properties of the solutions are generalizations of those for the archetypal equation (i.e. the one-dimensional wave equation), which is itself, in general, much easier to solve than is Laplace's equation. It is in fact possible, by generalizing the analysis of §3.2, to obtain, fairly easily, a considerable amount of semi-qualitative information concerning the solution of various types of problems for the linear and half-linear hyperbolic equations. This analysis is carried out in §5.1 and a formal existence and uniqueness theorem for the Cauchy problem is proved in §5.2. Green's functions again provide a useful method of solving problems for hyperbolic equations and, in §5.3, Green's functions associated with the one-dimensional wave equation are defined and used to obtain integral representations of the solution. In §5.4 the concept of Green's functions is generalized to derive integral representations of solutions of the general linear hyperbolic equation. Historically, integral representations of solutions of the general equation were first obtained by a method due to Riemann and which involves introducing a function known as the Riemann function. §5.4 concludes with a description of Riemann's method and the relationship between Riemann functions and Green's functions is also established. The possibility of obtaining weak solutions of linear hyperbolic equations is discussed in §5.5 where it is shown that there can exist weak solutions which are discontinuous across characteristics and that discontinuities in boundary data are propagated along characteristics.

5.1. General properties

The Cauchy–Kowalesky theorem (see §3.1) asserts that, provided the data is analytic, there exists, in the neighbourhood of a non-characteristic curve on which Cauchy conditions are prescribed, a power-series expansion of the solution of a linear hyperbolic equation. It can in fact be proved independently of the Cauchy–Kowalesky theory that, even for non-analytic data, if such conditions are prescribed on a smooth curve, which is nowhere parallel to a characteristic, then a solution exists in some neighbourhood of the curve. The detailed proof is given in §5.2 where it is also proved that the solution is unique. It is also possible to prove that the solution of a hyperbolic equation can be determined in some neighbourhood of two intersecting characteristics when the dependent variable is prescribed on these characteristics. In some senses hyperbolic equations are somewhat simpler than elliptic ones in that it is possible, assuming the existence of a solution, to obtain useful qualitative information concerning the nature of this solution by a slight extension of the method used in §3.2. We therefore in this section first consider some of the general features of solutions of hyperbolic equations and defer the formal proof of existence to the following section.

The general equation considered is

$$u_{xy} = F(x, y, u, u_x, u_y), \tag{5.1}$$

where F is some given function. Equation (5.1) is the canonical form of the general almost-linear hyperbolic equation and the characteristics are lines parallel to the coordinate axes. Imposing Cauchy conditions is equivalent to prescribing u and its first partial derivatives on a curve and we therefore attempt to determine u satisfying equation (5.1) and such that, on a curve which is nowhere characteristic,

$$u_x = p(x), \qquad u_y = q(y), \qquad u \equiv u_1(x) \equiv u_2(y).$$

Such a curve is defined by $y = y(x)$, (or equivalently $x = x(y)$); hence any function of x on the curve can also be expressed as a function of y and the above choices are the most convenient for the following discussion. The functions p, q, and u_1 are not independent since $u_1(x) = u(x, y(x))$ and differentiating this identity gives

$$\frac{\mathrm{d}u_1}{\mathrm{d}x} = p + qy_x.$$

We first look at a slight generalization of the initial value problem of §3.2, namely the determination at $P(x_0, y_0)$, for a given function $E(x, y)$, of that solution u of

$$u_{xy} = E(x, y), \tag{5.2}$$

which satisfies the above Cauchy conditions on an arbitrary curve Γ. The points $A(x_A, y_A)$ and $B(x_B, y_B)$ in Fig. 5.1 are the points of intersection of Γ with the characteristics drawn from P.

The solution can, in view of the linearity of equation (5.2), be written as the sum of the solution of

$$u_{xy} = 0, \tag{5.3}$$

with u satisfying the given conditions on Γ, and the solution of equation (5.2) which satisfies homogeneous Cauchy conditions on Γ. Integrating equation (5.3) from $y = y(x)$ to $y = y_0$ (i.e. along CD) gives

$$u_x(x, y_0) - u_x(x, y(x)) = 0,$$

i.e.

$$u_x(x, y_0) = p(x).$$

A further integration along BP now gives

$$u(x_0, y_0) - u_B = \int_{x_B}^{x_0} p(x) \, \mathrm{d}x. \tag{5.4}$$

It would have been equally permissible to integrate equation

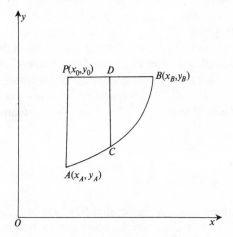

Fig. 5.1

(5.3) with respect to x first and then with respect to y and this would have given

$$u(x_0, y_0) - u_A = \int_{y_A}^{y_0} q(y)\, dy. \qquad (5.5)$$

We obtain, on adding equations (5.4) and (5.5),

$$u(x_0, y_0) = \tfrac{1}{2}(u_A + u_B) + \tfrac{1}{2}\int_{x_B}^{x_0} p(x)\, dx + \tfrac{1}{2}\int_{y_A}^{y_0} q(y)\, dy,$$

which is a more symmetric representation of the solution. The last two terms in the above equation are integrals of the given values of u_x and u_y on Γ and hence the solution can be rewritten as

$$u = \tfrac{1}{2}(u_A + u_B) + \tfrac{1}{2}\int_{\Gamma} (q\, dy - p\, dx), \qquad (5.6)$$

where the positive sense of integration is taken from A to B. It can now be verified by direct differentiation (i.e. reversing the above argument) that u defined by equation (5.6) satisfies equation (5.3) when u_1 is twice differentiable and p and q are differentiable. The form of equation (5.6) shows immediately that any discontinuity in u on Γ will be propagated along the characteristics through the point of discontinuity. Further, since $u_x(x_0, y_0) = p(x_0)$ and $u_y(x_0, y_0) = q(y_0)$, we see that discontinuities on Γ in p and q will be propagated along lines parallel to Oy and Ox respectively and passing through the point of discontinuity.

The solution of equation (5.2) satisfying homogeneous Cauchy conditions on Γ can be obtained by first integrating with respect to y from C to D and then integrating with respect to x. This is exactly the method used in §3.2 and for the configuration of Fig. 5.1 gives

$$u = -\int_{APB} E(x, y)\, dx\, dy. \qquad (5.7)$$

If y were a monotonically decreasing function of x on Γ then the integral in equation (5.7) would not have to be preceded by a minus sign. The complete solution of the Cauchy problem for equation (5.2) is therefore

$$u(x_0, y_0) = \tfrac{1}{2}(u_A + u_B) + \tfrac{1}{2}\int_{\Gamma} (q\,\mathrm{d}y - p\,\mathrm{d}x) - \int_{APB} E(x, y)\,\mathrm{d}x\,\mathrm{d}y,$$
(5.8)

and it can be verified by direct substitution and differentiation that u defined by equation (5.8) satisfies equation (5.2) (in the variables x_0, y_0) and the conditions on Γ.

The above verification is still valid even if E is replaced by $F(x, y, u, u_x, u_y)$ and in this case $u(x_0, y_0)$ defined by

$$u = \tfrac{1}{2}(u_A + u_B) + \tfrac{1}{2}\int_{\Gamma} (q\,\mathrm{d}y - p\,\mathrm{d}x) - \int_{APB} F(x, y, u, u_x, u_y)\,\mathrm{d}x\,\mathrm{d}y,$$
(5.9)

satisfies the conditions on Γ and is a solution of equation (5.1). F is dependent on u, u_x, and u_y and is not known explicitly so that equation (5.9) does not constitute a solution of our problem but is merely a method of representing the solution. Equation (5.9) is an integro-differential equation for u and forms the basis of the existence proof of §5.2. However it can, assuming the existence of a solution, also be used to obtain some useful qualitative information concerning the solution of equation (5.1). It should be remembered in the following discussion that, for non-linear equations, a solution of equation (5.9) can only be shown to exist in some neighbourhood of Γ but that, for linear equations, this restriction can be removed.

Equation (5.9) shows that the solution at a point P depends only on the conditions within the area APB so that the domain of dependence of P is the whole of APB, i.e. the region bounded by Γ and the characteristics through P. Also, for linear equations, the data on the arc AB of Γ determines the solution everywhere in, and on the boundary of, the sector APB and hence APB is part of the domain of determinacy of AB (the local nature of the existence theorem means that this statement can only hold for non-linear equations in a region near to Γ). There is also no *a priori* reason why P has to be on one particular side of Γ and for linear equations the complete domain of determinacy of the arc AB is the rectangular region shown in Fig. 5.2.

Similarly the domain of influence of AB is, for the linear equation, the shaded region of Fig. 5.3. In problems where one of

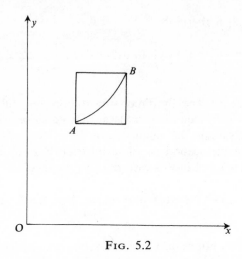

Fig. 5.2

the independent variables is time it is normally assumed that the future cannot affect the past and that initial conditions are only propagated in the sense of increasing time. This implies the existence of a preferential side to a curve on which conditions are imposed and in this case, as in §3.2, the domains of influence and determinacy lie on only one side of a given arc. In cases where time is one of the variables it is customary to define the positive

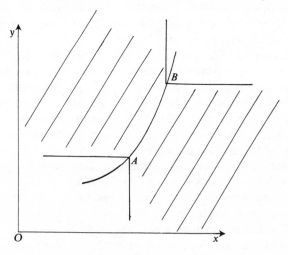

Fig. 5.3

sense along a charactersitic to be in the direction of increasing time. The positive senses of the characteristic coordinates $\phi(=x-ct)$ and $\psi(=x+ct)$ in Fig. 3.4 are thus in the direction of ϕ-decreasing and ψ-increasing respectively. Once such a convention has been established it is possible to classify curves in two dimensions as being either 'space-like' or 'time-like'. A 'space-like' curve is defined to be one such that the characteristics drawn in the positive sense from a point of it are on the same side of the curve, whilst for a 'time-like' curve they are on opposite sides. The line AB in Fig. 3.5 is thus 'space-like' whilst the line $\phi = -\psi$ (which corresponds to $x = 0$) is 'time-like'.

Examination of the form of the solution of equation (5.2) with different boundary conditions can also suggest other types of boundary-value problems that can be solved, and provide a method of reducing these boundary-value problems to an integro-differential equation similar to equation (5.9). Integrating equation (5.2) with respect to y from $y = 0$ to $y = y_0$ (i.e. along CD in Fig. 5.4) gives

$$u_x(x, y_0) - u_x(x, 0) = \int_0^{y_0} E(x, y) \, dy,$$

and after a further integration with respect to x we obtain

$$u(x_0, y_0) - u(0, y_0) - u(x_0, 0) + u(0, 0) = \int_0^{x_0} \int_0^{y_0} E(x, y) \, dx \, dy$$

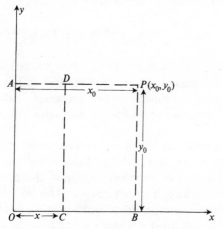

Fig. 5.4

or, in terms of the configuration shown in Fig. 5.4,

$$u_P = u_A + u_B - u_O + \int_{OBPA} E(x, y) \, dx \, dy, \qquad (5.10)$$

where the suffixes denote values evaluated at the corresponding points. Equation (5.10) shows that a solution of equation (5.2) can be obtained when u is prescribed on $x = 0$ and $y = 0$ (i.e. along two intersecting characteristics) and that the solution is defined in the region bounded by the arcs of the characteristics on which u is prescribed and by the characteristics through the end points of these arcs. It has been shown above (and is discussed in detail in §5.4) that discontinuities in boundary data are propagated along the characteristics through the point of discontinuity and such discontinuities can be avoided (so that u is a normal solution of a partial differential equation) be requiring that the prescribed values of u on the x- and y-axes are twice differentiable and are equal at O. Equation (5.10) also shows that the problem of solving equation (5.1), with u prescribed on $x = 0$ and $y = 0$, reduces to solving the integro-differential equation

$$u_P = u_A + u_B - u_O + \int_{OBPA} F(x, y, u, u_x, u_y) \, dx \, dy \qquad (5.11)$$

or, equivalently,

$$u(x_0, y_0) = u(0, y_0) + u(x_0, 0) - u(0, 0)$$
$$+ \int_0^{x_0} \int_0^{y_0} F(x, y, u, u_x, u_y) \, dx \, dy. \qquad (5.12)$$

The method used in §5.2 can be applied to demonstrate the existence and uniqueness of the solution of equation (5.12). The solution will be determined in the rectangle, two of whose adjacent sides are the segments of the axes on which u is prescribed.

A similar problem for which an explicit solution of equation (5.2) can be obtained is that of determining u in a region bounded by a characteristic and by an intersecting non-characteristic curve Γ when u is prescribed on both curves. This problem is often referred to as the Goursat problem and a possible configuration is shown in Fig. 5.5, where it is assumed that u is known on Γ and on the positive x-axis. Requiring that the prescribed values of u

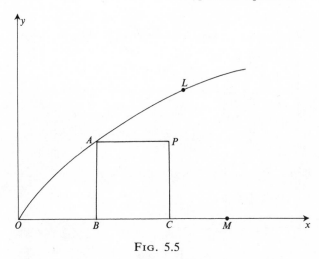

Fig. 5.5

on Γ and on the x-axis are twice differentiable is sufficient to ensure continuity of the solution. We have that

$$\int_{PABC} u_{xy}\, \mathrm{d}x\, \mathrm{d}y = u_P + u_B - u_A - u_C,$$

so that

$$u_P = u_A + u_C - u_B + \int_{PABC} E(x, y)\, \mathrm{d}x\, \mathrm{d}y \qquad (5.13)$$

which, as u_A, u_B, and u_C are known, provides the explicit solution of the Goursat problem. Equation (5.13) shows that, for equation (5.1), the Goursat problem reduces to the solution of

$$u_P = u_A + u_C - u_B + \int_{PABC} F(x, y, u, u_x, u_y)\, \mathrm{d}x\, \mathrm{d}y,$$

and existence of the solution of this equation can be demonstrated by means of the method described in §5.2. If u is prescribed on the arc OL of Γ and the segment OM of the x-axis (as shown in Fig. 5.5) then it will be determined everywhere in the rectangle whose adjacent sides are the y-characteristic LN drawn from L to Ox and ON. If M is to the left of L then the solution is determined in the rectangle whose two adjacent sides are OM and the y-characteristic drawn from M to Γ.

It is possible, assuming the existence of the solution of the

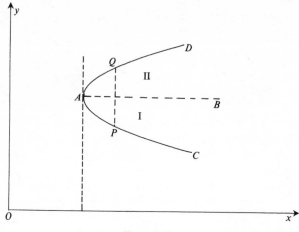

Fɪɢ. 5.6

Goursat problem, to determine the conditions that can be imposed on a curve of the form *DAC* shown in Fig. 5.6. The
rectilinear U-shaped boundary shown in Fig. 3.5, and associated
with the mixed initial-value boundary value problem for the wave
equation, is evidently a degenerate form of *DAC*.

In Fig. 5.6, *A* denotes the point at which the curve is tangent
to a characteristic and *AB* denotes the other characteristic
through the point of tangency. For a linear differential equation
the solution is determined completely in region I (including *AB*)
by Cauchy conditions on *AC*. Hence prescribing *u* on *AD* would
give a Goursat problem in region II and thus a solution can be
obtained in both regions. It would of course be equally possible
to have Cauchy conditions on *AD* and *u* prescribed on *AC*. In
general, both problems are reasonable but, for problems involving time as an independent variable, only one formulation is
physically sensible. In order to see which is the appropriate
problem to consider it is necessary to see qualitatively why it is
inappropriate to prescribe Cauchy conditions everywhere on
CAD. On any characteristic (e.g. any line $x =$ constant) the
differential equation effectively reduces to an ordinary one relating u, u_x and u_y, e.g.

$$(u_x)_y = F(x, y, u, u_x, u_y),$$

and thus, if u, u_x, u_y are known at one point of a characteristic, the equation will imply some relationship between these quantities at all points of the characteristic (the relations $cp \pm q =$ constant of §3.2 are particular cases of such a relationship). Therefore, when Cauchy conditions are prescribed on AC there will be some relationship between u and its first derivatives along any characteristic such as PQ, therefore prescribing all these quantities on AD would then lead to an inconsistency. Thus conditions on AC have an influence on AD (or vice versa) and, for problems involving time, the Cauchy conditions must be prescribed on the arc which is effectively earlier in time. If the positive x- and y-directions in Fig. 5.6 correspond to the direction of increasing time then the Cauchy conditions must be prescribed on the arc AC. Alternatively, using the nomenclature of 'space-like' and 'time-like' curves, the Cauchy conditions must be imposed on 'space-like' curves and u is to be prescribed on 'time-like' curves; this is also clearly the case in Fig. 3.5.

The above arguments are still valid if the slope of CAD is discontinuous at A, provided that CA and DA still remain on opposite sides of the line drawn from A parallel to the x-axis. A simple example of a problem of this type is the initial-value–boundary-value problem for the one-dimensional wave equation in $x \geq 0$ with $u(0, t) = 0$. The solution of this problem is obtained in §3.4 by integrating along the characteristics. The Cauchy conditions on AC determine the first derivatives of u at A and prescribing u on AD gives one relation between the first derivatives at A and therefore, in order to avoid discontinuities, the latter relation must be satisfied by the derivatives obtained from Cauchy conditions on AC.

It is also possible, by successive application of the above arguments, to demonstrate the existence of the solution of equation (5.1) in the U-shaped region bounded by a piecewise smooth curve $DACE$ of the general form shown in Fig. 5.7. The arcs CE, CA, and AD are assumed to be non-characteristic and it is assumed that Cauchy conditions are imposed on AC whilst u is prescribed on AD and CE. AH and CF are the characteristics drawn from A and C into the region and G denotes their point of intersection. The Cauchy conditions on AC determine u in the region AGC and on AG and CG. The problems for determining u in AGF and CGH are of Goursat-type and hence u can be

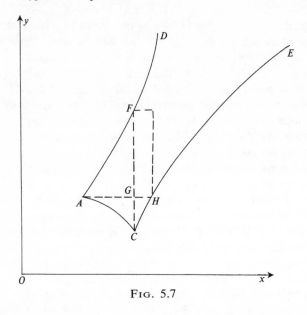

Fig. 5.7

found in these regions and, in particular, on *GH* and *GF*. The values of u on these two lines then determine u in the rectangle whose adjacent sides are *GF* and *GH*. The above process can now be repeated until the complete region in which u is defined has been determined. It should again be noted that consistency conditions will have to be satisfied by the data at *A* and *C*.

The derivatives u_{xx}, u_{yy} do not occur in the canonical form of the hyperbolic equation and hence these derivatives can in fact be discontinuous across the lines parallel to the coordinate axes (i.e. across the characteristics). Clearly u_{yy}, for example, can only be discontinuous across lines parallel to the x-axis since a discontinuity across lines parallel to the y-axis would imply that u_y is discontinuous and hence not differentiable. If u_{yy} were discontinuous across some characteristic $y = y_0$ then differentiating equation (5.1) with respect to y and subtracting the form of the equation on either side of the characteristic would give

$$[u_{yy}]_x = 0,$$

and therefore if there is a discontinuity in a second derivative at a point it is propagated along the appropriate characteristic through that point. A similar situation can, in a like manner, be shown to hold for all higher derivatives. An alternative method of deriving the characteristics would in fact be to seek those curves across which discontinuities in second derivatives may exist. If the curve $y = y(x)$ used in equation (3.2) is assumed to be such a curve then subtracting the appropriate form of equation (3.2) on either side gives a homogeneous equation relating the discontinuities $[u_{xx}]$, $[u_{xy}]$ in u_{xx} and u_{xy}. Similar homogeneous relationships can be obtained from equations (3.3) and (3.4) and, in order that there may be discontinuities, the determinant of the resulting equations must be zero and this then shows that any such curve must be a characteristic. It should be noted that, although the conditions imposed on the data or Γ in the above discussion are stronger than is necessary to ensure that existence of a solution, they are not sufficient to ensure the existence of continuous second derivatives. In this case, it is necessary for example, that prescribed values of u must be twice continuously differentiable.

It is worth noting that it is the existence for hyperbolic equations of curves across which second (or higher) derivatives of solutions can be discontinuous which explains why phenomena involving any kind of wave propagation lead to hyperbolic equations. A wave in its simplest sense is a disturbance propagating itself into a region in which the dependent variable is constant. Unless the dependent variable is entirely constant there will have to be a curve (the wave front) across which there must be some kind of discontinuity and for a smooth wave governed by a second-order equation (i.e. such that u and its first derivatives are continuous) there therefore have to be discontinuities in the second (or higher) derivatives.

Exercises 5.1

1. Use equation (5.6) to solve equation (5.3) when $u = x$, $u_x = 1 - 2x$ on $y = 2x$.
2. Use equation (5.6) to solve equation (5.3) when $u = x$, $\partial u / \partial n = 0$ on $y = x^2$, $x > 0$.
3. The equation

$$u_{xx} - u_{yy} = 0$$

is to be solved in the triangular region bounded by the lines $y = 0$, $y = x$, $y = 1 - x$. Determine, for each of the following sets of boundary conditions, whether or not a solution can be found.

(i) u given on $y = x$ and $y = 1 - x$.

(ii) u given on $y - x = 0$ and its normal derivative given on $y = 1 - x$.

(iii) u given on the boundary of the triangle.

4. u satisfies

$$u_{xx} - u_{yy} = 5e^{-2x}\sin y, \qquad x > 0, \qquad y > 0,$$

and one of the following sets of conditions

(i) $u = 0$, $y = 0$; $u_x = 0$, $x = 0$.

(ii) $u = 0$, $y = 0$; $u_x = 0$, $x = 0$, $u \to 0$ as $x \to \infty$.

(iii) $u = u_y = 0$, $y = 0$.

(iv) $u = u_y = 0$, $y = 0$; $u_x = 0$, $x = 0$.

Select, with reasons, the condition which you would expect to give a unique solution and find this solution.

5. Determine, without actually solving the equation, the conditions that must be satisfied by the twice differentiable functions ϕ_0, ϕ_1, ϕ_2 such that equation (3.17) has a solution defined in $x < ct$ with

$$u(x, 0) = \phi_0(x), \qquad u_t(x, 0) = \phi_1(x), \qquad u(x, 2x/c) = \phi_2(x), \qquad x \geq 0.$$

6. $u(x, t)$ satisfies the equation

$$c^{-2}u_{tt} + 2ku_t - u_{xx} = 0, \qquad x > 0, \qquad t > 0,$$

and the conditions

$$u(x, 0) = u_t(x, 0) = 0, \qquad x \geq 0; \qquad u(0, t) = \tfrac{1}{2}t^2, \qquad t \geq 0.$$

Show that $(u_t)_{t=(x/c)^+} = 0$ and $(u_{tt})_{t=(x/c)^+} = e^{-kcx}$.

5.2. Proof of existence of solution of the Cauchy problem

We wish to prove the existence of a function u satisfying

$$u_{xy} = F \tag{5.1}$$

and such that, on a non-characteristic curve Γ defined by $y = y(x)$,

$$u_x = p(x), \qquad u_y = q(y), \qquad u = u_1(x),$$

where u_1 is twice-differentiable and p and q are differentiable. We have already shown that the right-hand side of equation (5.6), namely

$$\tfrac{1}{2}(u_A + u_B) + \tfrac{1}{2}\int_\Gamma (q\,\mathrm{d}y - p\,\mathrm{d}x),$$

satisfies equation (5.1), with $F \equiv 0$, and the above Cauchy conditions on Γ. Therefore, if v is defined by

$$u = v + \tfrac{1}{2}(u_A + u_B) + \tfrac{1}{2}\int_\Gamma (q\,\mathrm{d}y - p\,\mathrm{d}x),$$

where u is the solution being sought, we see that v satisfies equation (5.1) and, on Γ, $v \equiv v_x \equiv v_y \equiv 0$. Therefore the problem under consideration can be transformed into one of solving equation (5.1) with homogeneous Cauchy conditions imposed on Γ. There is therefore no loss of generality in proving the existence theorem for this latter problem. Also taking new variables $x' = y(x)$, $y' = -y$ produces an equation similar to equation (5.1), in terms of these new variables, the curve Γ in the (x', y')-plane being the line $x' + y' = 0$ and therefore there is no loss of generality incurred by assuming Γ to be the line $x + y = 0$.

We therefore prove formally the existence theorem for solutions of equation (5.1) subject to the conditions $u = u_x = u_y = 0$ on $x + y = 0$. Equation (5.9) shows that (as y on the data curve is a monotonically decreasing function of x)

$$u(x_0, y_0) = \int_{APB} F(x, y, u, u_x, u_y)\,\mathrm{d}x\,\mathrm{d}y,$$

where A, B are the intersections with the line $x + y = 0$ of the characteristics drawn from $P(x_0, y_0)$, (see Fig. 5.8). This equation can be written in the two alternative forms

$$u(x_0, y_0) = \int_{-y_0}^{x_0} \int_{-x}^{y_0} F(x, y, u, u_x, u_y)\,\mathrm{d}y\,\mathrm{d}x \qquad (5.14)$$

$$= \int_{-x_0}^{y_0} \int_{-y}^{x_0} F(x, y, u, u_x, u_y)\,\mathrm{d}x\,\mathrm{d}y. \qquad (5.15)$$

We find, on differentiating equation (5.14) with respect to x_0 and

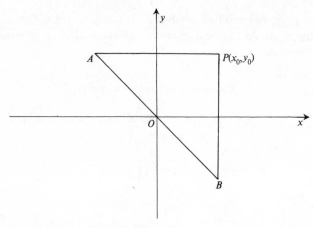

Fig. 5.8

differentiating equation (5.15) with respect to y_0 that, at P,

$$p = u_x = \int_{-x_0}^{y_0} F(x_0, y, u, p, q)\, \mathrm{d}y, \qquad (5.16)$$

$$q = u_y = \int_{-y_0}^{x_0} F(x, y_0, u, p, q)\, \mathrm{d}x. \qquad (5.17)$$

It will be assumed that, for $-2T \leq x + y \leq 2T$, F is a continuous function of all its arguments and satisfies a Lipschitz condition of the form

$$|F(x, y, u, p, q) - F(x, y, u', p', q')| \leq M\{|u - u'| + |p - p'| + |q - q'|\}. \qquad (5.18)$$

For linear equations, equation (5.18) will always be valid but in the non-linear case it will usually only hold when u, p, q, etc. are sufficiently small, i.e. when

$$|u| < \varepsilon, \qquad |u'| < \varepsilon, \qquad |p| < \varepsilon, \qquad |p'| < \varepsilon, \qquad |q| < \varepsilon, \qquad |q'| < \varepsilon,$$

for some ε. Existence will be demonstrated by showing that a sequence of successive approximations converges to the solution of the governing integro-differential equations, and the appropriate iterative scheme is defined by

$$u_0 = q_0 = p_0 = 0,$$

$$u_{n+1} = \int_{APB} F(x, y, u_n, p_n, q_n) \, dx \, dy, \qquad (5.19)$$

$$p_{n+1} = \int_{-x_0}^{y_0} F(x_0, y, u_n, p_n, q_n) \, dy, \qquad (5.20)$$

$$q_{n+1} = \int_{-y_0}^{x_0} F(x, y_0, u_n, p_n, q_n) \, dx. \qquad (5.21)$$

Our assumption concerning F means that, for $|x + y| \le 2T$, there exists a constant N such that $|F| \le N$ and therefore, for $|x_0 + y_0| \le 2T$, equations (5.19) to (5.21) give

$$|u_{n+1}| \le \left| \int_{APB} N \, dx \, dy \right| \le 2NT^2,$$

$$|p_{n+1}| \le \left| \int_{-x_0}^{y_0} N \, dy \right| \le 2NT,$$

$$|q_{n+1}| \le 2NT.$$

Thus by choosing T small enough all the iterates will be such that the Lipschitz condition on F will hold and, provided T is so chosen, we deduce from equations (5.19) to (5.21) that, for $|x_0 + y_0| < 2T$,

$$|u_{n+1} - u_n| \le M \int_{APB} \{|u_n - u_{n-1}| + |p_n - p_{n-1}| + |q_n - q_{n-1}|\} \, dx \, dy, \qquad (5.22)$$

$$|p_{n+1} - p_n| \le M \int_{-x_0}^{y_0} \{|u_n - u_{n-1}| + |p_n - p_{n-1}| + |q_n - q_{n-1}|\}_{x=x_0} \, dy, \qquad (5.23)$$

$$|q_{n+1} - q_n| \le M \int_{-y_0}^{x_0} \{|u_n - u_{n-1}| + |p_n - p_{n-1}| + |q_n - q_{n-1}|\}_{y=y_0} \, dx. \qquad (5.24)$$

It is convenient at this stage, in order to simplify the algebra, to assume that $x_0 + y_0 > 0$. This involves no loss of generality as the case $x_0 + y_0 < 0$ can be treated by replacing x_0 by $-x_0$ and y_0 by $-y_0$. The functions $|u_0|, |u_1|, |p_0|, |p_1|, |q_0|, |q_1|$ will be bounded for

$x_0 + y_0 < 2T$ and therefore a constant K exists such that

$$|u_2 - u_1| \leq MK\left|\int_{APB} dx\, dy\right| = \tfrac{1}{2}MK(x_0 + y_0)^2,$$

$$|p_2 - p_1| \leq MK\left|\int_{-x_0}^{y_0} dy\right| = MK(x_0 + y_0),$$

$$|q_2 - q_1| \leq MK\left|\int_{-y_0}^{x_0} dx\right| = MK(x_0 + y_0).$$

Therefore, when $x_0 + y_0 \leq 2T$,

$$|u_2 - u_1| + |p_2 - p_1| + |q_2 - q_1| \leq 2MK(x_0 + y_0) + \tfrac{1}{2}MK(x_0 + y_0)^2$$

$$\leq MK(T + 2)(x_0 + y_0). \tag{5.25}$$

Applying equation (5.25) to equations (5.22) to (5.24) with $n = 2$ gives

$$|u_3 - u_2| \leq M^2 K(T + 2)\left|\int_{APB} (x + y)\, dx\, dy\right|,$$

$$|p_3 - p_2| \leq MK^2(T + 2)\left|\int_{-x_0}^{y_0} (x_0 + y)\, dy\right| = \frac{M^2 K}{2!}(T + 2)(x_0 + y_0)^2,$$

$$|q_3 - q_2| \leq MK^2(T + 2)\left|\int_{-y_0}^{x_0} (x + y_0)\, dx\right| = \frac{M^2 K}{2!}(T + 2)(x_0 + y_0)^2.$$

In order to complete the calculation we need to evaluate expressions of the form

$$\int_{APB} (x + y)^n\, dx\, dy,$$

(the case $n = 1$ occurs in the above inequality for $|u_3 - u_2|$). These are most easily evaluated by introducing new variables ξ and η defined by $\xi = x + y$, $\eta = x - y$; we then obtain

$$\int_{APB} (x + y)^n\, dx\, dy = \tfrac{1}{2}\int_0^{x_0 + y_0} d\xi \int_{-2y_0}^{2x_0 - \xi} \xi^n\, d\eta = \frac{(x_0 + y_0)^{n+2}}{(n + 1)(n + 2)}.$$

Applying this result to the inequality for $|u_3 - u_2|$ gives

$$|u_3 - u_2| \leq \frac{M^2 K}{3!}(T + 2)(x_0 + y_0)^3.$$

We now prove, by induction, that for $n \geq 3$, when $0 < x + y < 2T$,

$$|u_n - u_{n-1}| \leq \frac{M^{n-1}K(T+2)^{n-2}(x+y)^n}{(n)!},$$

$$|p_n - p_{n-1}| \leq \frac{M^{n-1}K(T+2)^{n-2}(x+y)^{n-1}}{(n-1)!},$$

$$|q_n - q_{n-1}| \leq \frac{M^{n-1}K(T+2)^{n-2}(x+y)^{n-1}}{(n-1)!}.$$

The inequalities certainly hold for $n = 3$ and, assuming they hold for any specific value of n, gives, when $x + y < 2T$,

$$|u_n - u_{n-1}| + |p_n - p_{n-1}| + |q_n - q_{n-1}|$$

$$\leq \frac{M^{n-1}K(T+2)^{n-2}}{(n-1)!}(x+y)^{n-1}\left\{2 + \frac{(x+y)}{n}\right\}$$

$$\leq \frac{M^{n-1}K(T+2)^{n-1}}{(n-1)!}(x+y)^{n-1}.$$

Equations (5.22) to (5.24) now show that, at (x_0, y_0),

$$|u_{n+1} - u_n| \leq \frac{M^n K(T+2)^{n-1}}{(n-1)!}\int_{APB}(x+y)^{n-1}\,\mathrm{d}x\,\mathrm{d}y$$

$$= \frac{M^n K(T+2)^{n-1}(x_0+y_0)^{n+1}}{(n+1)!},$$

$$|p_{n+1} - p_n| \leq \frac{M^n K(T+2)^{n-1}}{(n-1)!}\int_{-x_0}^{y_0}(x_0+y)^{n-1}\,\mathrm{d}y$$

$$= \frac{M^n K(T+2)^{n-1}(x_0+y_0)^n}{n!},$$

$$|q_{n+1} - q_n| \leq \frac{M^N K(T+2)^{n-1}}{(n-1)!}\int_{-y_0}^{x_0}(y_0+x)^{n-1}\,\mathrm{d}x$$

$$= \frac{M^n K(T+2)^{n-1}(x_0+y_0)^n}{n!}.$$

It now follows by induction that $|u_n - n_{n-1}|$, etc. satisfy the inequalities suggested above.

Hence

$$\left| \sum_{n=0}^{\infty} (u_{n+1} - u_n) \right| \leq \sum_{n=0}^{\infty} |u_{n+1} - u_n|$$

$$\leq \sum_{n=0}^{\infty} \frac{M^n K (T+2)^{n-1} (2T)^n}{(n+1)!}$$

$$\leq K \sum_{n=0}^{\infty} \frac{[2MT(T+2)]^n}{n!}$$

$$= K \exp 2MT(T+2).$$

Thus, since $\sum_{n=0}^{\infty} (u_{n+1} - u_n)$ converges, u_n tends to a limit.

In order to prove uniqueness we assume that there are two independent solutions u and u' and hence in some neighbourhood of the initial line,

$$|u' - u| \leq \left| \iint_{APB} |F(x, y, u, p, q) - F(x, y, u', p', q')| \, \mathrm{d}x \, \mathrm{d}y \right|$$

$$\leq M \left| \iint_{APB} \{ |u' - u| + |p' - p| + |q' - q| \} \, \mathrm{d}x \, \mathrm{d}y \right| \quad (5.26)$$

and also

$$|p' - p| \leq M \left| \int_{-x_0}^{y_0} \{ |u' - u| + |p' - p| + |q' - q| \} \, \mathrm{d}y \right|, \quad (5.27)$$

$$|q' - q| \leq M \left| \int_{-y_0}^{x_0} \{ |u' - u| + |p' - p| + |q' - q| \} \, \mathrm{d}x \right|. \quad (5.28)$$

Applying the method used to derive the inequality for $|u_{n+1} - u_n|$ to equations (5.26) to (5.28) shows that there will be a constant K_1 such that for all n (Garabedian 1964, p. 117)

$$|u' - u| \leq \frac{K_1 (T+2)^{n-1} (2T)^n}{(n+1)!}$$

and hence $u' \equiv u$. It can also be proved that the Cauchy problem is well-posed.

The above iterative approach may also be used to demonstrate the existence of solutions of the Goursat problem and of the problem when conditions are prescribed on intersecting characteristics. In one case at least the iterative method can be used to derive the solution in a closed form, and this is illustrated in the following example.

Example 5.1

Find the solution of

$$u_{xy} + \lambda u = 0$$

with $u(x, 0) = u(0, y) = 1$, $x \geq 0$, $y \geq 0$.

Equation (5.10) shows that this problem reduces to the solution of

$$u = 1 - \lambda \int_0^{x_0} \int_0^{y_0} u \, dx \, dy,$$

and applying the iterative scheme

$$u_{n+1} = 1 - \lambda \int_0^{x_0} \int_0^{y_0} u_n \, dx \, dy$$

with $u_0 = 1$, gives $u_1 = 1 - \lambda x_0 y_0$, $u_2 = 1 - \lambda x_0 y_0 + \frac{1}{4} \lambda^2 x_0^2 y_0^2$. This suggests that

$$u_n = \sum_{r=0}^{n} (-1)^r \frac{\lambda^r x^r y^r}{(r!)^2} \qquad (5.29)$$

and substitution in the equation then gives

$$u_{n+1} = 1 - \lambda \sum_{r=0}^{n} (-1)^r \frac{\lambda^r x_0^{r+1} y_0^{r+1}}{(r+1)!^2} = \sum_{r=0}^{n+1} (-1)^r \frac{(\lambda x_0 y_0)^r}{(r!)^2}.$$

The validity of equation (5.29) for all n follows by induction and hence the required solution is

$$u = \sum_{r=0}^{\infty} (-1)^r \frac{(\lambda x_0 y_0)^r}{(r!)^2} = J_0[2(\lambda x_0 y_0)^{\frac{1}{2}}].$$

5.3. Fundamental solutions and Green's functions for the one-dimensional wave equation

Fundamental solutions and Green's functions for hyperbolic equations are defined, exactly as those for elliptic equations, to be solutions of an inhomogeneous equation with the right-hand side of the equation proportional to a delta function. The boundary conditions that have to be imposed on Green's functions for hyperbolic equations turn out to be very different from those appropriate to elliptic equations. Differences are of course to be expected as the typical boundary-value problems for the two classes of equations are completely dissimilar. There is also a further complication with practical problems in that, as mentioned earlier, they generally occur in a context where one of the variables is time so that solutions are only sought on one side of an initial curve. In order to emphasize some of the differences between Green's functions for hyperbolic equations and those for elliptic equations we first consider the special case of the one-dimensional wave equation for unit wave speed (i.e. $c \equiv 1$ in equation (3.17)) in some detail, and then generalize the analysis to cover the general hyperbolic equation.

Any solution $F(x, t, x', t')$ of

$$H_1' F \equiv F_{x'x'} - F_{t't'} = \delta(x' - x)\delta(t' - t), \tag{5.30}$$

is said to be a fundamental solution, and the convention is again adopted of using undashed variables to denote the coordinates of the point at which a solution of the wave equation will eventually be determined. The variables t and t' are used rather than y and y' in order to emphasize that they are time-like variables. Adjoint operators can be defined for hyperbolic equations in exactly the same way as for elliptic equations and the identity

$$vH_1'u - uH_1'v = (vu_{x'} - uv_{x'})_{x'} - (vu_{t'} - uv_{t'})_{t'}, \tag{5.31}$$

shows that the operator H_1' is self-adjoint. The analysis for elliptic equations suggests that, if fundamental solutions are useful for solving hyperbolic equations, there must be some suitable analogue of the Green's identity of equation (4.10). The appropriate identity is obtained by integrating equation (5.31) over a region D of the (x', t')-plane and applying the divergence theorem in the form

$$\int_D (P_{x'} + Q_{t'}) \, \mathrm{d}x' \, \mathrm{d}t' = \int_C (-Q \, \mathrm{d}x' + P \, \mathrm{d}t'),$$

where C is the boundary of D. This gives

$$\int_D (vH_1'u - uH_1'v) \, \mathrm{d}x' \, \mathrm{d}t' = \int_C (vu_{t'} - uv_{t'}) \, \mathrm{d}x' + \int_C (vu_{x'} - uv_{x'}) \, \mathrm{d}t'. \tag{5.32}$$

The simplest problem for the wave equation is the Cauchy one on the line $t = 0$ and we now determine the boundary conditions which define the Green's function appropriate for this problem. Thus we seek to solve

$$H_1 u = u_{xx} - u_{tt} = 0, \qquad (5.33)$$

subject to the conditions

$$u(x, 0) = f(x), \qquad u_t(x, 0) = g(x), \qquad -\infty < x < \infty. \quad (5.34)$$

In Fig. 5.9, P denotes the point (x, t) and *IJKL* is the rectangle $x_1 < x' < x_2$, $0 < t' < T$, with $x_1 < x < x_2$ and $T > t$. Taking D in equation (5.32) to be the rectangle *IJKL*, choosing u to satisfy equations (5.33) and (5.34) and v to be a solution F of equation (5.30) gives

$$-u(x, t) = \int_{x_1}^{x_2} (Fu_{t'} - uF_{t'})_{t'=0} \, dx' + \int_0^T (Fu_{x'} - uF_{x'})_{x'=x_2} \, dt'$$

$$+ \int_{x_2}^{x_1} (Fu_{t'} - uF_{t'})_{t'=T} \, dx' + \int_T^0 (Fu_{x'} - uF_{x'})_{x'=x_1} \, dt'. \quad (5.35)$$

Taking the limit as $x_1 \to -\infty$ and $x_2 \to \infty$ gives, assuming that u and F vanish as $|x| \to \infty$,

$$-u(x, t) = \int_{-\infty}^{\infty} (Fu_{t'} - uF_{t'})_{t'=0} \, dx' - \int_{-\infty}^{\infty} (Fu_{t'} - uF_{t'})_{t'=T} \, dx'.$$

$$(5.36)$$

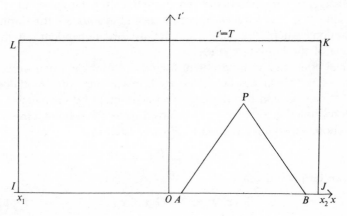

Fig. 5.9

In order to obtain $u(x, t)$ in terms of the known quantities $u(x, 0)$ and $u_t(x, 0)$ we see from equation (5.36) that F has to be chosen so that F and $F_{t'}$ vanish for $t' = T$. The only restriction made on T was that $T > t$ and hence the appropriate Green's function for solving the initial-value problem defined by equations (5.33) and (5.34) is the function $G(x, t, x', t')$ satisfying equation (5.30) and such that

$$G(x, t, x', t') = G_{t'}(x, t, x', t') = 0, \qquad t' > t; \qquad (5.37)$$

such a Green's function is said to be a '*causal*' Green's function. Equation (5.36) then becomes

$$u(x, t) = -\int_{-\infty}^{\infty} (Gu_{t'} - uG_{t'})_{t'=0} \, dx'. \qquad (5.38)$$

The right-hand side of equation (5.38) represents the effects of conditions at some value of t' (in this case $t' = 0$) on the value of u at $t(> t')$, and hence the representation is sensible in that 'future' values are given in terms of 'past' values (hence the adjective causal).

Taking D, in the above analysis, to be the region bounded by a curve C_1 which is completely below P (i.e. $t' < t$) and a curve C_2 completely above P together with connecting arcs at $x' = \pm\infty$ yields a representation for $u(x, t)$ in terms of G and the values of u and its derivatives on C_1. Therefore the function G defined by equations (5.30) and (5.37) is equally suitable for solving a Cauchy problem posed on any non-characteristic curve in the region $t' > t$. This differs significantly from the situation for elliptic equations where the Green's functions are dependent on the shape of the boundary curve.

The boundary-value problem for G is not of the usual initial-value type which can be solved by Laplace transform methods, but it can be transformed into such a problem by introducing a second function $G^*(x, t, x', t')$ referred to as the adjoint Green's function, satisfying equation (5.30) and such that

$$G^* = G^*_{t'} = 0 \quad \text{for} \quad t' < t. \qquad (5.39)$$

It can be proved that

$$G^*(x', t', x, t) = G(x, t, x', t') \qquad (5.40)$$

and hence G can be obtained immediately from G^*. Equation

(5.40) can be proved by applying equation (5.32) with D being the rectangle $IJKL$ $u = G(x_3, t_3, x', t')$, and $v = G^*(x_4, t_4, x', t')$, where the points (x_3, t_3) and (x_4, t_4) are both within the rectangle. The line integrals parallel to the t'-axis can again be assumed to vanish as $|x'| \to \infty$ and equations (5.37) and (5.39) ensure that the line integrals over the lines $t' = 0$ and $t' = T$ vanish. Hence

$$G^*(x_4, t_4, x_3, t_3) = G(x_3, t_3, x_4, t_4),$$

which is equivalent to equation (5.40).

It follows from equations (5.40) and (5.30) that

$$H_1 G(x, t, x', t') = H_1 G^*(x', t', x, t) = \delta(x - x')\delta(t - t');$$

the right-hand side of this equation represents some kind of disturbance applied at time t' at $x = x'$ (for a string the disturbance is an impluse applied at $x = x'$). Hence $G(x, t, x', t')$ represents the effect at (x, t) of the disturbance (or impulse) applied at (x', t').

The boundary-value problem for G^* is of a standard type which can be solved by Laplace transform (readers not yet familiar with the Laplace transform should omit the detailed analysis and go directly to equation (5.44), where the solution is stated), and we define

$$\bar{G}^*(x, t, x', s) = \int_0^\infty e^{-st'} G^*(x, t, x', t')\, dt'.$$

Equation (5.39) gives $G^* = G_{t'}^* = 0$ for $t' = 0$, and the standard results regarding the Laplace transforms of derivatives (cf. equations (7.13) and (7.14)) show the Laplace transform of $G_{t't'}^*$ to be $s^2 \overline{G}^*$. Taking the Laplace tranform of equation (5.30) gives

$$\int_0^\infty e^{-st'} G_{x'x'}^*\, dt' - s^2 \overline{G}^* = \int_0^\infty e^{-st'} \delta(t' - t)\delta(x' - x)\, dt',$$

and, if it is assumed that the order of differentiation and integration may be interchanged, we obtain

$$\overline{G_{x'x'}^*} - s^2 \overline{G}^* = \int_0^\infty e^{-st'} \delta(t' - t)\delta(x' - x)\, dt'.$$

This equation simplifies further, on using the basic property of the delta function (equation (4.24)), to

$$\overline{G^*_{x'x'}} - s^2\overline{G^*} = e^{-st}\delta(x'-x). \tag{5.4.1}$$

It follows from equation (5.41), exactly as in Example 4.4, that $\overline{G^*}$ is continuous for all x' and

$$(\overline{G^*_{x'}})_{x'=x+0} - (\overline{G^*_{x'}})_{x'=x-0} = e^{-st}. \tag{5.42}$$

Since $\overline{G^*}$ is a Laplace transform it must be such that $\lim_{s\to\infty} \overline{G^*} = 0$ (Theorem 7.1) and therefore the appropriate forms for $\overline{G^*}$ are given by

$$\overline{G^*} = \begin{cases} A e^{s(x'-x)}, & x' \le x \\ A e^{s(x-x')}, & x' \ge x. \end{cases}$$

Equation (5.42) now becomes

$$-2sA = e^{-st},$$

so that

$$\overline{G^*} = -\frac{e^{-st}}{2s} e^{-s|x'-x|}.$$

The Heaviside shift theorem (Table 7.2, (6)) states that the inverse of $e^{-as}\overline{f}(s)$ is $f(t'-a)H(t'-a)$, where $f(t')$ is the inverse of $f(s)$ and $H(t')$ is the unit step function which is equal to 1 for $t' \ge 0$ and zero otherwise. The inverse of s^{-1} is 1 so that

$$G^*(x, t, x', t') = -\tfrac{1}{2}H(t'-t-|x'-x|), \tag{5.43}$$

and

$$G(x, t, x', t') = -\tfrac{1}{2}H(t-t'-|x'-x|). \tag{5.44}$$

Equation (5.44) shows that the Green's function $G(x, t, x', t')$ is discontinuous across the half-lines $x' = x \pm (t-t')$, $t' < t$ (these are parts of the characteristics through P and are shown in Fig. 5.9,) and is only non-zero in the sector bounded by these half-lines. It is in fact a general feature of the fundamental solutions for hyperbolic equations that they are discontinuous and are only 'weak' solutions of the equation. (The notion of weak solutions is discussed further in §5.4.)

The fact that G vanishes except within the above sector can also be deduced directly from the representation of equation

(5.38). The analysis leading to this equation shows that $u(x, t)$ will be given by the integral on the right-hand side of equation (5.38) with the integrand evaluated at $t' = \tau$ for any τ such that $\tau < t$. By definition of the domain of dependence, however, there can be no contribution to $u(x, t)$ from those parts of $t' = \tau$ which are outside the domain of dependence of the point (x, t). Since u and $u_{t'}$ are completely arbitrary it follows that, on $t' = \tau$, G must be zero outside the domain of dependence of (x, t). This latter domain is the sector bounded by the lines $x' = x \pm (t - t')$, $t' < t$, and hence, as τ ranges over all possible values, we deduce that, when $t' < t$, G is non-zero only within this sector. Coupling this with equation (5.37) shows that G is identically zero outside the sector. Thus an alternative and completely equivalent method of defining G would have been to require it to vanish outside the sector rather than to satisfy equation (5.37). This additional restriction on G is not a necessary one as it will be automatically satisfied, in view of the hyperbolic nature of the equation, by any solution satisfying equation (5.37) but it does prove to be the most satisfactory method of defining the causal Green's function for the general hyperbolic equation.

The formal solution to the initial-value problem posed by equations (5.33) and (5.34) is, therefore,

$$u(x, t) = -\int_{-\infty}^{\infty} [(Gg(x') - f(x')G_{t'}] \, dx', \qquad (5.45)$$

where G is given by equation (5.44). Some care has to be taken in evaluating the integral in equation (5.45) since $G_{t'}$ is not defined at those points where G is discontinuous. The simplest method of evaluating the integral is to use the relationship between the delta function and the Heaviside function and we shall demonstrate both this method and one which is independent of the theory of the delta function.

We have that the derivative with respect to x of $H(x - a)$, for any a, vanishes except at $x = a$ where it is unbounded and this suggests that

$$\frac{d}{dx} H(x - a) = A\delta(x - a)$$

for some constant A. Integrating this equation from $x = a - \varepsilon$ to

$x = a + \varepsilon$ gives $A = 1$ so that

$$\frac{d}{dx} H(x-a) = \delta(x-a). \tag{5.46}$$

Hence

$$u(x, t) = \tfrac{1}{2}\int_{x-t}^{x+t} g(x')\,dx' + \tfrac{1}{2}\int_{-\infty}^{\infty} f(x')\delta[t - |x - x'|]\,dx',$$

and there will only be non-zero contributions to the second integral from the intervals $x' = x - t - \varepsilon$ to $x' = x - t + \varepsilon$ and $x' = x + t - \varepsilon$ to $x' = x + t + \varepsilon$. As $\varepsilon \to 0$ these give $\tfrac{1}{2}f(x-t) + \tfrac{1}{2}f(x+t)$ so that

$$u(x, t) = \tfrac{1}{2}[f(x-t) + f(x+t)] + \tfrac{1}{2}\int_{x-t}^{x+t} g(x')\,dx',$$

which is the result previously obtained as equation (3.25) ($c \equiv 1$).

An alternative method of deriving this result is to use the fact that the solution of the initial-value problem is given by

$$u = u^g + u_t^f, \tag{5.47}$$

where u^f and u^g are solutions of equation (5.33) with

$$u^f(x, 0) = 0, \qquad u_t^f(x, 0) = f,$$

and

$$u^g(x, 0) = 0, \qquad u_t^g(x, 0) = g, \quad \text{respectively.}$$

u, defined by equation (5.47), is clearly a solution of equation (5.33) and satisfies $u(x, 0) = f$, also

$$u_t(x, t) = u_t^g + u_{tt}^f = u_t^g + u_{xx}^f,$$

so that $u_t(x, 0) = g$. The problems for u^f and u^g can be solved using an appropriate form of equation (5.45), the integration in both cases now becomes trivial as the troublesome term involving $G_{t'}$ is absent. We have that

$$u^f = \tfrac{1}{2}\int_{x-t}^{x+t} f(x')\,dx', \qquad u^g = \tfrac{1}{2}\int_{x-t}^{x+t} g(x')\,dx',$$

and equation (3.25) is recovered.

The concept of Green's function can also be extended to obtain formal integral representations to the initial-value–boundary-value problem when, in addition to the initial conditions of equation (5.34), the solution u of equation (5.33) has to be such that $u(0, t)$ and $u(a, t)$ are both known.

In this case the appropriate rectangular region D to use in equation (5.32) is that obtained when the line LI in Fig. 5.9 becomes the line $x' = 0$ and the line KJ becomes the line $x' = a$. Inspection of equation (5.35) shows that u will be defined in terms of known functions if F is taken to be a Green's function satisfying equation (5.37) and the condition

$$G(x, t, 0, t') = 0, \qquad G(x, t, a, t') = 0.$$

G can be found fairly simply by using the Fourier series method of Example 4.4. The representation

$$G = \sum_{n=1}^{\infty} \sin \frac{n\pi x'}{a} G_n(t')$$

automatically satisfies the conditions on $x' = 0$ and $x' = a$, and substituting this representation into equation (5.30) gives

$$\sum_{n=1}^{\infty} \sin \frac{n\pi x'}{a} \left(-\frac{n^2\pi^2}{a^2} G_n - \frac{d^2 G_n}{dt'^2} \right) = \delta(x' - x)\delta(t' - t).$$

The delta function $\delta(x' - x)$ can, as in Example 4.4, be written as

$$\frac{2}{a} \sum_{n=1}^{\infty} \sin \frac{n\pi x'}{a} \sin \frac{n\pi x}{a},$$

so that

$$\frac{d^2 G_n}{dt'^2} + \frac{n^2\pi^2}{a^2} G_n = -\frac{2}{a} \sin \frac{n\pi x}{a} \delta(t' - t).$$

Equation (5.37) requires that $G_n = (G_n)_{t'} = 0$ for $t' > t$, so that

$$G_n = \begin{cases} 0, & t' > t, \\ A \cos \dfrac{n\pi}{a}(t' - t) + B \sin \dfrac{n\pi}{a}(t' - t), & t' < t. \end{cases}$$

The continuity of G_n at $t' = t$ implies that $A = 0$, and the condition

$$[(G_n)_{t'}]_{t'=t+0} - [(G_n)_{t'}]_{t'=t-0} = -\frac{2}{a} \sin \frac{n\pi x}{a}$$

gives

$$B = \frac{2}{n\pi} \sin \frac{n\pi x}{a}.$$

Hence

$$G = \frac{2}{\pi} \sum_{n=1}^{\infty} \frac{1}{n} \sin \frac{n\pi x}{a} \sin \frac{n\pi x'}{a} \sin \frac{n\pi}{a} (t' - t).$$

Replacing F in equation (5.35) by G as defined above will eventually result in the same formal Fourier series solution obtained in §3.3 and given by equations (3.33), (3.36), and (3.37).

The next step is to generalize the concept of Green's function to the more general hyperbolic equation. The analysis will be carried out using the general canonical form of the equation and it is helpful, in order to see the relationship between the special and general case, to rewrite equation (5.30) in terms of the characteristic coordinates $\phi' = x' - t'$ and $\psi' = x' + t'$. The concept of the delta function as a function vanishing except at one point gives

$$\delta(x' - x)\delta(t' - t) = A\delta(\phi' - \phi)\delta(\psi' - \psi),$$

where $\phi = x - t$, $\psi = x + t$, and A is a constant which can be found from the condition that

$$\int \delta(x' - x)\delta(t' - t) \, dx' \, dt' = 1,$$

where the integration is over any region containing the point (x, t). We have that $2 \, dx' \, dt' = d\phi' \, d\psi'$ so that

$$\int \delta(x' - x)\delta(t' - t) \, dx' \, dt' = \frac{A}{2} \int \delta(\phi' - \phi)\delta(\psi' - \psi) \, d\phi' \, d\psi' = 1$$

and hence $A = 2$. Thus

$$G_{\phi'\psi'} = \tfrac{1}{2}\delta(\phi' - \phi)\delta(\psi' - \psi). \tag{5.48}$$

Exercises 5.2

1. Find, in $x' > 0$, $t' > 0$ the causal Green's function $G(x, t, x', t')$ satisfying equations (5.30) and (5.37) and the additional condition

$$G(x, t, 0, t') = 0.$$

2. Find, in $0 < x' < a$, $t' > 0$, the causal Green's function satisfying equations (5.30) and (5.37) and such that

$$G(x, t, 0, t') = 0, \qquad G_x(x, t, a, t') = 0.$$

Use this function to obtain a series representation of the solution of

$$u_{xx} - u_{tt} = 0,$$

in $0 < x < a$, $t > 0$, with $u(0, t) = 0$, $u_x(0, t) = 0$, $u(x, 0) = f(x)$, $u_t(x, 0) = 0$.

3. Show that an integral representation of the solution of

$$u_{xx} - u_{tt} + k^2 u = 0, \qquad t > 0, \quad \text{all } x,$$

with u and u_t given on $t = 0$, can be found in terms of the function $G(x, t, x', t')$ satisfying

$$G_{x'x'} - G_{t't'} + k^2 G = \delta(x' - x)\delta(t' - t) \quad \text{and equation (5.37)}.$$

Use Laplace transforms to obtain G, given that the Laplace transform with respect to t of $I_0[a(t^2 - b^2)^{\frac{1}{2}}]H(t - b)$ is $s^{-1} \exp\{-b(s^2 - a^2)^{\frac{1}{2}}\}$.

5.4. Green's functions for the general hyperbolic equation

The concept of Green's functions as developed in the previous section can be extended to give a direct method (at least in principle) of solving the Cauchy problem for the equation

$$H_2 u = u_{xy} + h u_x + e u_y + f u = g. \tag{5.49}$$

The non-characteristic curve Γ on which the Cauchy conditions are prescribed is assumed to be such that, as shown in Fig. 5.10, y is a monotonically increasing function of x on Γ (the modifications necessary when y is a monotonically decreasing function of x on Γ will be indicated subsequently). It is also assumed that the points $P(x, y)$ at which the solution is sought should all lie on the

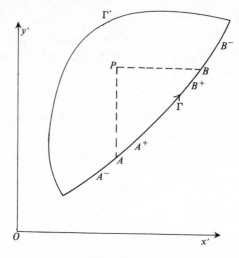

Fig. 5.10

same side of Γ and, as shown in Fig. 5.10, we take this side to be the left one to an observer looking in the direction of increasing y'. Our choice of the left side implies that the only allowable disturbances are those which, roughly speaking, propagate from right to left.

The previous discussions of Green's functions suggest expressing $vH_2u - uH_2v$, for all u and v, as a divergence and then applying the divergence theorem to obtain an identity similar to that of equation (5.32). The operator H_2, however, is not self-adjoint and in order to obtain an equation analogous to equation (5.32) it is necessary to introduce the adjoint operator H_3 defined by

$$H_3v = v_{xy} - (hv)_x - (ev)_y + fv. \qquad (5.50)$$

It can be verified by direct calculation that

$$vH_2u - uH_3v = U_x + V_y, \qquad (5.51)$$

where

$$U = \tfrac{1}{2}(vu_y - uv_y) + hvu, \qquad V = \tfrac{1}{2}(vu_x - uv_x) + evu. \quad (5.52)$$

We now have that, for any region D bounded by a closed curve C,

$$\int_D (vH_2'u - uH_3'v)\,dx'\,dy' = \int_C (U\,dy' - V\,dx'). \qquad (5.53)$$

Analogy with the case of the wave equation suggests that v has to be such that the coefficient of u on the left-hand side of equation (5.52) is proportional to a delta function. This, in turn, suggests that v should be taken to be a solution $G(x, y, x', y')$ of

$$H_3' G = \delta(x' - x)\delta(y' - y). \tag{5.54}$$

Therefore, as in the case of the general elliptic equation in §4.7, the Green's function appropriate for solving a particular equation is that associated with the adjoint operator. The differing factors on the right-hand sides of equations (5.48) and (5.54) mean that the Green's function defined by the former equation is one-half that defined by the latter. There is no obvious choice for the curve C, though clearly Γ must form some part of it and it must enclose P. We therefore take it to be any closed curve surrounding P and consisting of Γ and an arbitrary arc Γ' not intersecting the characteristics drawn from P towards Γ; one possible choice of Γ' is shown in Fig. 5.10. Equation (5.53) then reduces, when u satisfies equation (5.49), to

$$u(x, y) = \int_D vg \, dx' \, dy' - \int_\Gamma (U \, dy' - V \, dx') - \int_{\Gamma'} (U \, dy' - V \, dx').$$
$$\tag{5.55}$$

The appropriate form of boundary condition to be satisfied by G is suggested by the fact that the corresponding function for the wave equation vanishes outside the sector formed by the characteristics drawn from the point of observation towards the initial curve. Equation (5.55) reduces, on taking G to vanish outside the sector formed by the characteristics (which are lines parallel to Ox' and Oy') drawn from P towards Γ, to

$$u(x, y) = \int_{APB} Gg \, dx' \, dy' - \frac{1}{2} \int_A^B (Gu_{y'} - uG_{y'} + 2hGu) \, dy'$$
$$+ \frac{1}{2} \int_A^B (Gu_{x'} - uG_{x'} + 2eGu) \, dx', \tag{5.56}$$

where A and B are the points of intersection with Γ of the characteristics drawn from P. Equation (5.56) shows that $u(x, y)$ depends only on the values of u, u_x, and u_y on the arc of Γ which is within that part of the domain of dependence of P which is to the right of P. Thus the representation of equation (5.56) is appropriate for representing disturbances which are propagated from right to left.

The boundary-value problem for G appears to be an unusual one but, as will now be shown, it can be fairly easily transformed into a characteristic boundary-value problem of the type discussed in §5.1. Integrating equation (5.54) with respect to x' from $x' = x - \varepsilon$ to $x' = x + \varepsilon$ and letting $\varepsilon \to 0$ gives that, on AP (excluding P),

$$G_{y'}^+ - G_{y'}^- - h(G^+ - G^-) = 0,$$

where the affixes $+$ and $-$ denote the limiting values as AP is approached from inside and outside respectively. Similarly, integrating equation (5.54) with respect to y' gives

$$G_{x'}^+ - G_{x'}^- - e(G^+ - G^-) = 0.$$

G is identically zero outside ABP so that G^- etc. are all zero and

$$G_{y'} - hG = 0, \quad \text{on } AP, \tag{5.57}$$

$$G_{x'} - eG = 0, \quad \text{on } BP, \tag{5.58}$$

where the $+$ has now been dropped. Integrating equation (5.54) over the rectangle $x' = x \pm \varepsilon$, $y' = y \pm \varepsilon$ gives, as $\varepsilon \to 0$,

$$G(x, t, x, t) = -1. \tag{5.59}$$

G on AP is determined uniquely by equations (5.57) and (5.59) and G on BP is determined uniquely by equations (5.58) and (5.59). Within ABP, G satisfies the equation

$$H_3'G = 0, \tag{5.60}$$

and the boundary-vlaue problem for G is therefore a characteristic boundary-value problem for which there exists a unique solution. For the special case $h \equiv e \equiv f \equiv 0$, G is clearly -1 in agreement with the previous calculations for the wave equation (when the different factors on the right-hand sides of equations (5.48) and (5.54) are taken into account).

G, as defined by equations (5.57) to (5.59), is clearly discontinuous across AP and BP, confirming that Green's functions for hyperbolic equations are 'weak' solutions. Equation (5.56). involves the derivatives of G on Γ and, as G is discontinuous, some care has to be taken in evaluating the line integral in equation (5.56). It follows from the analysis leading to this equation that the line integral term has to be interpreted as

$$- \lim_{\substack{A^- \to A \\ B^- \to B}} \tfrac{1}{2} \int_{A^-}^{B^-} [Gu_y' - uG_{y'} + 2hGu] \, dy'$$

$$+ \lim_{\substack{A^- \to A \\ B^- \to B}} \tfrac{1}{2} \int_{A^-}^{B^-} [Gu_{x'} - uG_{x'} + 2eGu] \, dx',$$

where A^- and B^- are points adjacent to A and B but outside the sector APB. The only terms for which the limit cannot be taken are those involving $G_{y'}$ and $G_{x'}$ and we look at these terms separately. $G_{y'}$ is defined at A so that

$$\lim_{\substack{A^- \to A \\ B^- \to B}} \int_{A^-}^{B^-} uG_{y'} \, dy' = \int_{A}^{B^+} uG_{y'} \, dy' + \int_{B^+}^{B^-} uG_{y'} \, dy',$$

where B^+ is a point adjacent to B but within APB. Now

$$\int_{B^+}^{B^-} uG_{y'} \, dy' = \int_{B^+}^{B^-} [(uG)_{y'} - Gu_y] \, dy'$$

and this, in the limit at $B^+ \to B^-$, becomes $-(uG)_B$, where the suffix denotes values evaluated at B. Hence

$$- \lim_{\substack{A^- \to A \\ B^- \to B}} \tfrac{1}{2} \int_{A^-}^{B^-} [Gu_y' - uG_{y'} + 2hGu] \, dy'$$

$$= -(uG)_B - \tfrac{1}{2} \int_{A}^{B} [Gu_{y'} - uG_{y'} + 2hGu] \, dy',$$

where the integrand at the end points is to be evaluated in the limit as these points are approached from within. Similarly

$$\lim_{\substack{A^- \to A \\ B^- \to B}} \tfrac{1}{2} \int_{A^-}^{B^-} [Gu_{x'} - uG_{x'} + 2eGu] \, dx'$$

$$= -(uG)_A + \tfrac{1}{2} \int_{A}^{B} [Gu_{x'} - uG_{x'} + 2eGu] \, dx'.$$

Finally

$$u(x, y) = \int_{APB} Gg \, dx' \, dy' - (uG)_A - (uG)_B$$

$$- \tfrac{1}{2} \int_A^B [Gu_{y'} - uG_{y'} + 2hGu] \, dy'$$

$$+ \tfrac{1}{2} \int_A^B [Gu_{x'} - uG_{x'} + 2eGu] \, dx'. \quad (5.61)$$

It can be verified that for the case $h \equiv e \equiv f = 0$ equation (5.61), with $G = -1$ is equivalent to equation (5.8).

Equation (5.61) is still valid for points P to the right of Γ provided that A and B are still interpreted as the intersections with Γ of lines parallel to Oy' and Ox' respectively. If this interpretation of A and B is retained then, when Γ represents a curve on which y' is a monotonically decreasing function of x', the sense of integration has to be reversed for the divergence theorem to be valid. In this case the integral of $G_{x'y'}$ over the rectangle $x' = x \pm \varepsilon$, $y' = y \pm \varepsilon$ is equal, in the limit as $\varepsilon \to 0$, to $G(x, y, x, y)$ so that equation (5.59) has to be replaced by

$$G(x, y, x, y) = 1. \quad (5.62)$$

G^*, the Green's function adjoint to G, is defined to be the solution of

$$H_2'G^* = \delta(x' - x)\delta(y' - y), \quad (5.63)$$

which vanishes except in the section formed by the characteristic drawn from P in the direction away from Γ. It can be proved, by applying equation (5.52) with C equal to $\Gamma \cup \Gamma'$, that

$$G(x, y, x', y') = G^*(x', y', x, y). \quad (5.64)$$

Equation (5.61) demonstrates that Green's functions for hyperbolic equations reduce the solution of a Cauchy problem to evaluating integrals (provided G can be found) and therefore serve exactly the same purpose as Green's functions for elliptic equations. The main differences between the Green's functions for the two types of equations are that those for hyperbolic functions are non-singular but discontinuous, and are not dependent on the shape of the boundary curve but only on its general nature.

Green's functions were originally developed to deal with boundary-value problems for elliptic equations, in particular for Laplace's equation, and their use in problems for hyperbolic equations is comparatively new. In the past such problems were handled using a method first introduced by Riemann, which is entirely equivalent to the Green's function method. As the equivalence between the two methods is not, at first sight, entirely obvious it seems worthwhile to describe Riemann's method so that the relation between the methods can be seen.

The starting point for Riemann's method is equation (5.53), where D is now the region APB of Fig. 5.10 and C is the piecewise smooth curve APB. When u satisfies equation (5.49), we have that

$$\int_{APB} (vg - uH_3'v)\, dx'\, dy' = \int_{AB} (U\, dy' - V\, dx') - \int_B^P V\, dx'$$
$$+ \int_P^A U\, dy'. \quad (5.65)$$

Also,

$$\int_B^P V\, dx' = \tfrac{1}{2} \int_B^P [vu_{x'} - uv_{x'} + 2evu]\, dx'$$
$$= \tfrac{1}{2} \int_B^P [(vu)_{x'} - 2uv_{x'} + 2evu]\, dx'$$
$$= \tfrac{1}{2}[(vu)_P - (vu)_B] - \int_B^P u(v_{x'} - ev)\, dx',$$

where the suffixes mean the values at the corresponding points. Similarly

$$\int_P^A U\, dy' = \tfrac{1}{2}[(vu)_A - (vu)_P] - \int_P^A u(v_{y'} - hv)\, dy'.$$

Hence

$$\int_{APB} (vg - uH_3'v)\, dx'\, dy' = \tfrac{1}{2}[(vu)_A + (vu)_B] - (vu)_P$$
$$+ \int_{AB} (U\, dy' - V\, dx')$$
$$+ \int_B^P u(v_{x'} - ev)\, dx' - \int_P^A u(v_{y'} - hv)\, dy'.$$
$$(5.66)$$

Equation (5.66) can be considerably simplified by choosing v to be a function $R(x, y, x', y')$ such that

$$\left. \begin{array}{c} H'_3R = 0, \\ R_{y'} - hR = 0 \quad \text{on} \quad AP, \qquad R_{x'} - eR = 0 \quad \text{on} \quad BP, \\ R(x, y, x, y) = 1. \end{array} \right\} \quad (5.67)$$

Equation (5.66) then becomes

$$u(x, y) = \tfrac{1}{2}[(Ru)_A + (Ru)_B] - \int_{APB} Rg \, dx' \, dy'$$

$$+ \tfrac{1}{2} \int_{AB} (Ru_{y'} - uR_{y'} + 2hRu) \, dy'$$

$$- \tfrac{1}{2} \int_{AB} (Ru_{x'} - uR_{x'} + 2eRu) \, dx'. \quad (5.68)$$

R is defined to be the Riemann or Riemann–Green function for (or associated with) equation (5.49) (it should be noted that the Riemann function appropriate for solving a given equation satisfies the adjoint equation, cf. §4.9) and its definition is independent of the nature of the boundary curve and of the relative orientation of this curve and P. Comparison of the equations defining R with those defining G shows that, for the configuration of Fig. 5.10, $R = -G$ and making this substitution in equation (5.68) gives equation (5.61). For the case when y on Γ is a decreasing function of x we have $R = G$. Thus the Green's function is equal to $\pm R$, depending on the nature of the boundary curve, within the domain of dependence of P (R is only defined in this region), and is zero otherwise.

Since the Riemann function is independent of the orientation of the boundary curve it is simpler in illustrative examples to calculate it rather than the Green's function, the appropriate form of the latter can then be written down immediately in terms of R for any given curve.

The Riemann function for

$$u_{xy} + \tfrac{1}{4}u = 0$$

can be obtained directly from the results of Example 5.1. The Riemann function has, from equation (5.67), to be unity when $x' = x$ and $y' = y$. The operator in the above equation is self-adjoint and hence replacing x and y in Example 5.1 by $x' - x$ and $y' - y$ gives the Riemann function to be $J_0[(x' - x)^{\frac{1}{2}}(y' - y)^{\frac{1}{2}}]$.

There are no general methods available for determining Riemann (and hence Green's) functions though some useful results can be obtained for the equation

$$v_{xy} + N(1-N)(x+y)^{-2}v = 0, \qquad (5.69)$$

by seeking a series solution $\sum_{j=0}^{\infty} v_j(x'-x)^j(y'-y)^j$. Making this substitution in the equation eventually gives the Riemann function to be

$$F(N, 1-N; 1; -(x'-x)(y'-y)/[(x'+y')(x+y)]), \qquad (5.70)$$

where $F(a, b; c; z)$ is the hypergeometric function defined by

$$F(a, b; c; z) = 1 + \frac{ab}{c}z + \frac{a(a+1)b(b+1)}{c(c+1)2!}z^2 + \ldots .$$

For N a positive integer the series for the hypergeometric function terminates and a closed form can be obtained for the Riemann function. For integer N however it is just as easy to evaluate the Riemann function directly by noting that if v is written as $(x+y)^N u$, then u satisfies

$$u_{xy} + N(u_x + u_y)(x+y)^{-1} = 0, \qquad (5.71)$$

and the general solution of this is

$$u = \frac{1}{(x+y)}\left(\frac{\partial}{\partial x} + \frac{\partial}{\partial y}\right)^N [f(x) + g(y)], \qquad (5.72)$$

where f and g are arbitrary functions.

In cases where a change of dependent variable of the form $v = w(x, y)u$, with w known, is used to transform the differential equation for u to one for v it is possible to write down the Riemann function R_u, for determining u, directly in terms of the Riemann function R_v appropriate to the equation for v. The relationship is given by

$$R_u(x, y, x', y') = \frac{w(x', y')}{w(x, y)} R_v(x, y, x', y'), \qquad (5.73)$$

and can be proved by direct manipulation but a simpler proof is given by using equation (5.68). The equations for u and v will not be identical but if the equation for u is of the form

$$u_{xy} + \text{first-order derivatives} = E(x, y),$$

then that for v will be of the form

$$v_{xy} + \text{first-order derivatives} = E(x, y)w(x,y).$$

If u, u_x, and u_y are prescribed to vanish on a curve Γ, then v, v_x and v_y will also be zero on Γ and the solutions of the equations for u and v will be, using equation (5.68),

$$u(x, y) = -\int_{ABP} R_u(x, y, x', y')E(x', y')\,\mathrm{d}x'\,\mathrm{d}y'$$

and

$$v(x, y) = -\int_{ABP} R_v(x, y, x', y')E(x', y')w(x', y')\ \mathrm{d}x'\,\mathrm{d}y'.$$

However $v(x, y) = w(x, y)u(x, y)$ and therefore

$$R_u(x, y, x', y') = \frac{R_v(x, y, x', y')w(x', y')}{w(x, y)}.$$

As a particular example we use equation (5.73) to evaluate the Riemann function associated with equation (5.71) for $N = 1$. The corresponding Riemann function for equation (5.69) when $N = 1$ is obviously unity (the operator is self-adjoint), and hence the required Riemann function is $(x' + y')/(x + y)$.

Example 5.2

Find, when $N = 1$, the solution of equation (5.71) such that $u(x, x) = 0$ and $u_x(x, x) = f(x)$.

Differentiating the condition $u(x, x) = 0$ gives

$$u_x(x, x) + u_y(x, x) = 0,$$

so that $u_y(x, x) = -f(x)$. The point A in this case will be (y, y), B will be (x, x), and equation (5.68) gives, using the above Riemann function, that

$$u(x, y) = -\tfrac{1}{2}\int_y^x \frac{x'}{x+y}[-f(x')\,\mathrm{d}x' - f(x')\,\mathrm{d}x']$$

$$= \frac{2}{x+y}\int_y^x x'f(x')\,\mathrm{d}x'.$$

Example 5.3

Find the Riemann function appropriate to equation (5.71) with $N = 2$.

We first find the Riemann function $R(x, y, x', y')$ appropriate for solving equation (5.69); equation (5.67) shows that $R(x, y, x, y') = R(x, y, x', y) = 1$. It now follows from equation (5.72) that, as the operator in equation (5.69) is self-adjoint,

$$R_v = (x' + y')^2 \left[\frac{1}{x' + y'} \left(\frac{\partial}{\partial x'} + \frac{\partial}{\partial y'} \right)^2 [f(x') + g(y')] \right]$$

$$= (x' + y') \left(\frac{\partial}{\partial x'} + \frac{\partial}{\partial y'} \right) \frac{f'(x') + g'(y')}{x' + y'}$$

$$= f''(x') + g''(y') - \frac{2[f'(x') + g'(y')]}{x' + y'},$$

the dashes on the functions denoting derivatives with respect to the argument. Setting $R_v = 1$ when $x = x'$ gives

$$\frac{f''(x)}{(x + y')^2} - \frac{2f'(x)}{(x + y')^3} + \frac{\mathrm{d}}{\mathrm{d}y'} \cdot \frac{g'(y')}{(x + y')^2} = \frac{1}{(x + y')^2};$$

integrating this equation shows that g' is a quadratic in y' and it follows similarly that f' is a quadratic in x'. Setting $f' = ax'^2 + bx' + c$ and $g' = a'y'^2 + b'y' + c'$ gives R_v to be of the form

$$\frac{Ax'y' + B(x' - y') + C}{x' + y'},$$

where A, B, and C are arbitrary constants. Using the conditions on $x' = x$ and $y' = y$ gives

$$A = \frac{2}{(x + y)}, \qquad B = \frac{x - y}{x + y}, \qquad C = \frac{2xy}{x + y}.$$

Hence,

$$R_v(x, y, x', y') = \frac{2x'y' + (x - y)(x' - y') + 2xy}{(x + y)(x' + y')}$$

and, on using equation (5.73),

$$R_u(x, y, x', y) = (x' + y') \frac{[2x'y' + (x - y)(x' - y') + 2xy]}{(x + y)^3}.$$

In this case it was more convenient to work out the Riemann equation for v first as the operator in equation (5.69) is self-adjoint, and the absence of first derivative terms simplifies the determination of the form of the Riemann function on the lines $x' = x$ and $y' = y$.

Exercises 5.3

Solve the following Cauchy problems by first finding the appropriate Riemann function and then making the appropriate substitution in equation (5.68).

1. $(x - y)u_{xy} - (u_x - u_y) = 0$,

$$u = 0, \qquad u_x = 2x^2 \quad \text{on} \quad x + y = 0.$$

2. $u_{xy} - \dfrac{u_x}{y} + \dfrac{u_y}{x} - \dfrac{u}{xy} = 0$,

$$u = 0, \qquad u_x = \cos x, \qquad u_y = -\cos y \quad \text{on} \quad y = x.$$

(Set $R = xv/y$.)

3. $x^2 u_{xx} - y^2 u_{yy} = 0$, $u = 3x^2$, $u_y = 0$ on $y = 1$.

5.5. Weak solutions

The concept of a weak solution for second-order equations involving self-adjoint operators was introduced in §4.6 but it can be extended very easily to equations involving more general operators. For the general linear operator H_2 of equation (5.49) we have that

$$\int_D (vH_2u - uH_3v)\,\mathrm{d}x\,\mathrm{d}y = \int_C [\tfrac{1}{2}(vu_y - uv_y) + hvu]\,\mathrm{d}y$$
$$- [\tfrac{1}{2}(vu_x - uv_x) + evu]\,\mathrm{d}x, \quad (5.74)$$

where C is a closed contour bounding a domain D. Thus, when v and its derivatives vanish on C,

$$\int_D (vH_2u - uH_3v)\,\mathrm{d}x\,\mathrm{d}y = 0$$

and a weak solution of equation (5.49) (i.e. $H_2u = g$) is defined to be one such that

$$\int_D uH_3v \, \mathrm{d}x \, \mathrm{d}y = \int_D vg \, \mathrm{d}x \, \mathrm{d}y \qquad (5.75)$$

for all functions v which, together with their derivatives, vanish on C. It follows by reversing all the above arguments that a twice differentiable weak solution is a solution in the normal sense.

Hyperbolic equations differ from elliptic equations in that not all weak solutions are solutions in the normal sense. This can be seen by considering the simple counter-example of $H(x-y)$, where H is the unit step function, for the equation

$$u_{xx} - u_{yy} = 0$$

This equation is self-adjoint and hence equation (5.75) requires that

$$\int_{D'} (v_{xx} - v_{yy}) \, \mathrm{d}x \, \mathrm{d}y = 0,$$

where D' is the intersection of the region $x - y > 0$ and the region D. The integral is most easily evaluated using characteristic coordinates $\xi = x - y$ and $\eta = x + y$ and we need to show that

$$\int_{D'} v_{\xi\eta} \, \mathrm{d}\xi \, \mathrm{d}\eta = 0.$$

If the integration with respect to ξ is carried out then, as $v \equiv v_\xi \equiv v_\eta \equiv 0$ on the boundaries of D' other than $\xi = 0$, the left-hand side of the equation becomes

$$\int_{C'} v_\eta \mathrm{d}\eta,$$

where C' is a segment (or segments) of the line $\xi = 0$, with end points on C (see Fig. 5.11). The integration with respect to η gives zero as the end points of the interval lie on C; hence $H(x-y)$ is a weak solution of the equation and it can be shown similarly that $H(x+y)$ is another weak solution.

It was also proved in §4.8 that, for the class of equations considered, the only curves of discontinuity are the characteristics, and it will now be proved that this is also true for the general linear hyperbolic equation. To demonstrate this it is convenient

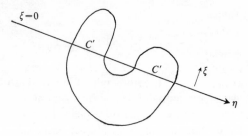

FIG. 5.11

to rewrite equation (5.74) in a slightly different form. We have
that

$$\int_C vu_y \, dy = \int_C (vu)_y \, dy - \int_C uv_y \, dy = -\int_C uv_y \, dy,$$

$$\int_C uv_x \, dx = \int_C (uv)_x \, dx - \int_C vu_x \, dx = -\int_C vu_x \, dx,$$

so that equation (5.74) is equivalent to

$$\int_D (vH_2u - uH_3v) \, dx \, dy = \int_C \{(hvu - uv_y) \, dy - (evu + vu_x) \, dx\}.$$

$$(5.76)$$

If it is assumed that a given solution u is a weak solution with
discontinuities across a curve Γ but nowhere else, then it can be
proved, as in §4.8, by using equation (5.76) that

$$\int_\Gamma \{(hv[u] - [u]v_y) \, dy - (ev[u] + v[u_x]) \, dx\} = 0,$$

where [ ] denotes a discontinuity. In general v and v_y are independent and hence

$$\int_\Gamma [u]v_y \, dy = 0 \quad \text{and} \quad \int_\Gamma hv[u] \, dy - \int_\Gamma \{ev[u] + v[u_x]\} \, dx\} = 0,$$

Thus, as v is arbitrary, in order that $[u] \neq 0$ we must have $dy = 0$
(i.e. Γ is a characteristic) so that

$$e[u] + [u_x] = 0,$$

and, as Γ is now parallel to the x-axis, this is equivalent to

$$e[u] + [u]_x = 0,$$

which is a differential equation for $[u]$ on a characteristic. If u is continuous but u_x is discontinuous (i.e. $[u] = 0$, but $[u_x] \neq 0$) then we require that $dx = 0$ so that Γ is again a characteristic. The above assumes that v_y and v are independent and this is true unless Γ is parallel to the y-axis (i.e. $dx = 0$) when

$$\int_\Gamma (hv[u] - [u]v_y)\, dy = 0$$

$$= \int_\Gamma \{hv[u] + [u]_y v - ([u]v)_y\}\, dy.$$

The last term can be integrated and vanishes as $v = 0$ at the end points of Γ and hence

$$h[u] + [u]_y = 0,$$

giving a differential equation for $[u]$ and showing that u can have a discontinuity. Thus weak solutions can only be discontinuous across characteristics. Also, since the discontinuity across a characteristic satisfies a first-order differential equation, any discontinuity in boundary data is propagated along the characteristics through the point of discontinuity. This is in contrast to the situation for elliptic equations where any such discontinuity is not propagated into the interior of a region.

Bibliography

Garabedian, P. R. (1964). *Partial differential equations*. Wiley, New York.

6 Parabolic equations

SOME aspects of the existence and uniqueness of solutions of a general class of parabolic equations in two independent variables are discussed in §6.1 and the particular case of the heat-conduction equation is considered in the subsequent two sections. The general properties of solutions of this equation are established in §6.2 where it is shown that there exists a maximum–minimum principle similar to that for Laplace's equation and in §6.3 it is shown that Green's functions can be defined for the heat-conduction equation and used to obtain integral representations of its solutions. The extension of some of the results for the heat-conduction equation to a more general class of equations is described in §6.4.

6.1. Preliminary remarks on existence and uniqueness

The canonical form of the general linear parabolic equation in two independent variables is

$$a(x, y)u_{xx} + d(x, y)u_x + e(x, y)u_y + f(x, y)u = g(x, y), \quad (6.1)$$

and, as for the general elliptic equation, it is relatively complicated to describe the kind of problems for which solutions are known to exist. It is possible, however, to state, in a fairly concise form, existence and uniqueness theorems for the slightly less general equation

$$P_1 u = au_{xx} + du_x - u_y + fu = g, \quad (6.2)$$

when $a > 0$ and $f \leq 0$ in any region in which a solution is being sought. It can be shown that, when the coefficients are continuous, there is a unique solution of equation (6.2) in $y > 0$, $0 < x < b$, such that

$$u(x, 0) = \phi_1(x), \qquad u(0, y) = \phi_2(y), \qquad u(b, y) = \phi_3(y), \quad (6.3)$$

where the functions $\phi_i (i = 1, 2, 3)$ are continuous bounded functions. The proof of existence is technically complicated and will not be given, but a proof of uniqueness for the special case $a \equiv 1$, $d \equiv f \equiv 0$, will be given in the following section and a proof of the general uniqueness theorem given in §6.4. The problem posed by equations (6.2) and (6.3) is obviously a generalization of the initial-value–boundary-value problem for the heat-conduction equation (§1.2(v)).

It is also possible to prove existence and uniqueness theorems when the last two conditions in equation (6.3) are replaced by ones prescribing u on two non-intersecting curves C_1 and C_2, completely in $y \geq 0$, intersecting $y = 0$ at $x = 0$ and $x = b$ respectively, and which are never parallel to $y = 0$ (Fig. 6.1). In this case the solution is defined in that part of $y > 0$ bounded by C_1 and C_2 and the segment $0 \leq x \leq b$ of the line $y = 0$.

Another problem for which existence and uniqueness theorems may be proved is the Cauchy problem on $y = 0$ (i.e. when $u(x, 0) = \phi_1(x), |x| < \infty$) for equation (6.2). The general conditions under which such theorems are valid are fairly elaborate (Friedman 1964), but from these theorems it can be deduced that, if the coefficients in equation (6.2) and the function $\phi_1(x)$ are bounded continuous functions satisfying a Lipschitz condition of order α $(0 < \alpha < 1)$ in x for $0 \leq x \leq b$, $0 \leq y \leq c$, then there exists only one solution of the Cauchy problem in this region, satisfying the

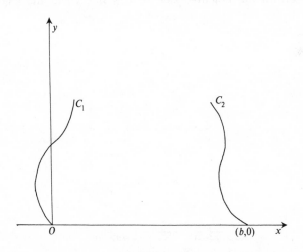

F<small>IG</small>. 6.1

condition

$$\int_{-\infty}^{\infty} \int_0^c |u(x, y)| e^{-kx^2} \, dy \, dx < \infty \qquad (6.4)$$

for some constant k. This inequality is effectively a boundedness condition on u.

The line $y = 0$ in equation (6.2) is a characteristic and the above results can obviously be rephrased in terms of conditions imposed on characteristics ($y = 0$) and on non-characteristic curves (e.g. C_1 and C_2).

Analogy with the wave equation suggests that, as time is often one of the independent variables in a parabolic equation, it may be necessary to specify on which side of a given curve a solution is being sought. This turns out to be the case but the situation is more restrictive for parabolic equations than for hyperbolic ones in that a problem is only well-posed when a solution is sought on a particular side of a given curve. The crucial factor in determining on which side of a line $y = $ constant the problem is well-posed is the relation between the signs of the coefficients of u_y and u_{xx} in equation (6.2). If the signs of these coefficients are different then the problem is well-posed on the positive side of a line $y = $ constant on which conditions are imposed; the problem is well-posed on the negative side of this line when the signs of

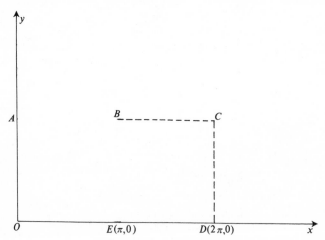

Fig. 6.2

the coefficients of u_{xx} and u_y are the same. Thus, for example, the problem of solving

$$u_{xx} - u_y \sin x = 0, \qquad 0 \le x \le 2\pi, \qquad 0 \le y \le 1,$$

will only be well-posed if u is prescribed on the lines OA, BC, CD, OE of Fig. 6.2.

It is of course a consequence of the Cauchy–Kowalesky theorem that solutions may be obtained, in principle, for parabolic equations when Cauchy conditions are prescribed on a non-characteristic curve. In general such problems are not well-posed as can be seen by solving

$$u_{xx} = u_y$$

for the two separate sets of conditions

   (i) $u(0, y) = y$, $u_x(0, y) = 0$,
   (ii) $u(0, y) = y$, $u_x(0, y) = \varepsilon$,

where ε is a constant. The solution to the first problem is $u = y + \frac{1}{2}x^2$, whilst that to the second problem is $u = y + \frac{1}{2}x^2 + \varepsilon x$. Thus, however small a value is taken for ε, the difference between these two solutions will be unbounded when x is large, even though the difference between the two sets of conditions may be very small.

In some ways parabolic equations are more complicated than either elliptic or hyperbolic ones as there is no single equation which is entirely typical of the class. It is in fact possible to construct parabolic equations for which Dirichlet problems on a closed boundary are well-posed but we shall not pursue such topics.

The remainder of the chapter is devoted to examining the properties of the solution of the heat-conduction equation with the boundary conditions defined by equation (6.3), and to extending some of the results to the more general equation (6.2). A detailed account of the properties of the heat-conduction equation is available in Widder (1975).

6.2. The heat-conduction equation

Most of the general features of solutions of equation (6.2) are exemplified in the simpler heat-conduction equation (when $d \equiv f \equiv 0$), and in this and the following section we concentrate entirely on the properties of solutions of this latter equation. The extension of the analysis to the more general equation is sketched briefly in §6.4.

We consider first the initial-value–boundary-value problem, i.e. the solution of

$$P_2 u = u_{xx} - u_t = g(x, t), \qquad t > 0, \qquad 0 < x < a, \qquad (6.5)$$

with

$$u(x, 0) = \phi_1(x), \qquad u(0, t) = \phi_2(t), \qquad u(a, t) = \phi_3(t). \quad (6.6)$$

In view of the physical significance of the heat-conduction equation, independent variables x and t are used rather than x and y so as to emphasize that one of the variables is 'time-like'. It would appear at first sight to be more sensible to consider first, as was done for the wave equation, the apparently simpler Cauchy problem when the last two conditions of equation (6.6) are omitted. This problem, however, turns out to be more difficult to handle than the mixed problem as it is necessary to impose the condition of equation (6.4) in order to cope with the fact that the range of x is unbounded. This problem was avoided for the wave equation as the existence of a finite speed of wave propagation ensures that conditions do not need to be imposed as $|x| \to \infty$. Parabolic equations, though similar to hyperbolic equations in that one of the variables in a practical context is usually time, are not such that the domain of dependence of a point is finite and a small change in initial conditions has an effect on the solution for all x and t. In this sense they are more like elliptic equations than hyperbolic ones and therefore effectively model phenomena in which disturbances are propagated with infinite speed. An infinite speed of propagation can, however, never be associated with a real physical phenomenon and the heat-conduction equation can only be regarded as giving a reasonable description, under most circumstances, of the mechanism of heat conduction. The wave equation and the heat-conduction equation are, for obvious reasons, often referred to as equations of evolution.

Physical sense suggests that it is only sensible to seek a solution of equations (6.5) and (6.6) for $t > 0$ and, as has been mentioned earlier, the nature of the equation is such that the problem considered is only well-posed for $t > 0$. Equation (6.5) is, unlike the wave equation, not invariant under the transformation $t' = -t$, and it is therefore to be expected that the behaviour of the solution for $t < 0$ could be radically different from that for $t > 0$.

The general theory and analysis of parabolic equations is very similar to that of elliptic equations and it will, for example, be shown that there is a maximum principle for equation (6.5) which is very similar to that for Laplace's equation, and one of the proofs of uniqueness that we give is also similar to that for Laplace's equation. This similarity is to be expected on mathematical grounds, since the problem posed by equations (6.5) and (6.6) is a Dirichlet one on a U-shaped region and it is only the absence of a condition on some line $t = T(T > 0)$ which stops the problem from being the typical problem for an elliptic equation. The absence of such a condition is of course a consequence of the heat equation being an equation of evolution.

The first step in proving uniqueness theorems for equation (6.5) is the derivation of an integral identity similar to the Green's identity of equation (4.11), and the appropriate identity is an almost immediate consequence of the result that

$$vP_2u - uP_3v = (vu_x - uv_x)_x - (uv)_t, \qquad (6.7)$$

where

$$P_3v = v_{xx} + v_t.$$

P_3 is therefore the adjoint operator to P_2 and its form shows that the heat equation, unlike the wave equation and Laplace's equation, is not self-adjoint. Setting $v \equiv 1$ and replacing u by u^2 in equation (6.7) gives

$$P_2u^2 = 2(uu_x)_x - (u^2)_t$$

and, as

$$P_2u^2 = 2uP_2u + 2u_x^2,$$

we have that

$$uP_2u + \tfrac{1}{2}(u^2)_t + u_x^2 = (uu_x)_x. \qquad (6.8)$$

Integration of equation (6.8) over the region D defined by $0 \le x \le a$, $0 < t < T$, then gives

$$\int_D (uP_2u + \tfrac{1}{2}(u^2)_t + u_x^2) \, \mathrm{d}x \, \mathrm{d}t$$
$$= \int_0^T (uu_x)_{x=a} \, \mathrm{d}t - \int_0^T (uu_x)_{x=0} \, \mathrm{d}t, \quad (6.9)$$

which is the required integral identity.

198 *Parabolic equations*

The following uniqueness theorem can now be deduced almost immediately from equation (6.9).

Theorem 6.1

There is at most one solution of equation (6.5) with $u(x, 0)$ given and either u or u_x prescribed for $x = 0$ and $x = a$.

Proof

If it is assumed that there are two such solutions u_1 and u_2 then, as equation (6.5) is linear, their difference $u = u_1 - u_2$ will satisfy

$$P_2 u = 0, \qquad (6.10)$$

and vanish when $t = 0$, further, either u or u_x will vanish at $x = 0$ and $x = a$. It then follows from equation (6.9) that

$$\int_D [\tfrac{1}{2}(u^2)_t + u_x{}^2] \, dx \, dt = 0,$$

and hence that u is constant within D; the condition $u(x, 0) = 0$ then gives $u \equiv 0$, thus showing uniqueness.

The above uniqueness proof is equally valid for positive and negative t and gives no indication of the difficulties that can arise in the latter case. It therefore seems worthwhile to give a second proof in which some of the difficulties associated with negative t become more apparent. If $I(t)$ is defined by

$$I(t) = \frac{1}{2} \int_0^a u^2 \, dx$$

then, when u satisfies equation (6.10),

$$\frac{dI}{dt} = \int_0^a u u_t \, dx = \int_0^a u u_{xx} \, dx = \int_0^a [(u u_x)_x - u_x{}^2] \, dx$$

$$= (u u_x)_{x=a} - (u u_x)_{x=0} - \int_0^a u_x{}^2 \, dx.$$

When either u or u_x vanish for $x = 0$ and $x = a$ we obtain

$$\frac{dI}{dt} = -\int_0^a u_x{}^2 \, dx, \qquad (6.11)$$

showing that I is a non-increasing function of t. For the function $u = u_1 - u_2$ of Theorem 6.1 we have that $I(0) = 0$ and hence, as I is positive and non-increasing, it follows that $I \equiv 0$, $t > 0$, and $u \equiv 0$, again showing uniqueness. In this case it is only possible to infer uniqueness for $t > 0$ as $I(t)$ could be positive for $t < 0$. In the practical context of heat conduction $I(t)$ is a measure of the thermal energy and the above proof illustrates that, in general, I may decrease with time so that the heat-conduction process is a dissipative one.

It is also possible, using the variational approach introduced in §1.3, and discussed in more detail in §10.3, to give some quantitative estimate of the rate of decay of I for the case when u vanishes for $x = 0$ and $x = a$. The basic result required is that obtained immediately preceding Example 10.1 namely that, for all functions u vanishing for $x = 0$ and $x = a$,

$$\int_0^a u_x^2 \, \mathrm{d}x \geq \frac{\pi^2}{a^2} \int_0^a u^2 \, \mathrm{d}x.$$

It follows from this inequality and equation (6.11) that

$$\frac{\mathrm{d}I}{\mathrm{d}t} \leq -\frac{\pi^2}{a^2} I$$

or

$$\frac{\mathrm{d}}{\mathrm{d}t} [I e^{\pi^2 t/a^2}] \leq 0,$$

so that

$$I \leq I(0) e^{-\pi^2 t/a^2}. \tag{6.12}$$

It can now be inferred directly from equation (6.12) that the mean square integral of the difference of any two solutions of equation (6.5) which take the same values for $x = 0$ and $x = a$ will tend to zero exponentially with time.

A third method of proving a uniqueness theorem for equation (6.5) is one based on a maximum–minimum principle and the appropriate principle in this case is given by

Theorem 6.2

The values within the rectangle R defined by $0 \leq x \leq a$, $0 \leq t \leq T$, (Fig. 6.3) of a continuous function $u(x, t)$ satisfying the

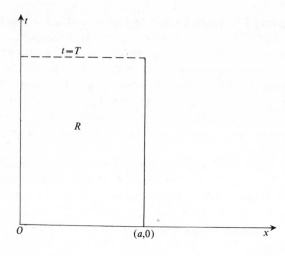

FIG. 6.3

homogeneous equation (6.10) do not exceed the maximum value of u on the lines $t = 0$, $x = 0$, $x = a$.

Proof

The simplest proof is by contradiction and we therefore assume that, when the maximum value of u in R is M, its maximum value on the lines $x = 0$, $x = a$ and $t = 0$ is $M - \varepsilon$ where $\varepsilon > 0$, and let (x_0, t_0) be the point at which the maximum is attained.

The function v defined by

$$v = u + \frac{\varepsilon}{4a^2}(x - x_0)^2$$

satisfies

$$v_t - v_{xx} = -\frac{\varepsilon}{2a^2}, \tag{6.13}$$

with $v(x_0, t_0) = M$, and on the lines $x = 0$, $x = a$, $t = 0$,

$$v < M - \varepsilon + \tfrac{1}{4}\varepsilon = M - \tfrac{3}{4}\varepsilon.$$

Thus the maximum value of v is not attained on any of the above lines and this maximum will therefore be at a point (x_1, t_1) such that $0 < x_1 < a$, $0 < t_1 \leq T$. Since v is a maximum at (x_1, t_1) we must have that, at this point, $v_t = 0$ if $t_1 < T$, or $v_t \geq 0$ if $t_1 = T$, and $v_{xx} \leq 0$. Hence at (x_1, t_1)

$$v_t - v_{xx} \geq 0, \qquad (6.14)$$

which contradicts equation (6.13), showing that our initial assumption regarding u was false.

It is possible by replacing u by $-u$ to deduce that the values of u within R cannot be less than the minimum value of u on the lines $x = 0$, $x = a$, $t = 0$.

Theorem 6.2 does not preclude the maximum values of u within R being equal to the maximum values of u on the lines $x = 0$, $x = a$, and $t = 0$, but by more delicate arguments than those above we can deduce

Theorem 6.3

The values within R of a continuous solution of equation (6.10) which does not take the same constant value on the three lines $x = 0$, $x = a$, and $t = 0$ is always strictly less than its maximum value on those lines, and its minimum value is strictly greater than its minimum value on those lines. In the case when u takes the same constant value on these lower boundaries of R it will be constant throughout R.

It is also worth noting that it is not possible to replace the line $t = 0$ in Theorem 6.2 by the line $t = T$. t_1 could then possibly be zero and this would mean $v_t \leq 0$ for $t = 0$ and it would no longer be possible to deduce that equation (6.14) holds.

If we consider the slab $0 \leq x \leq a$ whose initial temperature is a constant U and whose ends are subsequently kept at a lower temperature than U, then the maximum principle asserts that the temperature within the slab is always less than U. This, of course, makes physical sense in that it conforms with the principle that heat flows from hot to cool areas.

Uniqueness follows immediately by applying the maximum–minimum principle to the function $u = u_1 - u_2$ of Theorem 6.1. This function can be neither greater than, or less than, zero within R and hence is identically zero. (A more general maximum principle associated with the operator P_2 can also be established (cf. Exercises 6.1, 4)).

Continuous dependence of the solution of equation (6.5) on

initial data (i.e. well-posedness) can be deduced from the maximum–minimum principle by considering the difference $u = u_1 - u_2$ of two solutions u_1 and u_2 of equation (6.5) and such that

$$u_1(x, 0) = \phi_1(x), \qquad u_1(0, t) = \phi_2(t), \qquad u_1(a, t) = \phi_3(t),$$

$$u_2(x, 0) = \psi_1(x), \qquad u_2(0, t) = \psi_2(t), \qquad u_2(a, t) = \psi_3(t).$$

u satisfies equation (6.10) and

$$u(x, 0) = \phi_1 - \psi_1, \qquad u(0, t) = \phi_2 - \psi_2, \qquad u(a, t) = \phi_3 - \psi_3.$$

It now follows from the maximum principle that the maximum value of $|u|$ within R is less than the maximum of $|\phi_1 - \psi_1|$, $|\phi_2 - \psi_2|$, $|\phi_3 - \psi_3|$ showing that the problem is well-posed.

It has already been pointed out that it is not possible to modify the proof of the maximum principle so as to be able to deduce that u, within R, is less than its value on $t = T$, and we cannot deduce that the problem when $u(x, 0)$ in equation (6.6) is replaced by $u(x, T)$ is well-posed. We can in fact go further and actually show by a counter-example that it is sometimes not well-posed. If $u(x, T)$ is taken to be $\varepsilon \sin m\pi x/a$ for some integer m then it can be verified by direct differentiation that a solution of equation (6.5), vanishing for $x = 0$ and $x = a$ and taking the correct form at $x = T$ is $\varepsilon e^{-(m^2\pi^2/a^2)(t-T)} \sin m\pi x/a$. Hence

$$u(x, 0) = \varepsilon e^{(m^2\pi^2/a^2)T} \sin \frac{m\pi x}{a}$$

and, by making m arbitrarily large, we can make $u(x, 0)$ as large as we choose for all ε, however small; therefore the 'backward-time' problem is not well-posed in this particular case.

The general form of solution of equation (6.10) with $u(x, 0)$ prescribed and u vanishing on $x = 0$ and $x = a$ can be obtained in the form of a Fourier series and from this representation it is possible to produce a more general result concerning the solution of this 'backward time' problem. We consider first the general form of the Fourier series solution.

Writing u as $X(x)T(t)$ and substituting into equation (6.10) gives, as in §3.3, Example 4.1 and as described in the general case in §7.1,

$$\frac{X''(x)}{X(x)} = -\lambda^2, \qquad \frac{T'(t)}{T(t)} = -\lambda^2,$$

where λ is a constant. The general solution for X is $A \cos \lambda x + B \sin \lambda x$, the conditions on u at $x = 0$ and $x = a$ give $X(0) = X(a) = 0$. Thus, as in Example 4.1, $\lambda = n\pi/a$, where n is an integer, and X is then proportional to $\sin n\pi x/a$. Substituting into the equation for $T(t)$ shows the general solution to be

$$u = \sum_{n=1}^{\infty} e^{-n^2\pi^2 t/a^2} A_n \sin \frac{n\pi x}{a}. \tag{6.15}$$

Setting $t = 0$ in equation (6.15) gives

$$u(x, 0) = \sum_{n=1}^{\infty} A_n \sin \frac{n\pi x}{a},$$

and hence

$$A_n = \frac{2}{a} \int_0^a u(x', 0) \sin \frac{n\pi x'}{a} \, dx'.$$

The solution is therefore given by

$$u = \int_0^a \left[u(x', 0) \frac{2}{a} \sum_{n=1}^{\infty} \exp\left\{ -\frac{n^2\pi^2 t}{a^2} \right\} \sin \frac{n\pi x'}{a} \sin \frac{n\pi x}{a} \right] dx'. \tag{6.16}$$

If $u(x, 0)$ is integrable, the exponential factor in equation (6.16) ensures the uniform convergence of the series for all t such that $t \geq t_0 > 0$, even if $u(x, 0)$ is discontinuous, showing, that, as for Laplace's equation, discontinuities in data are smoothed immediately within the interior of a region.

We can use the property that a solution of equation (6.10) vanishing on $x = 0$ and $x = a$, is infinitely differentiable to prove that the backward-time problem does not, under these conditions, always have a solution. We assume that there exists a solution in $0 \leq t \leq T$ with $u(x, T) = X(x)$ for some arbitrary function X. It would then be possible to find $u(x, 0)$, and this quantity is denoted by $X^*(x)$. The forward-time problem with $u(x, 0) = X^*(x)$ can now be solved and in view of the existence of a uniqueness theorem for this problem we deduce that $X(x)$ will be equal to the solution of this forward-time problem evaluated at $t = T$. The fact that solutions of this problem are infinitely differentiable leads to the conclusion that $X(x)$ must also be infinitely differentiable and this was not an assumption originally made about $X(x)$ so that there is a contradiction. Thus for non-infinitely-differentiable initial data there does not exist a solution of the 'backward-time' problem.

It is possible to deduce from Theorem 6.3 a maximum–minimum principle valid when the range of x is infinite, and this principle can be used to prove a uniqueness theorem for the Cauchy problem and to show that this problem is well-posed. The basic result needed is

Theorem 6.4

Every continuous bounded solution of equation (6.10) in $t \geq 0$, $|x| < \infty$, satisfies the inequalities

$$m \leq u \leq M,$$

where

$$M = \sup_{|x| < \infty} u(x, 0) \quad \text{and} \quad m = \inf_{|x| < \infty} u(x, 0).$$

($u(x, 0)$ may not attain its maximum or minimum values on the infinite interval and hence the supremum and infimum have to be used rather than maximum or minimum values.)

Proof

We prove that $u \leq M$, the other inequality following by applying the same argument to $-u$.

The function $w(x, t) = x^2 + 2t$ is a solution of equation (6.10), as is the function

$$v = \varepsilon \frac{w(x, t)}{w(x_0, t_0)} + M - u,$$

where (x_0, t_0) denotes some arbitrary point with $t_0 \geq 0$, $|x| < \infty$ and ε is some positive number. Therefore

$$v \geq \frac{\varepsilon x^2}{w(x_0, t_0)} + M - u$$

and, if N denotes the supremum of u for $t \geq 0$, $|x| < \infty$, it follows that $v \geq 0$ for $t = 0$ and for

$$|x| = \left\{ \frac{(N - M) w(x_0, t_0)}{\varepsilon} \right\}^{\frac{1}{2}} + |x_0|.$$

Thus by Theorem 6.3 we have that $v \geq 0$ within the strip

$$|x| \leq \left\{ \frac{(N-M)}{\varepsilon_0} w(x_0, t_0) \right\}^{\frac{1}{2}} + |x|, \qquad 0 \leq t < T,$$

and therefore

$$u \leq M + \frac{\varepsilon w(t, x)}{w(t_0, x_0)}, \tag{6.17}$$

in the above rectangle which contains the point (x_0, t_0), so that $u(x_0, t_0) \leq M + \varepsilon$.

However (x_0, t_0) and ε are both arbitrary so that $u \leq M$ for all $t > 0$, $|x| < \infty$, thus proving the theorem.

In the above proof it should be noted that v would be positive for

$$|x| = \left\{ \frac{(N-M)}{\varepsilon} w(x_0, t_0) \right\}^{\frac{1}{2}},$$

but it was necessary to add an extra $|x_0|$ so that equation (6.16) can be established in a region containing the point (x_0, t_0). The assumption that u is bounded is necessary in order that N be finite.

The proof of the uniqueness theorem for a bounded solution of the Cauchy problem now follows from Theorem 6.4 by arguments almost identical to those used for the initial-value–boundary-value problem. The well-posedness of the problem follows in exactly the same way as before.

The necessity for u to be bounded in order to be able to prove uniqueness can be seen by the following counter-example. It can be formally verified that

$$u = \sum_{n=0}^{\infty} f^{(n)}(t) \frac{x^{2n}}{(2n)!},$$

where $f^{(n)}(t)$ denotes the nth derivative of e^{-1/t^2}, satisfies the homogeneous equation (6.10) and vanishes when $t = 0$. Thus, if a boundedness condition were not imposed, this function could be added to any solution of the Cauchy problem, so that such solutions could not be unique.

The fact that the backward-time Cauchy problem is not, in general, well-posed can be seen by considering the solution of equation (6.10) with $u(x, T) = \varepsilon \sin \alpha x$. It can be verified that the solution is

$$u = \varepsilon e^{\alpha^2(T-t)} \sin \alpha x,$$

with

$$u(x, 0) = \varepsilon e^{\alpha^2 T} \sin \alpha x.$$

Thus, by taking α arbitrarily large, $u(x, 0)$ can become unbounded for any ε. It is also possible to derive an integral representation for the solution of the Cauchy problem which shows that it is infinitely differentiable for $t > 0$, and this can be used to show that the backward-time Cauchy problem does not have a solution when the initial data is not differentiable. The precise integral representation of the solution will be given in the following section.

Exercises 6.1

1. Prove that there cannot be two solutions of equation (6.5) in $0 < x < a$, $t > 0$, with $u_x - \alpha u$ prescribed on $x = 0$, $u_x + \beta u$ prescribed on $x = a$, where $\alpha > 0$, $\beta > 0$, and u prescribed on $t = 0$.

2. A solution u of the heat equation in $0 < x < a$, $t > 0$ is such that on $x = 0$, $x = a$, and $t = 0$, $A < u < B$, where A and B are constants. Show that $A < u < B$ for $0 < x < a$, $t > 0$.

3. u is a solution of equation (6.5) in the region $0 < x < 2$ and satisfies the conditions

$$u(0, t) = f(t), \qquad u(2, t) = 0, \qquad u(x, 0) = 0, \qquad 0 \le x \le 2,$$

and also $f(0) = 0$ and $0 \le f(t) \le 1$ for all t.
   Show that

$$0 \le u \le 1 - \tfrac{1}{2}x.$$

4. Prove that if $P_2 u \ge 0 (\le 0)(0 < x < a, t > 0)$ then the maximum (minimum) value of u is attained when $x = 0$, $x = a$, or $t = 0$.

5. If $f(t)$ in Exercise 3 above is such that $f > 1 - e^{-kt}$, where $0 < k \le \tfrac{1}{16}\pi^2$, show that

$$u \ge \left(1 - \tfrac{1}{2}x - e^{-kt} \cos\frac{\pi x}{4}\right).$$

6. Solve equation (6.5) in $0 < x < a$, $t > 0$, with

$$u_x(0, t) = 0, \qquad u(a, t) = 0, \qquad u(x, 0) = \sin\frac{\pi x}{a}.$$

6.3. Fundamental solutions and Green's functions for the heat-conduction equation

Fundamental solutions and Green's functions can again be used to derive integral representations, at a point (x, t), of solutions of equation (6.5) in terms of the values of u on lines $t = $ constant and of the values of u(or u_x) on lines $x = $ constant. The first step in deriving these representations is to integrate equation (6.7) (i.e. the equation relating P_2 and its adjoint P_3), with x and t replaced by x' and t', over the region D defined by $-b < x' < c(b, c > 0)$, $0 \le t' \le T(T > t)$ (Fig. 6.4) and then apply the divergence theorem to the resulting identity.

This gives

$$\int_D (vP_2'u - uP_3'v)\, \mathrm{d}x'\, \mathrm{d}t' = \int_0^T (vu_{x'} - uv_{x'})_{x'=c}\, \mathrm{d}t'$$
$$- \int_0^T (vu_{x'} - uv_{x'})_{x'=-b}\, \mathrm{d}t' + \int_{-b}^c (uv)_{t'=0}\, \mathrm{d}x'$$
$$- \int_{-b}^c (uv)_{t'=T}\, \mathrm{d}x'. \quad (6.17)$$

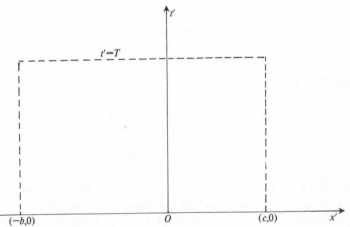

Fig. 6.4

An integral representation for $u(x, t)$ in terms of its values on the boundaries of D can therefore be found by taking v in equation (6.17) to be a solution $G(x, t, x', t')$ of

$$P_3' G = \delta(x' - x)\, \delta(t' - t), \tag{6.18}$$

giving

$$u(x, t) = \int_D Gg \, dx' \, dt' - \int_0^T (Gu_{x'} - uG_{x'})_{x'=c} \, dt'$$
$$+ \int_0^T (Gu_{x'} - G_{x'}u)_{x'=-b} \, dt' - \int_{-b}^c (uG)_{t'=0} \, dx'$$
$$+ \int_{-b}^c (uG)_{t'=T} \, dx'. \tag{6.19}$$

In order to solve initial-value problems it is necessary for there to be no contribution to $u(x, t)$ for values of $t' > t$ and it therefore follows that the last term on the right-hand side of equation (6.19) has to be zero for any $T > t$, i.e.

$$G \equiv 0, \qquad t' > t. \tag{6.20}$$

This is exactly the same condition as that obtained for the wave equation.

A representation of u suitable when u is prescribed on $x = 0$ and $x = a$ can be obtained by setting $b = 0$ and $c = a$ in equation (6.19) and, in order that this representation only involves u on $x = 0$ and $x = a$, the appropriate choice of Green's function is the function G_1 satisfying equation (6.20) and such that

$$G_1(x, t, 0, t') = G_1(x, t, a, t') = 0. \tag{6.21}$$

Equation (6.19) then becomes

$$u(x, t) = \int_D Gg \, dx' \, dt' + \int_0^t [u(G_1)_{x'}]_{x'=a} \, dt'$$
$$- \int_0^t [u(G_1)_{x'}]_{x'=0} \, dt' - \int_0^a (uG_1)_{t'=0} \, dx', \tag{6.22}$$

in equation (6.22) the upper limit of integration has been replaced by t as G_1 satisfies equation (6.20) and the range of t' in the double integral will be from 0 to t. Different boundary conditions on $x = 0$ and $x = a$ require Green's functions satisfying different conditions on $x' = 0$ and $x' = a$; for example, the problem when u_x is prescribed for $x = 0$ and $x = a$ requires the use of the Green's function G_2 whose derivative with respect to x' vanishes for $x' = 0$ and $x' = a$.

The appropriate representation for the Cauchy problem is obtained by letting $b = c = \infty$. In this case u will be bounded at infinity and in order that the integrals at $x' = \pm\infty$ are zero, the appropriate Green's function G_3 has to be such that

$$\lim_{|x'| \to \infty} G_3 = 0. \tag{6.23}$$

In this case

$$u = \int_D G_3 g \, dx' \, dt' - \int_{-\infty}^{\infty} (uG_3)_{t'=0} \, dx'. \tag{6.24}$$

The condition of equation (6.20) is best handled, like that for the wave equation, by introducing the adjoint Green's function $G^*(x, t, x', t')$, where

$$P_2' G^* = \delta(x' - x)\, \delta(t' - t) \tag{6.25}$$

and

$$G^* \equiv 0, \qquad t' < t. \tag{6.26}$$

The adjoint G_i^* of the functions $G_i (i = 1 \text{ to } 3)$ is a solution of equations (6.25) and (6.26) which satisfies the same condition as G_i on lines $x' = \text{constant}$ or as $|x'| \to \infty$ as appropriate. It can be proved by setting $u = G_i(x_3, t_3, x', t')$, $v = G_i^*(x_4, t_4, x', t')$ in the appropriate form of equation (6.17) (i.e. with $b = 0$ and $c = a$ for $i = 1, 2$ and $b = c = \infty$ for $i = 3$) that

$$G_i(x_3, t_3, x_4, t_4) = G_i^*(x_4, t_4, x_3, t_3),$$

i.e.

$$G_i(x, t, x', t') = G_i^*(x', t', x, t). \tag{6.27}$$

In this case it should be noted that, as for the general hyperbolic equation, the adjoint Green's function satisfies the equation adjoint to that satisfied by the Green's function and that the Green's function appropriate for solving a given differential equation is that associated with the adjoint equation.

We are now in a position to calculate the various Green's

functions, and we first evaluate G_3. Since t is positive it follows from equation (6.24) that $G_3^* = 0$ for $t' = 0$ and hence, as for the wave equation, an appropriate method of solving equation (6.23) would be by means of Laplace transforms. Readers not yet familiar with Laplace transform methods should omit the detailed analysis and go directly to equation (6.29), where the formal solution to the Cauchy problem is stated. The Laplace transform $\bar{G}^*$ of G^* with respect to t' is defined by

$$\bar{G}_3^*(x, t, x', s) = \int_0^\infty e^{-st'} G_3^* \, dt',$$

and taking the Laplace transform of equation (6.25) gives (cf. §7.2)

$$\frac{d^2}{dx'^2} \bar{G}_3^* - s\bar{G}_3^* = e^{-st} \delta(x' - x).$$

$\bar{G}_3^*$ has, as in the other problems considered where the right-hand side is a delta function, to be continuous at $x' = x$ and also equation (6.23) implies that $G_3^* \to 0$ as $|x'| \to \infty$. Hence the appropriate form of $\bar{G}_3^*$ is given by

$$\bar{G}_3^* = \begin{cases} Ae^{s^{\frac{1}{2}}(x-x')}, & x < x', \\ Ae^{-s^{\frac{1}{2}}(x-x')}, & x > x', \end{cases}$$

where $s^{\frac{1}{2}}$ is that branch of the square root whose real part is positive. We also have, by comparison with previous calculations, that

$$\left(\frac{d\bar{G}_3^*}{dx'} \right)_{x'=x+0} - \left(\frac{d\bar{G}_3^*}{dx'} \right)_{x'=x-0} = e^{-st},$$

giving $-2As^{\frac{1}{2}} = e^{-st}$ so that

$$\bar{G}_3^* = -\frac{1}{2} e^{-st} s^{-\frac{1}{2}} e^{-s^{\frac{1}{2}}|x-x'|}.$$

Table 7.2(12) gives the inverse of $s^{-\frac{1}{2}} e^{-s\,|x-x'|}$ to be $\pi^{-\frac{1}{2}} \exp\{-(x-x')^2/4t\}$ and by applying the Heaviside shift theorem we obtain

$$G_3^*(x, t, x', t') = -\frac{1}{2} H(t'-t) \pi^{-\frac{1}{2}} (t'-t)^{-\frac{1}{2}} \exp\left\{ -\frac{(x-x')^2}{4(t'-t)} \right\}.$$

$$(6.28)$$

The quantity to be substituted into equation (6.24) is

$G_3(x, t, x', t')$ which is, from equations (6.27) and (6.28), given by

$$G_3(x, t, x', t') = -\tfrac{1}{2}\pi^{-\frac{1}{2}}H(t-t')(t-t')^{-\frac{1}{2}}\exp\left\{-\frac{(x-x')^2}{4(t-t')}\right\}.$$
(6.29)

Thus the solution of the homogeneous equation (6.10) which vanishes as $|x| \to \infty$ is

$$u = \frac{(\pi t)^{-\frac{1}{2}}}{2}\int_{-\infty}^{\infty} u(x', 0)\exp\left\{-\frac{(x-x')^2}{4t}\right\}dx'.$$
(6.30)

It can be proved rigorously that, even when $u(x', 0)$ is discontinuous, equation (6.30) defines a function which is infinitely differentiable for $t > 0$, illustrating how data is smoothed immediately away from the boundary. It is possible to deduce immediately, as in the previous section, that the solution of the backward-time problem does not exist for an initial form of u which is not infinitely differentiable. We also see from equation (6.30) that if $u(x', 0)$ is taken to be positive for $-\infty < x < \infty$ then, for any $t > 0$, u will be positive, showing once again the infinite speed of propagation of phenomena associated with the heat equation.

In general it is not possible to give a simple characterization of the singularity associated with the Green's functions for the heat equation and all that can be deduced about the singularity is that, by equation (6.29), any fundamental solution behaves as $-\tfrac{1}{2}\pi^{-\frac{1}{2}}(t-t')^{-\frac{1}{2}}\exp-\{(x-x')^2/4(t-t')\}$ near $t' = t$, $x' = x$. This function, with x, t replaced by x_0, t_0, is a solution of the homogeneous equation (6.10) whenever $(x_0-x')^2 + (t_0-t')^2 \neq 0$ and this property can be used, in combination with the method of images, to obtain the Green's function suitable for solving the initial value boundary-value problem in $x > 0$ with u prescribed on $x = 0$ and bounded as $x \to \infty$. The required Green's function will have to vanish on $x' = 0$ and tend to zero as $x' \to \infty$, and the function

$$-\tfrac{1}{2}\pi^{-\frac{1}{2}}(t-t')^{-\frac{1}{2}}\left[\exp\left\{-\frac{(x-x')^2}{4(t-t')}\right\} - \exp\left\{-\frac{(x+x')^2}{4(t-t')}\right\}\right]H(t-t')$$

satisfies all these conditions.

The most direct method of calculating G_1 is by a Fourier series method, virtually identical to that used in calculating the Green's function for the mixed initial-value–boundary-value problem for the wave equation (§5.3). The conditions on $x' = 0$ and $x' = a$ are satisfied by setting

$$G_1(x, t, x', t') = \sum_{n=1}^{\infty} G^{(n)}(t') \sin \frac{n\pi x'}{a},$$

and substitution of this result into equation (6.18) gives, on using the representation of $\delta(x'-x)$ as a Fourier series and equating the coefficients of $\sin(n\pi x'/a)$ to zero,

$$-\frac{n^2\pi^2}{a^2} G^n + G_{t'}^{(n)} = \frac{2}{a} \sin \frac{n\pi x}{a} \delta(t'-t). \tag{6.31}$$

$G^{(n)}$ will, from equation (6.20), be zero for $t' > t$ and equation (6.31) shows that

$$(G^{(n)})_{t'=t+0} - (G^{(n)})_{t'=t-0} = \frac{2}{a} \sin \frac{n\pi x}{a}.$$

Thus

$$G^{(n)} = \begin{cases} -\dfrac{2}{a} \sin \dfrac{n\pi x}{a} \exp\left\{\dfrac{n^2\pi^2}{a^2} (t'-t)\right\}, & t' < t, \\ 0 & t' > t, \end{cases}$$

and

$$G_1 = -\frac{2}{a} \sum_{n=1}^{\infty} \exp\left\{\frac{n^2\pi^2}{a^2} (t'-t)\right\} \sin \frac{n\pi x}{a} \sin \frac{n\pi x'}{a}, \qquad t' < t.$$

Equation (6.22) then shows that a solution of equation (6.10) vanishing for $x = 0$ and $x = a$ is given by

$$u(x, t) = \frac{2}{a} \int_0^a u(x', 0) \sum_{n=1}^{\infty} \exp\left\{-\frac{n^2\pi^2 t}{a^2}\right\} \sin \frac{n\pi x}{a} \sin \frac{n\pi x'}{a} \, dx',$$

which is the result previously obtained in equation (6.16).

Exercises 6.2

1. Find in $x' > 0$, $t' > 0$, the Green's function $G(x, t, x', t')$ satisfying equations (6.8) and (6.20) and the additional condition

$$G_{x'}(x, t, 0, t') = 0.$$

2. Find in $0 < x' < a$, $t' > 0$, the Green's function $G(x, t, x', t')$ satisfying equations (6.18) and (6.20) and the additional conditions

$$G_{x'}(x, t, a, t') = G(x, t, 0, t') = 0.$$

6.4. The general parabolic equation

We now look briefly at some extensions of the preceding analysis to the more general equation (6.2), i.e. to

$$P_1 u = a u_{xx} + d u_x - u_t + f u = g, \qquad (6.31)$$

where y has now been replaced by t.

One important result that can be generalized to the operator P_1 is the maximum principle and we have

Theorem 6.5

Suppose that $a \geq 0$, $f < 0$, for $0 \leq x \leq b$, $0 \leq t \leq T$, then a continuous function $u(x, t)$ satisfying the homogeneous equation

$$P_1 u = 0, \qquad (6.23)$$

can only attain its positive maximum (or negative minimum) on the lines $x = 0$, $x = b$, $t = 0$. [The result is in fact valid for $f \leq 0$ but the proof then becomes more complicated, cf. Chapter 2 of Friedman (1964).]

Proof

The proof is by contradiction and follows more or less the same pattern as that of Exercise 4.3, 1. If it is assumed that u attains its positive maximum at (x_0, t_0) with $0 < x_0 < b$, $0 < t \leq T$ then it follows as in Exercies 4.3, 1 that, at (x_0, t_0), $P_1 u < 0$, which is a contradiction. The proof that a negative minimum cannot be attained, except on $t = 0$, $x = 0$, $x = b$ is carried out by considering the function $-u$.

It is possible to prove, when $f \equiv 0$, the more delicate result that the maximum and minimum values, irrespective of sign, are attained on one or other of $x = 0$, $x = b$, $t = 0$.

Some information can be obtained from Theorem (6.5) concerning the solutions of equation (6.32) with $f > 0$ (and $a \leq 0$) which vanish for $x = 0$ and $x = b$. To obtain this we make the substitution $u = v e^{kt}$, for some k, and so obtain

$$a v_{xx} + d v_x - v_t + (f - k) v = 0.$$

It is possible, for any bounded function f, to find k such that the coefficient of v is negative, so that Theorem 6.5 may be applied to v. It then follows from this that $|v|$ is less than the maximum value attained by $|v|$ on $t = 0$ and hence $|v| < M$, where M is the maximum value of $|u|$ on the line $t = 0$. This result suggests that solutions of the initial-value–boundary-value problem for u will tend to infinity exponentially with t, and such problems are unlikely to occur in a physical context.

The proof of uniqueness of solutions of the initial-value–boundary-value problem for equation (6.31) follows directly from Theorem 6.5 in exactly the same fashion as the corresponding theorem for the heat equation followed from Theorem 6.2. The 'well-posedness' of the initial-value–boundary-value problem is again a direct consequence of the existence of a maximum–minimum principle.

The identity

$$vP_1u - uP_5v = [vau_x - uav_x + uv(d - a_x)]_x - (uv)_t, \quad (6.33)$$

where

$$P_5v = (av)_{xx} - (dv)_x + fv + v_t,$$

shows that the adjoint of P_1 is the operator P_5. The fundamental solutions appropriate for deriving integral representations of solutions of (6.31) therefore satisfy

$$P_5'v = \delta(x' - x)\,\delta(t' - t). \quad (6.34)$$

The relevant integral representations can be obtained by integrating equation (6.33), in the variables x' and t', with v being a solution of equation (6.34). The Green's function suitable for solving the initial-value–boundary-value problem with u prescribed on $x = 0$ and $x = b$ will vanish for $x' = 0$ and $x' = b$. The corresponding Green's function for the problem when u_x is prescribed has, because of the existence of the term $av(d - a_x)$ in equation (6.33), to satisfy a more complicated condition on $x' = 0, b$. In the case when $d = a_x$, however, the condition that has to be satisfied by this second Green's function becomes $G_{x'} = 0$ for $x' = 0$ and $x' = b$.

The formal definition given in equation (6.34) for a fundamental solution is not always the most convenient one and it is

possible to show that an equivalent definition is that any fundamental solution must satisfy the homogeneous form of equation
(6.34), except at $x' = x$, $t' = t$ where it behaves like

$$\frac{1}{2}\frac{(t-t')^{-\frac{1}{2}}}{[a(x, t)\pi]^{\frac{1}{2}}}\exp\left\{-\frac{a(x, t)(x-x')^2}{4(t-t')}\right\}.$$

Equation (6.33) can, as for hyperbolic and elliptic equations,
be used to define a weak solution of equation (6.31). It follows as
for hyperbolic equations that weak solutions can only be discontinuous across characteristics, i.e. lines $x = $ constant.

Bibliography

Friedman, A. (1964). *Partial differential equations of parabolic type.*
Prentice-Hall, New Jersey.
Widder, D. (1975). *The heat equation.* Academic Press New York.

7 Analytical methods of solution

IN THIS chapter an account is given of some of the analytical methods of solving linear partial differential equations. The methods described all have a common basis in that the solution is obtained in the form of a linear superposition (as an infinite series or as an integral) of products of functions of the separate independent variables. Particular problems in the previous chapters have been solved by using infinite series representations of the above type (Fourier series in all cases) and in §7.1 the approach used in previous examples (e.g. Example 4.1) is generalized to more general situations where expansions other than trigonometric ones are appropriate. The most convenient method of obtaining solutions to problems which are best solved by using integral representations is by means of integral transforms. The most common transforms are the Laplace and Fourier transforms and their use in solving boundary-value problems is described in §7.2 and §7.3, respectively. The essence of an integral transform method of solution is the reduction of the partial differential equation for the dependent variable u say, to an ordinary differential equation for the appropriate transform of u. It often proves reasonably easy to calculate this function but recovering u from its transform can prove more difficult. A particular feature of Laplace and Fourier transforms, however, is that it is often possible to obtain, directly from the transform of u, some useful information (often of an asymptotic nature) about the behaviour of u for a particular range of one of the independent variables. The methods for extracting this kind of information are also described in §7.2 and §7.3. The chapter concludes in §7.4 with a discussion of the method of obtaining the most appropriate transform to use for solving a particular problem for a given differential equation.

7.1. Separated series solutions of partial differential equations

The method used in §3.3 and Examples 4.1 and 4.4 of representing the solution of a partial differential equation as an infinite series of products of the form $X(x)Y(y)$, where x and y are independent variables, is a useful and powerful one. It seems, therefore, worthwhile to try and establish the features of a particular problem which suggest that this method is a feasible one to use and, with this end in view, we now examine the basic procedure involved in solving the above examples.

The basic steps in the solutions are

(1) *Determining whether or not product solutions exist for the particular equation considered.* In any particular case this can always be established by direct substitution and manipulation similar to that used in Example 4.1. It is, however, worth noting that, at least in principle, such solutions can always be found for the equation

$$a_1 u_{xx} + a_2 u_x + a_3 u + b_1 u_{yy} + b_2 u_y + b_3 u = 0, \qquad (7.1)$$

where the a_i's are functions of x only and the b_i's are functions of y only. That this is the case can be seen by setting $u = X(x)Y(y)$ in equation (7.1) and then dividing by XY, giving

$$\frac{(a_1 X_{xx} + a_2 X_x + a_3 X)}{X} + \frac{(b_1 Y_{yy} + b_2 Y_y + b_3 Y)}{Y} = 0.$$

The first term is a function of x only whilst the second is a function of y only; thus both must be separately constant so that

$$a_1 X_{xx} + a_2 X_x + a_3 X = \lambda X, \qquad (7.2)$$

$$b_1 Y_{yy} + b_2 Y_y + b_3 Y = -\lambda Y, \qquad (7.3)$$

where λ is a constant. X and Y satisfy separate equations and therefore can be found, at least in principle, though they will both depend on λ. The equations previously solved are all particular cases of equation (7.1).

(2) *Determining, from the boundary conditions, an infinitely denumerable set of values of λ $(\lambda_1, \lambda_2, \ldots)$ such that the solution u can be represented by*

$$u = \sum_{n=1}^{\infty} X_n(x) Y_n(y). \qquad (7.4)$$

In equation (7.4) one of the sets of functions $X_n(x)$, $Y_n(y)$ $(n = 1, 2, \ldots \infty)$ can be specified uniquely and the representation is such that some of the boundary conditions on u are satisfied automatically. (A representation, like that of equation (7.4), of the solution of a partial differential equation as a sum of independent solutions is only possible for linear equations, and is an illustration of the principle of superposition.) We assume that $X_n(x)$ are the known functions; they will be particular solutions of equation (7.2) with $\lambda = \lambda_n$. In the examples considered previously $\lambda_n = n\pi/a$, $X_n \equiv \sin n\pi x/a$, and the representation used vanished automatically at $x = 0$ and $x = a$.

(3) *Determining, from properties of u (and/or u_y) on lines y = constant, sufficient information about Y_n (and/or $(Y_n)_y$) on such lines for a unique solution to be found for Y_n.* This effectively requires solving, for $Y_n(y_0)$, equations like

$$u(x, y_0) = \sum_{n=1}^{\infty} X_n(x) Y_n(y_0),$$

where y_0 is a constant and $u(x, y_0)$ is a known function of x, and the solution has to be unique. Explicit values can be obtained for $Y_n(y_0)$ if the X_n form a complete orthogonal set, and in the previous examples this is known to be the case from the theory of Fourier series.

(4) *Determining Y_n and hence the complete series solution.*

The best method of tackling any given problem is to systematically work through the above steps and see whether the whole calculation can be completed. Steps (2) and (4) can, in many circumstances, lead to problems in ordinary differential equations which can be resolved only by either detailed knowledge of particular special functions (e.g. Bessel functions, Legendre functions), or by numerical methods. In the latter case it is generally best to abandon trying to find series solutions and to attempt to solve the original problem by direct numerical methods. It is worth noting, however, that all the above steps can, in principle, be carried out when:

(a) the equation has the general form in equation (7.1),

(b) one of the variables (which we will always denote by x) has bounded range, i.e. $a \leq x \leq b$, where a and b are finite,

(c) the differential equation for X can be rearranged in the Sturm–Liouville form (§1.5),

(d) the boundary conditions on the lines $x = a$ and $x = b$ are one of

$$\left.\begin{array}{l} \text{(i)} \quad \alpha u(a, y) + \beta u_x(a, y) = 0, \ \gamma u(b, y) + \delta u_x(b, y) = 0, \\ \text{where } \alpha, \beta, \gamma, \delta \text{ are constants,} \\ \text{(ii)} \quad u(a, y) = u(b, y), \ u_x(a, y) = u_x(b, y). \end{array}\right\} \quad (7.5)$$

Equations (7.5)(i) and (ii) will be satisfied when X satisfies the same conditions on $x = a$ and $x = b$ as u and this, together with (c), means that the problem for X is a non-singular Sturm–Liouville eigenvalue problem. The results quoted in §1.5 then ensure that steps (2) and (3) are feasible.

The method may still be suitable when the coefficient of u_{xx} (i.e. p of equation (1.30)) vanishes at $x = a$ or $x = b$, or when one (or both) of a and b are unbounded, provided that the resulting singular Sturm–Liouville problem is known to have a denumerable infinity of solutions. Success in this case will, however, depend on what is known, or can be determined, about a particular differential equation and in this chapter we only consider problems which lead to non-singular Sturm–Liouville problems or to singular Sturm–Liouville problems described in §1.5 (i.e. problems leading to expansions in terms of Bessel functions or Legendre polynomials).

It follows from the general form of equations (7.2) and (7.3) that if it is possible to guess suitable X_n for the representation of equation (7.4), then the equation for Y_n can be found directly by substituting the representation into equation (7.1), interchanging summation and differentiation, and equating the coefficients of X_n to zero. In general it is better to go formally through the above steps than to attempt to guess the correct representation, though for the particular case $a_2 = 0$, a_1, a_3 constant (which often occurs) it is possible to bypass some of the steps. In this case the equation for X becomes

$$a_1 X_{xx} + (a_3 - \lambda) X = 0,$$

whose solutions are trigonometric functions. Requiring X to vanish when $x = a$ and $x = b$ will lead, as in Examples 4.1 and 4.4, to $X_n = \sin\{n\pi(x-a)/(b-a)\}$. Similarly $X = 0$ at $x = a$ and $X_x = 0$ at $x = b$ gives $X_n = \sin\{(2n-1)\pi(x-a)/2(b-a)\}$. The possibilities for the most commonly occurring boundary conditions are summarized in Table 7.1.

TABLE 7.1

(i) $u = 0$, $x = a, b$,	$\displaystyle\sum_{n=1}^{\infty} Y_n(y)\sin\frac{n\pi(x-a)}{(b-a)}$
(ii) $u_x = 0$, $x = a, b$	$\displaystyle\sum_{n=0}^{\infty} Y_n(y)\cos\frac{n\pi(x-a)}{(b-a)}$
(iii) $u = 0$, $x = a$, $u_x = 0$, $x = b$	$\displaystyle\sum_{n=1}^{\infty} Y_n(y)\sin\frac{(2n-1)\pi(x-a)}{2(b-a)}$
(iv) $u_x = 0$, $x = a$, $u = 0$, $x = b$	$\displaystyle\sum_{n=1}^{\infty} Y_n(y)\cos\frac{(2n-1)\pi(x-a)}{2(b-a)}$
(v) $u(a, y) = u(b, y)$ $u_x(a, y) = u_x(b, y)$	$\displaystyle\sum_{n=1}^{\infty} Y_n(y)\sin\frac{2n\pi(x-a)}{(b-a)}$ $+$ $\displaystyle\sum_{n=0}^{\infty} Z_n(y)\cos\frac{2n\pi(x-a)}{(b-a)}$

$a_2 = 0$, a_1, a_3 constant.

Unless $b_1 \equiv 0$, the equation for Y_n will be of the second order and two conditions will have to be imposed on Y_n for particular values of y. These can either be prescribing Y_n and $(Y_n)_y$ on $y = c$ (arising from prescribing u and u_y on $y = c$) or prescribing Y_n (or $(Y_n)_y$) on $y = c$ and $y = d$ (arising from prescribing u (or u_y) on $y = c$ and $y = d$). In the first case the problem is a Cauchy one on $y = c$ so that the equation has to be hyperbolic (i.e. $a_1 b_1 < 0$) and the solution is then determined in $a < x < b$, $y > c$ or $y < c$, depending on which side of $y = c$ it is sensible, in a given context, to obtain a solution. In the second case the boundary conditions are those appropriate to an elliptic equation (i.e. $a_1 b_1 > 0$) and the solution is then determined in $a < x < b$, $c < y < d$.

When $b_1 = 0$ the equation is parabolic and prescribing u on $y = c$ determines $Y_n(c)$ and hence $Y_n(y)$. In this case the solution is determined in $a < x < b$, $y > c$ ($y < c$) according, as discussed in Chapter 6, as to whether a_1/b_2 is negative or positive. In discussing hyperbolic and parabolic equations we shall adopt the convention of using x for the 'space-like' variable and y for the 'time-like' variable.

Straightforward application of separation of variables is only effective when the equation does not change its character (e.g.

from elliptic to hyperbolic) within a region and this means that the coefficients a_1 and b_1 (and also b_2 if $b_1 \equiv 0$) must always remain the same sign in that region.

It should be noted that the method of deriving series solutions is a purely formal one, which assumes the legitimacy of interchanging summation and differentiation, and it is necessary to examine whether or not the final solution justifies the assumptions made in deriving it.

Examples 7.1 to 7.4 provide some additional illustrations of deriving solutions of boundary-value problems and further examples may be found in Williams (1973).

Example 7.1

Solve

$$u_{xx} = u_y,$$

in $0 < x < a$, $y > 0$, with

$$u_x(0, y) = 0, \qquad u(a, y) = 0, \qquad u(x, 0) = x^2(x - a).$$

(If y is taken to be time this problem can be interpreted as determining the temperature in the slab $0 < x < a$ with the face $x = 0$ insulated and that at $x = a$ kept at zero temperature, the initial temperature in the slab being $x^2(x - a)$.)

In this case the coefficient of u_{xx} is constant and that of u is zero so that the representation can be obtained immediately from Table 7.1(iv). Substituting the appropriate series into the equation gives

$$(Y_n)_y = -\frac{(2n-1)^2 \pi^2}{4a^2} Y_n,$$

so that

$$u = \sum_{n=1}^{\infty} C_n \cos \frac{(2n-1)}{2a} \pi x \exp\left\{-\frac{(2n-1)^2}{4a^2} \pi^2 y\right\}.$$

The condition on $y = 0$ gives

$$x^2(x - a) = \sum_{n=1}^{\infty} C_n \cos \frac{(2n-1)}{2a} \pi x,$$

and it then follows from equation (1.29) that

$$C_n = \frac{2}{a} \int_0^a x^2(x-a)\cos \frac{(2n-1)}{2a} \pi x \, dx$$

$$= \frac{(-1)^n 64a^3}{(2n-1)^3 \pi^3} - \frac{192a^3}{(2n-1)^4 \pi^4}.$$

Example 7.2

Determine, within the circle $r = 1$, the solution of Laplace's equation equal to $\cos^2\theta$, where (r, θ) are polar coordinates with origin at the centre of the circle.

Laplace's equation in polar coordinates is

$$r^2 u_{rr} + r u_r + u_{\theta\theta} = 0, \quad \text{where} \quad 0 \le r \le 1, \quad 0 \le \theta \le 2\pi,$$

and the boundary condition is $u(1, \theta) = \cos^2\theta$.

At first sight it appears that there are insufficient boundary conditions but the fact that the equation is defined for $0 \le \theta \le 2\pi$ means that $u(r, 0) = u(r, 2\pi)$, $u_\theta(r, 0) = u_\theta(r, 2\pi)$. If we take θ to correspond to x then the above problem is one for which the representation can be found from Table 7.1(v) and we obtain

$$u = \sum_{n=1}^{\infty} Y_n(r)\sin n\theta + \sum_{n=0}^{\infty} Z_n(r)\cos n\theta.$$

Substituting this representation into Laplace's equation and equating the coefficients of $\cos n\theta$ and $\sin n\theta$ to zero implies that Y_n and Z_n are both solutions of

$$r^2 V_{rr} + r V_r - n^2 V = 0.$$

Again it appears that there are not enough conditions to find V but it should be noted that the fact that the equation holds in $0 \le r < 1$ means that V has to be bounded at $r = 0$. The independent solutions for V are $r^{\pm n}$ ($n > 0$), 1 and $\log r$ ($n = 0$) and the boundedness criterion then enables us to reject r^{-n} and $\log r$, so that

$$u = \sum_{n=1}^{\infty} A_n r^n \sin n\theta + \sum_{n=0}^{\infty} B_n r^n \cos n\theta.$$

Setting $r = 1$ gives

$$\cos^2 \theta = \sum_{n=1}^{\infty} A_n \sin n\theta + \sum_{n=0}^{\infty} B_n \cos n\theta.$$

A_n and B_n can, therefore, be found from equation (1.19) but in this case it is much simpler to write $\cos^2\theta$ as $\frac{1}{2}(1 + \cos 2\theta)$, giving

$$A_n \equiv 0, \qquad B_n \equiv 0, \qquad n \neq 0, 2. \qquad B_0 = B_2 = \tfrac{1}{2}.$$

The solution is therefore $u(r, \theta) = \frac{1}{2} + \frac{1}{2}r^2 \cos 2\theta$.

The validity of the representation used here is a consequence of the existence of the Fourier series expansion of $u(r, \theta)$ as a function of θ. We can therefore deduce that the general solution of Laplace's equation, in a region excluding the origin, which is bounded at infinity is of the form

$$\text{constant} + \sum_{n=1}^{\infty} r^{-n}(A_n \cos n\theta + B_n \sin n\theta).$$

Thus u_r is $O(r^{-2})$ as $r \to \infty$ and this is the result quoted in §4.2 in proving Theorem 4.2 for unbounded regions.

Example 7.3

Solve

$$u_{xx} + x^{-1}u_x + u_{yy} = 0, \qquad 0 \leq x < 1, \qquad y > 0,$$

with

$$u(1, y) = 0, \qquad u(x, 0) = 1, \qquad \lim_{y \to \infty} u = 0.$$

(If x is taken to be the radial coordinate of a cylindrical polar coordinate system then the equation is the axi-symmetric form of Laplace's equation, with y measured along the axis of the coordinate system. The problem is, therefore, that of solving Laplace's equation within the semi-infinite cylinder, $0 \leq r < 1$, $y > 0$, with u vanishing on the curved surface of the cylinder, tending to zero at infinity and equal to unity on the base of the cylinder.)

Seeking a product form of solution gives

$$x^2 X_{xx} + x X_x + \lambda x^2 X = 0,$$

$$Y_{yy} - \lambda Y = 0.$$

As x is the bounded variable it seems better to try and solve the equation for X to give values for λ. The equation for X is Bessel's equation of order zero with, in a usual notation, the general solution $X = AJ_0(\lambda^{\frac{1}{2}}x) + BY_0(\lambda^{\frac{1}{2}}x)$. Since the equation holds in $0 \le x < 1$, u must be bounded when $x = 0$, so that $B \equiv 0$. In order that u vanishes when $x = 1$ we must have $J_0(\lambda^{\frac{1}{2}}) = 0$, it is known from the properties of Bessel functions (Watson 1966) that there exist an infinite number of such real values of λ denoted, in the convention defined in §1.5, by j_{n0}^2 $(n = 1, \ldots)$. Hence

$$u = \sum_{n=1}^{\infty} (A_n e^{-j_{n0}y} + B_n e^{j_{n0}y}) J_0(j_{n0}x),$$

and the condition as $y \to \infty$ gives $B_n = 0$; further, setting $x = 1$ gives

$$1 = \sum_{n=1}^{\infty} A_n J_0(j_{n0}x).$$

We therefore deduce, using equations (1.34) and (1.35), that

$$A_n = \frac{2}{J_1^2(j_{n0})} \int_0^1 x J_0(j_{n0}x) \, dx,$$

A_n can be evaluated using results quoted in Watson (1966) to give

$$A_n = \frac{2}{j_{n0}} J_1(j_{n0}),$$

and hence

$$u = 2 \sum_{n=1}^{\infty} \frac{e^{-j_{n0}y} J_0(j_{n0}x)}{j_{n0} J_1(j_{n0})}.$$

Example 7.4

Solve

$$(r^2 u_r)_r + \operatorname{cosec} \theta (\sin \theta u_\theta)_\theta = 0$$

in $r > a$, $0 \le \theta \le \pi$, with

$$u(a, \theta) = f(\theta), \qquad \lim_{r \to \infty} u = 0.$$

(This equation is Laplace's equation in the usual spherical coordinate system (r, θ, ϕ) when axial symmetry is assumed. The boundary-value problem is thus that of solving the axi-symmetric Laplace equation in the region exterior to a sphere of radius a; with u taking given values on the sphere and vanishing at infinity.)

Assuming product solutions of the form $R(r)\Theta(\theta)$ gives
$$(r^2 R_r)_r - \lambda R = 0,$$

$$\operatorname{cosec} \theta [\sin \theta \Theta_\theta]_\theta + \lambda \Theta = 0,$$

where λ is a constant. On writing $x = \cos \theta$ and setting $\Theta(\cos^{-1} x) = X(x)$, the equation for Θ becomes

$$(1 - x^2)X_{xx} - 2xX_x + \lambda X = 0. \tag{7.6}$$

The boundary conditions on $r = a$ are inhomogeneous and this suggests that, if separation of variables is to work, there must be a denumerable infinity of eigenvalues of equation (7.6). A difficulty arises, however, in that there are no obvious boundary conditions imposed on lines $x = \text{constant}$, the only requirement being that u is differentiable in $0 \le \theta \le \pi$ (i.e. in $|x| \le 1$). This, however, is sufficient to determine values of λ since equation (7.6) is Legendre's equation, and it is known from the general properties of this equation that it only has a solution bounded for $|x| \le 1$ when $\lambda = n(n+1)$, where n is a positive integer. This solution is proportional to the Legendre polynomial $P_n(x)$, and hence

$$u = \sum_{n=0}^{\infty} R_n(r)P_n(\cos \theta),$$

where

$$(r^2(R_n)_r)_r - n(n+1)R_n = 0.$$

The general solution of this equation is

$$R_n = A_n r^n + B_n r^{-n-1},$$

and the condition as $r \to \infty$ gives $A_n \equiv 0$. Imposing the condition on $r = a$ gives

$$f(\theta) = \sum_{n=0}^{\infty} A_n a^{-n-1} P_n(\cos \theta)$$

and it now follows from equations (1.36) and (1.37) that

$$A_n = \frac{(2n+1)a^{n+1}}{2} \int_0^\pi f(\theta) P_n(\cos\theta) \sin\theta \, d\theta.$$

If the given function $f(\theta)$ is reasonably simple then it may be possible to use the properties of Legendre polynomials (Ritt 1970) to find a closed form for A_n.

For the case when $f(\theta)$ is a polynomial in $\cos\theta$ and of low degree it is possible to avoid evaluating the integral by using the known expressions for the Legendre polynomials. We have that

$$P_0(x) = 1, \quad P_1(x) = x, \quad P_2(x) = \tfrac{1}{2}(3x^2 - 1), \quad P_3(x) = \tfrac{1}{2}(5x^3 - 3x),$$

and in general $P_n(x)$ is determined by Rodrigues formula

$$P_n(x) = \frac{1}{2^n n!} \frac{d^n}{dx^n} (x^2 - 1)^n.$$

For the particular case of $f(\theta) = \cos\theta$ we therefore have

$$A_n \equiv 0, \quad n \neq 1, \quad A_1 = a^2,$$

so that

$$u = \frac{a^2}{r^2} \cos\theta.$$

The last two examples illustrate the point that, in particular cases, successful completion of a solution may require extensive knowledge of special function theory.

By a generalization of the approach used to calculate Green's functions in Example 4.4, the general method of obtaining product solutions can be used to solve inhomogeneous equations of the form

$$a_1 u_{xx} + a_2 u_x + a_3 u + b_1 u_{yy} + b_2 u_y + b_3 u = g(x, y). \qquad (7.7)$$

If the X_n corresponding to equation (7.2) and the relevant homogeneous boundary conditions can be found then, since the X_n are assumed to form a complete set, there will exist unique functions G_n such that

$$g = \sum_{n=1}^\infty G_n(y) X_n(x). \qquad (7.8)$$

Substituting from equations (7.4) and (7.8) into equation (7.7), interchanging the orders of summation and integration and equating the coefficients of X_n to zero, gives

$$b_1(Y_n)_{yy} + b_2(Y_n)_y + b_3 Y_n = \lambda_n Y_n + G_n,$$

which is a differential equation for the Y_n.

It is also possible to relax the restriction that the boundary conditions in equation (7.5)(i) have to be homogeneous. If a function v is found which satisfies the inhomogeneous boundary conditions (but not the partial differential equation) then, when u is a solution of equation (7.7), u_1 defined by $u = v + u_1$ will, in view of the linearity of the boundary conditions and of equation (7.7), satisfy the corresponding homogeneous boundary condition and be a solution of an equation of the same type as equation (7.7). If, for example, $u(a, y) = k_1(y)$ and $u(b, y) = k_2(y)$, then a possible form for v is

$$\frac{(x-b)}{(a-b)} k_1(y) - \frac{(x-a)}{(a-b)} k_2(y).$$

Worked examples relating to the above points may be found in Williams (1973).

The extension to cover inhomogeneous equations and boundary conditions can be achieved in many cases by introducing finite transforms. One simple such transform is described in §7.4 and the use of finite transforms is effectively another method of presenting the derivation of series solutions obtained using separation of variables.

In cases where the solution of a partial differential equation cannot be represented as a series of discrete products of the form $X(x)Y(y)$ an integral transform method often proves useful. Such methods still rely on the existence of product solutions and are applicable to cases where the solution satisfying some of the conditions can only be expressed in the form

$$\int X(s, x) Y(s, y) \, ds,$$

where s is effectively the separation constant. They can be used for equation (7.7) and for

$$u_{xy} + a_2 u_x + a_3 u + b_2 u_y + b_3 u = g(x, y), \qquad (7.9)$$

where the *a*'s and *b*'s are as in equation (7.1). In the subsequent sections the use of the more commonly occurring integral transforms will be described.

Exercises 7.1

1. Solve
$$u_{xx} + u_{yy} = 2, \qquad 0 < x < 1, \qquad 0 < y < 1,$$
with
$$u(0, y) = u(1, y) = u(x, 0) = u_y(x, 1) = 0.$$

2. Solve
$$u_{xx} - u_{yy} = e^{-y}, \qquad y > 0, \qquad 0 < x < \pi,$$
with
$$u(0, y) = u(\pi, y) = u(x, 0) = u_y(x, 0) = 0.$$

3. Solve
$$u_{xx} = u_y, \qquad 0 < x < 1, \qquad y > 0,$$
with
$$u(0, y) = y, \qquad u(1, y) = 0, \qquad u(x, 0) = 0.$$

4. Solve
$$u_{rr} + \frac{1}{r} u_r = u_{yy}, \qquad 0 \le r < 1, \qquad y > 0,$$
with
$$u(1, y) = 0, \qquad u(r, 0) = 1 - r^2, \qquad u_y(r, 0) = 0.$$

5. Solve
$$\sin\theta (r^2 u_r)_r + (\sin\theta u_\theta)_\theta = 0, \qquad r > a, \qquad 0 \le \theta \le \pi,$$
with
$$u(a, \theta) = \cos^2\theta, \qquad u \to 0 \quad \text{as} \quad r \to \infty.$$

7.2. The Laplace transform

The Laplace transform which is often used to reduce initial-value problems for certain classes of ordinary differential equations to the solution of algebraic equations, also proves to be a powerful method of solving initial-value problems for certain types of partial differential equations. In this case the Laplace transform reduces the partial differential equation to an ordinary differential equation rather than to an algebraic one.

The Laplace transform $\bar{f}(s)$ of a function $f(y)$ is defined by

$$\bar{f}(s) = \int_0^\infty e^{-sy} f(y) \, dy \qquad (7.10)$$

and hence the Laplace transform with respect to y of a function $u(x, y)$ is the function $\bar{u}(x, s)$ given by

$$\bar{u}(x, s) = \int_0^\infty e^{-sy} u(x, y) \, dy. \qquad (7.11)$$

In general the Laplace transform turns out to be most useful in problems where one of the variables is 'time-like', and the Laplace transform is generally taken with respect to this variable. Equations (7.10) and (7.11) anticipate this and perpetuate our convention of denoting, where such a distinction is possible, the 'time-like' variable by y. We shall not pursue at all the question of the existence of the Laplace transform; this is discussed in detail in Sneddon (1972) and we simply quote the basic results proved by Sneddon.

Theorem 7.1

If

(a) $f(y)$ is integrable over any finite interval $0 < a \leq y \leq b$,

(b) there exists a real number c such that, for any arbitrary $b > 0$, $\int_b^\lambda e^{-cy} f(y) \, dy$ tends to a finite limit as $\lambda \to \infty$,

(c) for arbitrary $a > 0$, $\int_\varepsilon^a |f(y)| \, dy$ tends to a finite limit as $\varepsilon \to 0$,

then the Laplace transform of f exists and is an analytic function of s for $\mathrm{Re}\, s > c$ and tends to zero as $s \to \infty$. Further, if a given function $\bar{f}(s)$ is such that $\lim_{s \to \infty} \bar{f} \neq 0$ then it is not the Laplace transform of a function satisfying the above conditions and $\lim_{s \to \infty} \bar{f} = 0$, is a necessary condition for $\bar{f}$ to be the Laplace transform of a function for which such a transform can be defined.

Theorem 7.2

If $\bar{f}(s)$ is an analytic function of the complex variable s in some half-plane $\operatorname{Re} s \geq \gamma$ and is $O(s^{-k})$ as $|s| \to \infty$, where k and γ are real with $k > 1$, then

$$\frac{1}{2\pi i} \lim_{w \to \infty} \int_{c-iw}^{c+iw} e^{sy}\bar{f}(s) \, ds,$$

where $c > \gamma$, converges independently of c, to a function $f(y)$ whose Laplace transform is $\bar{f}(s)$. Furthermore the function $f(y)$ is continuous for each $y \geq 0$ and is $O(e^{\gamma y})$ as $y \to \infty$.

Theorem 7.2 can be stated more crudely as implying that the 'inverse' $f(y)$ of the Laplace transform $\bar{f}(s)$ is given by

$$f(y) = \frac{1}{2\pi i} \int_C e^{sy}\bar{f}(s) \, ds, \tag{7.12}$$

where C is a line parallel to the imaginary s-axis and to the right of all singularities of $\bar{f}$.

Theorem 7.1 gives sufficient conditions for the existence of a Laplace transform and Theorem 7.2 provides a method of obtaining the original function from its Laplace transform.

The value of the Laplace transform in solving differential equations lies in the simple relationship between the Laplace transform of f and that of $f_y, f_{yy}, \ldots$ etc. We have, on integrating by parts, that for continuous functions f,

$$\int_0^\infty e^{-sy}f_y \, dy = [e^{-sy}f]_0^\infty + s\int_0^\infty fe^{-sy} \, dy.$$

A necessary condition for the existence of the Laplace transform is that $\lim_{y \to \infty} e^{-sy}f = 0$ (otherwise the integral does not converge); hence

$$\int_0^\infty e^{-sy}f_y \, dy = -f(0) + s\bar{f}. \tag{7.13}$$

Repeated integration by parts gives

$$\int_0^\infty e^{-sy}f_{yy} \, dy = s^2\bar{f} - sf(0) - f_y(0), \tag{7.14}$$

and similar expressions can be obtained for the higher derivatives. It also follows from equation (7.13) and (7.14) that

$$\int_0^\infty e^{-sy} u_y \, dy = s\bar{u} - u(x, 0), \qquad (7.15)$$

$$\int_0^\infty e^{-sy} u_{yy} \, dy = s^2\bar{u} - su(x, 0) - u_y(x, 0). \qquad (7.16)$$

Taking the Laplace transform with respect to y of equation (7.7) (i.e. multiplying by e^{-sy} and integrating with respect to y from 0 to ∞) gives, on using equations (7.15) and (7.16) and assuming that b_1 to b_3 are constant,

$$\int_0^\infty (a_1 u_{xx} + a_2 u_x + a_3 u)e^{-sy} \, dy + b_1[s^2\bar{u} - su(x, 0) - u_y(x, 0)]$$
$$+ b_2[s\bar{u} - u(x, 0)] + b_3\bar{u} = \bar{g}.$$

This equation simplifies further if it is assumed that the order of differentiation and integration may be interchanged, and we then obtain

$$a_1 \bar{u}_{xx} + a_2 \bar{u}_x + a_3 \bar{u} + b_1[s^2\bar{u} - su(x, 0) - u_y(x, 0)]$$
$$+ b_2[s\bar{u} - u(x, 0)] + b_3\bar{u} = \bar{g}. \quad (7.17)$$

Equation (7.17) is, when $u(x, 0)$ and $u_y(x, 0)$ are known, effectively an ordinary differential equation for $\bar{u}$, and the problem is solved if $\bar{u}$ can be found and u then obtained from it. The Laplace transform therefore provides a method of solving equation (7.7) (with b_1 to b_3 constant) when Cauchy conditions are imposed on $y = 0$, and analogy with the wave equation and the heat-conduction equation implies that y is a time-like variable.

Similarly taking the Laplace Transform of equation (7.9) (with b_1 to b_3 constant) gives

$$s\bar{u}_x - u_x(x, 0) + a_2\bar{u}_x + a_3\bar{u} + b_2[s\bar{u} - u(x, 0)] + b_3\bar{u} = \bar{g}, \quad (7.18)$$

so that, when $u(x, 0)$ is known, a differential equation is obtained for $\bar{u}$. Equation (7.9) is hyperbolic with characteristics parallel to the axes and equation (7.18) shows that the Laplace transform can therefore be used to solve some characteristic boundary-value problems for hyperbolic equations.

Once it has been decided that it is appropriate to use the

Laplace transform, the detailed calculation can be regarded as consisting of three separate stages:

(1) Obtaining the ordinary differential equation for $\bar{u}$.

(2) Solving this equation to determine $\bar{u}$ uniquely (this requires there being enough boundary conditions on lines $x =$ constant, if u or u_x are known for specific values of x then u or u_x can be found for these values of x).

(3) Determining u from $\bar{u}$.

The first stage is carried out using equations (7.15) and (7.16) and leads to an equation of the same general form as equations (7.17) and (7.18). The most difficult part is likely to be calculating $\bar{g}$; there are however many tables of Laplace transforms and use of these can often help to avoid tedious calculation. One such table is available in Volume 1 of Erdelyi, Magnus, and Oberhettinger (1954) which will be referred to as IT1. For ease of reference some of the simpler and more frequently occurring Laplace transform relations are displayed in Table 7.2. The second stage can only be carried out when the equation for $\bar{u}$ is such that its general solution can be found, and this places a restriction on the method. The final stage of the calculation is most easily carried out by using tables, but if appropriate tables are not available it then becomes necessary to calculate the contour integral in equation (7.12). Example 7.8 and the analysis following this example illustrate the general methods that can be used to invert Laplace transforms. The determination of the Green's functions in §5.3 and §6.3 followed the above pattern, the most complicated part in each case being the final inversion, this was effected in both cases by using tables.

It is however often possible to obtain some useful information about u for large and small values of y directly from properties of $\bar{u}$ without carrying out the explicit inversion. This kind of information can be very valuable in practical circumstances, particularly when the solution cannot be obtained in closed form. It is convenient to state the basic results in the form of theorems and we have

Theorem 7.3 (**Watson's lemma**)

If the function $f(y)$ is such that, for some α, $|f(y)|e^{-\alpha y}$ is bounded and, in a neighbourhood of $y = 0$, $f(y)$ has the expan-

TABLE 7.2

	Function	Laplace transform
1	$f(y)$	$\bar{f}(s)$
2	$e^{-ay}f(y)$	$\bar{f}(s+a)$
3	$f(ay), \quad a>0$	$a^{-1}\bar{f}(s/a)$
4	$\displaystyle\int_0^y f(t)\,dt$	$\bar{f}(s)/s$
5	$\displaystyle\int_0^y f(t)g(y-t)\,dt$	$\bar{f}(s)\bar{g}(s)$

(This is known as the convolution theorem)

6	$f(y-a)H(y-a)$, where $H(z)$ is Heaviside's unit function $H(z)=1, \ z\geq 0, \ H(z)=0, \ z<0.$	$e^{-as}\bar{f}(s)$

(This result is often referred to as Heaviside's shift theorem)

	Function	Laplace transform
7	1	s^{-1}
8	e^{-ay}	$(s+a)^{-1}$
9	$\cos ay$	$s(s^2+a^2)^{-1}$
10	$\sin ay$	$a(s^2+a^2)^{-1}$
11	y^n	$\Gamma(n+1)s^{-n-1} \, n>-1$
12	$y^{-\frac{1}{2}}e^{-c/y}$	$(\pi/s)^{\frac{1}{2}}e^{-2(cs)^{\frac{1}{2}}}$
13	$y^{\frac{1}{2}n}J_n[2(ay)^{\frac{1}{2}}]$	$a^{\frac{1}{2}n}s^{-n-1}e^{-a/s}$
14	$\mathrm{Erfc}(ay^{-\frac{1}{2}})$	$s^{-1}e^{-2as^{\frac{1}{2}}}$

where $\mathrm{Erfc}\, z = \dfrac{2}{\pi^{\frac{1}{2}}}\displaystyle\int_z^\infty e^{-t^2}\,dt$

Most of the above results are proved in Sneddon (1972).

sion $\sum_{n=0}^\infty a_n y^{\lambda+n}$ then, as $|s|\to\infty$, $\bar{f}$ has the asymptotic expansion $\sum_{n=0}^\infty a_n\Gamma(n+1+\lambda)s^{-(\lambda+n+1)}$.

This lemma is proved in most books on asymptotic expansions, e.g. Murray (1974) or Copson (1965).

Watson's lemma thus relates the form of $f(y)$, for y small, to the form of $\bar{f}$ for large s and, in particular, shows that the behaviour for large s is determined by substituting into the defining integral the series expansion for f and integrating term by term. It is not strictly correct to attempt to deduce, from the asymptotic expansion of $\bar{f}(s)$, the behaviour of f near $y=0$ as there is no *a priori* reason to expect f to satisfy the conditions of

Watson's lemma. In practical circumstances in partial differential equations, however, the expectation is that u would satisfy these conditions and Watson's lemma can be treated as being reversible. In the particular case when an absolutely convergent expansion in inverse powers of s can be obtained for $\bar{f}$, then it can be proved rigorously that term-by-term inversion is permissible. More generally, if $\bar{f}$ can be written as an uniformly convergent series of the form $\sum_{n=0}^{\infty} a_n \phi_n(s)$, where $\lim_{s \to \infty} \phi_{n+1}/\phi_n = 0$, then term-by-term inversion of this expansion is permissible and can lead to a useful representation of f for small values of y. This procedure is illustrated in Example 7.8.

Theorem 7.4

If (i) $\bar{f}(s)$ is $O(1/|s|)$ as $|s| \to \infty$, (ii) the singularity of $\bar{f}$ with the largest real part is at $s = s_1$, and (iii) in some neighbourhood of s_1,

$$\bar{f}(s) = (s - s_1)^{\lambda} \sum_{n=0}^{\infty} a_n (s - s_1)^n, \qquad -1 < \lambda \le 0,$$

then

$$f(y) \sim -\frac{1}{\pi} \frac{e^{s_1 y}}{y^{\lambda+1}} \sin \lambda \pi \sum_{n=0}^{\infty} a_n (-1)^n y^{-n} \Gamma(\lambda + n + 1) \quad \text{as} \quad y \to \infty.$$

If the singularity at $s = s_1$ is a pole of residue r, then

$$f(y) \sim r e^{s_1 y} \quad \text{as} \quad y \to \infty.$$

Proof

We assume first that the only singularity of $\bar{f}(s)$ is at $s = s_1$, $\bar{f}(s)$ will have a branch point at $s = s_1$ and the corresponding branch line is drawn from $s = s_1$, as shown in Fig. 7.1.

The condition on $\bar{f}(s)$ as $|s| \to \infty$ and the absence of singularities other than at $s = s_1$ ensure that the contour in the inversion integral of equation (7.12) may be deformed into the loop Γ shown in Fig. 7.1 so that

$$f(y) = \frac{1}{2\pi i} \int_{\Gamma} e^{sy} \bar{f}(s) \, ds.$$

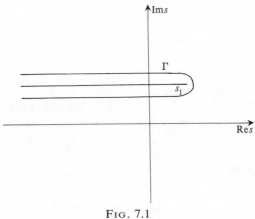

Fig. 7.1

On the lower side of the branch cut we have that $s - s_1 = re^{-i\pi}$, $0 \le r \le \infty$ and on the upper side $(s - s_1) = re^{i\pi}$ so that

$$f(y) = \frac{1}{2\pi i} \int_\infty^0 e^{(s_1 - r)y} \bar{f}(s_1 + re^{-i\pi}) e^{-i\pi} \, dr$$

$$+ \frac{1}{2\pi i} \int_0^\infty e^{(s_1 - r)y} \bar{f}(s_1 + re^{i\pi}) e^{i\pi} \, dr$$

$$= -\frac{1}{2\pi i} \int_0^\infty e^{(s_1 - r)y} [\bar{f}(s_1 + re^{+i\pi}) - \bar{f}(s_1 + re^{-i\pi})] \, dr. \quad (7.18)$$

This integral is of the general form which can be evaluated asymptotically by Watson's lemma on replacing $\bar{f}(s_1 + re^{\pm i\pi})$ by series expansions in powers of r and integrating term by term, carrying out the necessary algebra gives the required result.

In the particular case when $\bar{f}$ just has a pole as $s = s_1$, the contour in the inversion formula can be deformed into a small closed loop round $s = s_1$, and it then follows from Cauchy's theorem that the inverse is $r_1 e^{s_1 y}$, where r_1 is the residue at the pole at $s = s_1$.

In the general case when $\bar{f}$ has singularities at $s = s_i$ the contour can be deformed into a sum of loops round the various branch cuts and the presence of the factor $e^{s_i y}$ shows that the dominant contribution to the asymptotic expansion comes from the value of s_i with the largest real part.

Example 7.5

Solve

$$u_{xy} + \tfrac{1}{4}u = 0, \qquad x > 0, \qquad y > 0,$$

with $u(x, 0) = u(0, y) = 1$.

Taking the Laplace transform with respect to y and using equation (7.15) and the condition $u(x, 0) = 1$ gives

$$s\bar{u}_x + \tfrac{1}{4}\bar{u} = 0,$$

whose solution is $\bar{u} = A e^{-x/4s}$. The condition $u(0, y) = 1$ gives

$$\bar{u}(0, s) = \int_0^\infty e^{-sy} u(0, y)\,\mathrm{d}y = \frac{1}{s},$$

and hence $A = 1/s$ so that

$$\bar{u} = \frac{1}{s}\,e^{-x/4s}.$$

The inversion can be effected using Table 7.2(13) (with $n = 0$) giving

$$u = J_0[(xy)^{\frac{1}{2}}],$$

which is the result previously found in Example 5.1 (with $\lambda = \tfrac{1}{4}$).

Example 7.6

Solve

$$u_{xx} = u_y \quad \text{in} \quad x > 0, \qquad y > 0,$$

with

$$u(x, 0) = 0, \qquad u(0, y) = 1, \qquad y > 0, \quad \text{and} \quad \lim_{x \to \infty} u(x, y) = 0.$$

(With y taken to be proportional to a time variable this problem reduces to the Rayleigh problem of determining the motion of a viscous fluid in the half-space immediately above an impermeable plane which is jerked into motion parallel to itself.)

Taking the Laplace transform with respect to y gives

$$\bar{u}_{xx} = s\bar{u},$$

and the solution satisfying the condition as $x \to \infty$ is

$$\bar{u} = A e^{-s^{\frac{1}{2}}x},$$

where $s^{\frac{1}{2}}$ is that branch with positive real part. Also

$$\bar{u}(0, s) = \int_0^\infty u(0, y) e^{-sy} \, dy = \frac{1}{s},$$

so that

$$\bar{u} = \frac{e^{-s^{\frac{1}{2}}x}}{s}.$$

This can be inverted using Table 7.2 (14) to give

$$u = \text{Erfc}(\tfrac{1}{2}xy^{-\frac{1}{2}}).$$

Example 7.7

Find the behaviour as $y \to \infty$, for fixed x, of the solution of

$$u_{yy} = u_{xx} - u, \qquad x > 0, \qquad y > 0,$$

such that

$$u(x, 0) = u_y(x, 0) = 0, \qquad u(0, y) = 1.$$

The Laplace transform, $\bar{u}$, of u with respect to y, satisfies

$$(1 + s^2)\bar{u} = \bar{u}_{xx}$$

and the two independent solutions of this are $e^{-(1+s^2)^{\frac{1}{2}}x}$ and $e^{(1+s^2)^{\frac{1}{2}}x}$. The latter does not tend to zero as $s \to \infty$ and hence does not represent a Laplace transform (Theorem 7.1). Thus the appropriate solution satisfying the condition on $x = 0$ is

$$\bar{u} = \frac{1}{s} e^{-(1+s^2)^{\frac{1}{2}}x}.$$

It is not possible to obtain the inverse of this transform in a closed form but nevertheless the required information can be extracted from $\bar{u}$ by using Theorems 7.3 and 7.4. $\bar{u}$ has a pole at $s = 0$ and branch points at $s = \pm i$, the real parts of s being equal in all three cases. Thus all have to be taken into account in evaluating u as $y \to \infty$. We have that

$$\text{near } s = i, \qquad \bar{u} \sim \frac{1}{i} \{1 - (2i)^{\frac{1}{2}}(s-i)^{\frac{1}{2}}x + O[(s-i)^2]\},$$

$$\text{near } s = -i, \qquad \bar{u} \sim -\frac{1}{i} \{1 - (-2i)^{\frac{1}{2}}(s+i)^{\frac{1}{2}}x + O[(s+i)^2]\},$$

and $\bar{u}$ has residue e^{-x} at $s = 0$.

Theorem 7.4 then gives that, for fixed x, as $y \to \infty$

$$u(x, y) \sim e^{-x} + \frac{1}{\pi y^{\frac{3}{2}}} [2^{\frac{1}{2}} x \Gamma(\tfrac{3}{2}) e^{i[y+(\pi/4)]} + 2^{\frac{1}{2}} x \Gamma(\tfrac{3}{2}) e^{-i[y+(\pi/4)]}]$$

$$= e^{-x} + \frac{2^{\frac{1}{2}} x}{\pi^{\frac{1}{2}} y^{\frac{3}{2}}} \cos\left(y + \frac{\pi}{4}\right).$$

Example 7.8

Solve

$$u_{xx} = u_y \quad \text{in} \quad y > 0, \qquad 0 < x < a,$$

subject to $u(x, 0) = 1$, $u(0, y) = u(a, y) = 0$.

The Laplace transform of u with respect to y satisfies

$$\bar{u}_{xx} = s\bar{u} - 1,$$

giving

$$\bar{u} = \frac{1}{s} + A \cosh s^{\frac{1}{2}} x + B \sinh s^{\frac{1}{2}} x.$$

The boundary conditions at $x = 0$ and $x = a$ imply that $A = -s^{-1}$ and $B = s^{-1} \tanh \tfrac{1}{2} s^{\frac{1}{2}} a$; $\bar{u}$ is then given by

$$\bar{u} = \frac{1}{s} - \frac{1}{s} \frac{\cosh s^{\frac{1}{2}}(x - \tfrac{1}{2}a)}{\cosh \tfrac{1}{2} s^{\frac{1}{2}} a}.$$

The inverse of $\bar{u}$ is not available in closed form in tables of transforms and we therefore have to use the formula

$$u = \frac{1}{2\pi i} \int_C \frac{e^{sy}}{s} \left\{1 - \frac{\cosh s^{\frac{1}{2}}(x - \tfrac{1}{2}a)}{\cosh \tfrac{1}{2} s^{\frac{1}{2}} a}\right\} ds,$$

where C is to the right of all singularities of the integrand. The only singularities of $\bar{u}$ are poles at $s = s_n = -(2n+1)^2 \pi^2/a^2$, $n = 0, 1, \ldots$ and hence u is equal to the sum of the residues at the poles of $e^{sy}\bar{u}(x, s)$. The residue at $s = s_n$ is given by

$$\frac{-a^2}{(2n+1)^2\pi^2} e^{s_n y} \frac{\cosh s_n^{\frac{1}{2}}(x-\frac{1}{2}a)}{\left[\dfrac{d}{ds}\cosh \frac{1}{2}s^{\frac{1}{2}}a\right]_{s=s_n}}$$

$$= \frac{4 \exp\left(\dfrac{-(2n+1)^2\pi^2 y}{a^2}\right)}{\pi(2n+1)} \sin(2n+1)\frac{\pi x}{a},$$

Hence

$$u = \frac{4}{\pi} \sum_{n=0}^{\infty} \frac{\exp\left(\dfrac{-(2n+1)^2\pi^2 y}{a^2}\right)}{(2n+1)} \sin(2n+1)\frac{\pi x}{a},$$

which agrees with the result obtained from equation (6.16) by setting $u(x', 0) \equiv 1$ and replacing t by y.

The above series is fairly slowly convergent for small y and, for such values of y, a more useful representation can be obtained by expanding $\bar{u}$ in powers of $\exp - s^{\frac{1}{2}}$ and inverting term by term. We have that

$$\bar{u} = \frac{1}{s}\{1 - e^{-s^{\frac{1}{2}}(a-x)} - e^{-s^{\frac{1}{2}}x} + e^{-s^{\frac{1}{2}}(a+x)} + e^{-s^{\frac{1}{2}}(2a-x)} + O(e^{-2us^{\frac{1}{2}}})\}.$$

This series is uniformly convergent for $\text{Re } s^{\frac{1}{2}} > 0$ and can be inverted term by term to give

$$u = 1 - \text{Erfc}\left(\frac{a-x}{2y^{\frac{1}{2}}}\right) - \text{Erfc}\left(\frac{x}{2y^{\frac{1}{2}}}\right) + \text{Erfc}\left(\frac{a+x}{2y^{\frac{1}{2}}}\right)$$

$$- \text{Erfc}\left(\frac{2a-x}{2y^{\frac{1}{2}}}\right) + \text{Order}\left(\text{Erfc}\frac{2a}{y^{\frac{1}{2}}}\right).$$

$\text{Erfc}(z)$ is of $O(e^{-z^2})$ as $z \to \infty$ and therefore, for small y, the error made in truncating the series obtained by term-by-term inversion is extremely small.

This example illustrates one general method used to obtain inverses of transforms, namely to express them as infinite series, and this method will always work provided that the only singularities of $\bar{u}$ are poles. This approach will of course not work when $\bar{u}$ has branch points but in the case when $\bar{u}$ has a simple branch point some progress can often be made by deforming the

240   *Analytical methods of solution*

contour C into a loop round the branch cut. The resulting integral over a real line can often be evaluated by using tables of infinite integrals or by using real variables. The basic steps in this kind of calculation can be illustrated by looking at the particular case of $\bar{f} = s^{-\frac{1}{2}}e^{-2as^{\frac{1}{2}}}$. The contour of integration in the inversion integral can be deformed into the contour Γ of Fig. 7.1 (with $s_1 = 0$) and f is then given by equation (7.12) with $s_1 = 0$ and $\bar{f} = s^{-1}e^{-2as^{\frac{1}{2}}}$. Carrying out the necessary substitutions gives

$$f(y) = \frac{1}{\pi} \int_0^\infty \frac{e^{-ry} \cos 2ar^{\frac{1}{2}}}{r^{\frac{1}{2}}} \, dr$$

$$= \frac{2}{\pi} \int_0^\infty e^{-v^2y} \cos 2av \, dv, \quad \text{where} \quad v = r^2.$$

This integral can be found either from tables or by obtaining a simple differential equation for f as a function of a, we use the latter method as it is a useful one for evaluating infinite integrals. We have that

$$\frac{df}{da} = -\frac{4}{\pi} \int_0^\infty v e^{-v^2y} \sin 2av \, dv$$

$$= -\frac{4}{\pi} \int_0^\infty \frac{a}{y} e^{-v^2y} \cos 2av \, dv, \quad \text{on integration by parts,}$$

$$= -\frac{2a}{y} f.$$

Hence $f = Ae^{-a^2/y}$ and as $f = (\pi y)^{-\frac{1}{2}}$ when $a = 0$ we have that $A = (\pi y)^{-\frac{1}{2}}$. Thus $f = (\pi y)^{-\frac{1}{2}}e^{-a^2/y}$, a result which agrees with that of Table 7.2(12) with $c = a^2$.

Exercises 7.2

1. u satisfies the equation

$$au_{xx} - bu_x - u_t = 0, \quad x > 0, \quad t > 0, \quad a > 0, \quad b > 0,$$

and the conditions

$$u(x, 0) = 0, \quad u(0, t) = 1, \quad u \to 0 \text{ as } x \to \infty, \quad t > 0.$$

Find the Laplace transform of u with respect to t.

2. Find u satisfying

$$u_{xx} = u_t, \qquad x, t \geq 0,$$

and the conditions

$$u(0, x) = 0, \qquad u(0, t) = t, \qquad \lim_{x \to \infty} u = 0.$$

3. Show that if u satisfies

$$u_{tt} = u_{xx} + u_{xxt}, \qquad x, t \geq 0,$$

and

$$u(x, 0) = u_t(x, 0) = 0, \qquad u(0, t) = 1, \qquad \lim_{x \to \infty} u = 0, \quad \text{then, on} \quad x = 0,$$

$$u_x = -(\pi t)^{-\frac{1}{2}} e^{-t}.$$

4. u satisfies

$$u_{xx} - u_{tt} + u = 0,$$

$$u(x, 0) = u_t(x, 0) = 0, \qquad u(0, t) = t.$$

Show that, for fixed x,

$$u \sim \frac{x e^t}{(2\pi t^3)^{\frac{1}{2}}}, \qquad \text{as} \quad t \to \infty.$$

5. Find, for $x = 0$, the Laplace transform with respect to t of the function u satisfying

$$u_{xx} = u_t, \; x > 0, \; t > 0,$$

$$u(x, 0) = 0, \; -u_x(0, t) = \sin t, \; \lim_{x \to \infty} u = 0.$$

Find the first two terms in the expansion of $u(0, t)$ for small t.

6. Find the Laplace transform with respect to t of u satisfying

$$u_{xx} = u_t, \; t > 0, \; 0 < x < 1,$$

and the conditions

$$u(0, t) = 0, \; u(1, t) = 1, \; u(x, 0) = 1.$$

Obtain an approximation for u which is suitable for evaluating it for small values of t.

7.3. Fourier transforms

The exponential Fourier transform $F(w)$ of a function $f(x)$ is defined by

$$F(w) = \frac{1}{(2\pi)^{\frac{1}{2}}} \int_{-\infty}^{\infty} f(x)e^{iwx}\, dx, \quad \text{where } w \text{ is real.} \quad (7.19)$$

The exponential Fourier transform is defined for real w provided that $\int_{-\infty}^{\infty} |f(x)|\, dx$ exists and it can be shown (Sneddon 1972) that this latter condition is sufficient, when $f(x)$ is continuous, to ensure the validity of the inversion formula

$$f(x) = \frac{1}{(2\pi)^{\frac{1}{2}}} \int_{-\infty}^{\infty} F(w)e^{-iwx}\, dw. \quad (7.20)$$

The choice of the factor multiplying the integral in equation (7.19) is somewhat arbitrary and different conventions are adopted in different texts. The advantage of taking it to be $(2\pi)^{-\frac{1}{2}}$ is that it produces a useful similarity between the inversion formula and the equation defining the transform. Equation (7.20) demonstrates that, over an infinite interval, an arbitrary function can be represented as a superposition of oscillations of all possible frequencies. This representation is the analogue, for functions defined over an infinite interval, of the general Fourier series representation (cf. equation (1.19)) of a function defined over a finite interval. The relationship between the two representations is discussed in most texts relating to Fourier transforms (see, for example, Carslaw 1930).

We shall always use capital letters to denote Fourier transforms so that the exponential Fourier transform with respect to x of the function $u(x, y)$ is $U(w, y)$ given by

$$U(w, y) = \frac{1}{(2\pi)^{\frac{1}{2}}} \int_{-\infty}^{\infty} u(x, y)e^{iwx}\, dx. \quad (7.21)$$

In general it is the exponential Fourier transform with respect to a 'space-like' variable which turns out to be useful in solving boundary-value problems and we continue, where appropriate, to use x as the 'space-like' variable. In some contexts, however, taking the Fourier transform with respect to time can prove to be more convenient. The use of such a transform implies representing a function as a superposition of harmonic oscillations and such a representation is often very useful in analysing the response of linear electric circuits.

The utility of the exponential Fourier transform in solving boundary value problems lies in the simple relationship that exists between the Fourier transform of f and that of f_x, f_{xx} etc. We have by integration by parts that, when f is continuous,

$$\int_{-\infty}^{\infty} f_x e^{iwx} \, dx = [fe^{iwx}]_{-\infty}^{\infty} - iw \int_{-\infty}^{\infty} fe^{iwx} \, dx.$$

A necessary condition for the existence of F is that $\lim_{|x| \to \infty} f = 0$ so that the integrated part in the above expression is zero and therefore

$$\int_{-\infty}^{\infty} e^{iwx} f_x \, dx = -iwF.$$

Similarly, when f and f_x are both continuous,

$$\int_{-\infty}^{\infty} e^{iwx} f_{xx} \, dx = (-iw)^2 F,$$

and the Fourier transform of the nth derivative is obviously $(-iw)^n F$. Thus

$$\int_{-\infty}^{\infty} e^{iwx} u_x \, dx = -iwU, \qquad (7.22)$$

$$\int_{-\infty}^{\infty} e^{iwx} u_{xx} \, dx = -w^2 U, \qquad (7.23)$$

and it follows from these equations that taking the exponential Fourier transform with respect to x of equation (7.7),

$$a_1 u_{xx} + a_2 u_x + a_3 u + b_1 u_{yy} + b_2 u_y + b_3 u = g(x, y), \qquad (7.7)$$

gives, when a_1 to a_3 are constant, the ordinary differential equation

$$-a_1 w^2 U - iw a_2 U + a_3 U + b_1 U_{yy} + b_2 U_y + b_3 U = G. \qquad (7.24)$$

The existence of U presumes that $\lim_{|x| \to \infty} u = 0$ and hence the exponential Fourier transform would appear to be well suited to problems in which $\lim_{|x| \to \infty} u = 0$. Such conditions at either end of an infinite interval normally occur when x is a 'space-like' variable.

The stages involved in solving a problem by means of the exponential Fourier transform are virtually the same as those occurring in the Laplace transform method. The most difficult stage is again the inversion from U to u, and tables of transforms are again available to help to do this; one such table is available in IT1. Some of the simpler exponential Fourier transform relations are displayed in Table 7.3.

The convolution theorem (Table 7.3(4)) is a particularly useful result in that, if a given Fourier transform is a product of the Fourier transforms of relatively simple functions, then the inverse can at least be obtained in terms of a real integral.

In view of the difficulties that can be associated with inverting a transform it is again fortunate that, as for Laplace transforms, it often turns out to be possible to deduce the asymptotic properties of f directly from properties of F. (There is a basic result referred to as the Riemann–Lebesgue lemma which states that the Fourier transform of a continuous function tends to zero as $|w| \to \infty$, but it gives no precise information concerning the behaviour of the transform as $|w| \to \infty$.) The method of calculation is not quite as direct as for Laplace transforms and it is more convenient to use one method when F can be analytically continued into a region

TABLE 7.3

	Function	Exponential Fourier transform
1.	$f(x)$	$F(w)$
2.	$f(ax+b)$	$a^{-1}e^{-ibw/a}F(w/a)$
3.	$e^{iax}f(x)$	$F(w+a)$
4.	$(2\pi)^{-\frac{1}{2}}\displaystyle\int_{-\infty}^{\infty} f(t)g(x-t)\,dt$	$F(w)G(w)$

(This is known as the convolution theorem)

	Function	Exponential Fourier transform
5.	$\delta(x-x')$	$(2\pi)^{-\frac{1}{2}}e^{iwx'}$
6.	$e^{-a|x|}$	$(2/\pi)^{\frac{1}{2}}a(a^2+w^2)^{-1}$
7.	$(a^2+x^2)^{-1}$	$(\pi/2)^{\frac{1}{2}}a^{-1}e^{-a|w|}$
8.	$\left(\dfrac{\pi}{2}\right)^{\frac{1}{2}} b^{-1} \dfrac{\sinh \pi a/b}{\cosh \pi x/b + \cos \pi a/b}$	$\sinh aw \operatorname{cosech} bw$
9.	$e^{-a^2x^2}$	$2^{-\frac{1}{2}}a^{-1}e^{-w^2/4a^2}$
10.	$K_0[a(x^2+b^2)^{\frac{1}{2}}]$	$(\pi/2)^{\frac{1}{2}}(w^2+a^2)^{-\frac{1}{2}}e^{-b(w^2+a^2)^{\frac{1}{2}}}$

including the real axis in the complex w-plane and another method when F cannot be continued analytically off the real axis in the w-plane. An example of the first situation is when $F(w) = (w^2+1)^{-1}$ and an example of the second is $f = e^{-|w|}$. In this second case F could of course be written as $\lim_{\varepsilon\to 0} e^{-(w^2+\varepsilon^2)^{\frac{1}{2}}}$, with an appropriate choice of the square root, but it is easier to try and calculate its asymptotic form directly rather than by a limiting process using the method which will be developed when F can be analytically continued off the real axis. In this latter case we have that

Theorem 7.5

Suppose that $F(w) = O(1/|w|)$ as $|w| \to \infty$. We have that

(i) If the singularity with the largest negative imaginary part (i.e. in the lower half of the complex w-plane but nearest the real axis) is at $w = w_1$ and, in some neighbourhood of $w = w_1$,

$$F(w) = (w - w_1)^{\gamma_1} \sum_{n=0}^{\infty} a_n (w - w_1)^n, \qquad -1 < \gamma_1 < 0,$$

then

$$f(x) \sim -\left(\frac{2}{\pi}\right)^{\frac{1}{2}} \exp\left\{-i\left(w_1 x - \gamma_1 \frac{\pi}{2}\right)\right\} x^{-(\gamma_1+1)} \sin \gamma_1 \pi$$

$$\times \sum_{n=0}^{\infty} a_n \left(\frac{-i}{t}\right)^n \Gamma(\gamma_1 + n + 1) \quad \text{as} \quad x \to \infty. \quad (7.25)$$

If F has a pole with residue r_1 at $w = w_1$ then equation (7.25) has to be replaced by

$$f \sim -i(2\pi)^{\frac{1}{2}} r_1 e^{-iw_1 x} \quad \text{as} \quad x \to \infty.$$

(ii) If the singularity with the smallest positive imaginary part is at $w = w_2$ and, in some neighbourhood of $w = w_2$,

$$F(w) = (w - w_2)^{\gamma_2} \sum_{n=0}^{\infty} b_n (w - w_2)^n, \qquad \gamma_2 > -1,$$

then

$$f(x) \sim -\left(\frac{2}{\pi}\right)^{\frac{1}{2}} e^{-i(w_2 x + \gamma_2 \frac{1}{2}\pi)} (-x)^{-(\gamma_2+1)} \sum_{n=0}^{\infty} b_n \left(\frac{i}{t}\right)^n \Gamma(\gamma_2 + n + 1)$$

$$\text{as} \quad x \to -\infty. \quad (7.26)$$

When F has a pole with residue r_2 at $w = w_2$ then equation (7.26) has to be replaced by

$$f \sim i(2\pi)^{\frac{1}{2}} r_2 e^{-iw_2 x}.$$

The proof can be completed more or less as in Theorem 7.4, the main difference now being that the contour of integration in equation (7.20) has to be deformed into the upper (lower) half-plane for $x < 0$ ($x > 0$). A detailed proof is given in Murray (1974, Chapter 5).

If there are several singularities equally near the real axis then the asymptotic expansion is the sum of the contributions from the separate singularities.

In the case when the Fourier transform cannot be continued analytically off the real axis the asymptotic form of f is determined by the behaviour of F near those points at which it is not infinitely differentiable and we shall refer to such points as singularities. The basic result is

Theorem 7.6

If the only singularity of $F(w)$, defined only on the real axis, is at $w = w_1$ and, in a neighbourhood of $w = w_1$,

$$F(w) = \sum_{n=0}^{\infty} a_n \, |w - w_1|^{\beta+n} + \sum_{n=0}^{\infty} b_n \, |w - w_1|^{\beta+n} \, \mathrm{sgn}(w - w_1),$$

where

$$\mathrm{sgn}(w - w_1) = \begin{array}{l} 1, \; w > w_1, \\ -1, \; w < w_1, \end{array}$$

then, as $|x| \to \infty$,

$$f(x) \sim -\left(\frac{2}{\pi}\right)^{\frac{1}{2}} e^{-ixw_1} \Bigg[\sum_{n=0}^{\infty} a_n \, \sin\{\tfrac{1}{2}\pi(\beta+n)\} \frac{\Gamma(\beta+n+1)}{|x|^{\beta+n+1}}$$
$$-i\left(\frac{2}{\pi}\right)^{\frac{1}{2}} \mathrm{sgn} \, x \sum_{n=0}^{\infty} b_n \, \cos\{\tfrac{1}{2}\pi(\beta+\eta)\} \frac{\Gamma(\beta+n+1)}{|x|^{\beta+n+1}} \Bigg]. \quad (7.27)$$

The proof is carried out most easily using the theory of generalized functions, and is given in Lighthill (1962). (The convention used by Lighthill in defining the Fourier transform is slightly different from that adopted here.) Lighthill also gives the asymptotic behaviour when $F(w)$ has a logarithmic type of singularity near $w = w_1$ and also shows that, when F has a finite number of singularities, the asymptotic form of f is the sum of the contributions of the individual singularities as defined in equation (7.27).

Example 7.9

Solve

$$u_{xx} = u_y, \qquad |x| < \infty, \qquad y > 0,$$

with

$$u \to 0 \quad \text{as} \quad |x| \to \infty \quad \text{and} \quad u(x, 0) = e^{-x^2}.$$

$U(w, y)$, the Fourier transform of $u(x, y)$ with respect to x, satisfies

$$-w^2 U = U_y,$$

and hence

$$U = A e^{-w^2 y}.$$

The condition on u on $y = 0$ gives $U(x, 0)$ to be the Fourier transform of e^{-x^2} and this, using Table 7.3(9), is $2^{-\frac{1}{2}} e^{-\frac{1}{4} w^2}$ so that

$$U = 2^{-\frac{1}{2}} e^{-w^2(y + \frac{1}{4})}.$$

This result can be inverted using Table 7.3(9) (with $a = (4y + 1)^{-\frac{1}{2}}$) to give

$$u = \frac{1}{(4y + 1)^{\frac{1}{2}}} e^{-x^2/(4y + 1)}.$$

Example 7.10

Solve

$$u_{xx} + u_{yy} - k^2 u = \delta(x - x') \, \delta(y - y'), \quad k \text{ real},$$

with $u \to 0$ as $x^2 + y^2 \to \infty$. (This is effectively the Green's function obtained in Example 4.5.)

Taking the Fourier transform with respect to x gives

$$U_{yy} - (k^2 + w^2) U = \frac{1}{(2\pi)^{\frac{1}{2}}} e^{iwx'} \delta(y - y').$$

The solution of this equation has to vanish as $|y| \to \infty$ so that

$$U = \begin{cases} A e^{-(k^2+w^2)^{\frac{1}{2}}y}, & y > y', \\ B e^{+(k^2+w^2)^{\frac{1}{2}}y}, & y < y', \end{cases}$$

where that branch of the square root is taken which has positive real part. We have, as in Example 4.4, that U is continuous at $y = y'$ and

$$(U_y)_{y=y'+0} - (U_y)_{y=y'-0} = \frac{1}{(2\pi)^{\frac{1}{2}}} e^{iwx'};$$

hence $A = B$ and

$$U = \frac{-1}{2(2\pi)^{\frac{1}{2}}} e^{iwx'} (k^2 + w^2)^{-\frac{1}{2}} e^{-(k^2+w^2)^{\frac{1}{2}}|y-y'|}.$$

The inversion can be carried out using Table 7.3(2) and (10) to give

$$u = -\frac{1}{2\pi} K_0[k\{(x-x')^2 + (y-y')^2\}^{\frac{1}{2}}],$$

which is the result previously obtained.

Example 7.11

Solve

$$u_{xx} + u_{yy} = 0, \qquad y > 0,$$

with $u \to 0$ as $(x^2 + y^2)^{\frac{1}{2}} \to \infty$ and $u(x, 0) = 1$, $|x| \le 1$, $u(x, 0) = 0$, $|x| > 1$.

U, the Fourier transform of u with respect to x, satisfies

$$U_{yy} - w^2 U = 0$$

and the solution satisfying the condition at infinity is therefore $U = A e^{-|w|y}$. Also

$$U(w, 0) = \frac{1}{(2\pi)^{\frac{1}{2}}} \int_{-1}^{1} e^{iwx} \, dx = \left(\frac{2}{\pi}\right)^{\frac{1}{2}} \frac{\sin w}{w}$$

so that

$$U = \left(\frac{2}{\pi}\right)^{\frac{1}{2}} \frac{\sin w}{w} e^{-|w|y}.$$

This expression is the product of the Fourier transforms of $u(x, 0)$ and $(2/\pi)^{\frac{1}{2}}y(x^2 + y^2)^{-1}$ (Table 7.3(7)) and hence, by the convolution theorem,

$$u = \frac{y}{\pi} \int_{-1}^{1} \frac{u(t, 0)\, dt}{(x - t)^2 + y^2}$$

$$= \frac{y}{\pi} \int_{-1}^{1} \frac{dt}{(x - t)^2 + y^2}$$

$$= \frac{1}{\pi} \left(\tan^{-1} \frac{x - 1}{y} + \tan^{-1} \frac{x + 1}{y} \right).$$

This is a particular case of the general result of equation (4.8) which was first obtained by complex variable methods and re-derived in Example 4.2 by using a Green's function.

If it had not been possible to invert U directly we could still have obtained its form for large x, keeping y fixed, by using Theorem 7.6. U has a singularity at $w = 0$ and can be written as $(2/\pi)^{\frac{1}{2}}\{1 - |w|\, y + O(w^2)\}$. Equation (7.27) shows that there is no contribution from the first term to the asymptotic expansion of u whilst the second term gives

$$u \sim \frac{2y}{\pi\, |x|^2} + O(|x|^{-3}),$$

which agrees with the result obtained by direct expansion of the exact solution.

Example 7.12

Find a solution $u(x, y)$ of Laplace's equation in $0 < y < a$, all x, satisfying the conditions

$$u(x, 0) = 0, \qquad u(x, a) = f(x), \qquad \lim_{|x| \to \infty} u = 0,$$

where f is a given function.

We have that U, the Fourier transform of u with respect to x, satisfies

$$U_{yy} - w^2 U = 0,$$

and the conditions on u when $y = 0$ and $y = a$ give

$$U(w, 0) = 0, \qquad U(w, a) = F(w),$$

where F is the Fourier transform of f. The function U satisfying all these conditions is

$$U = \frac{\sinh wy}{\sinh wa} F(w).$$

This expression can be inverted, using the convolution theorem and Table 7.3(8), to give

$$u = \frac{1}{2a} \int_{-\infty}^{\infty} \frac{f(t)\sin \pi y/a \, dt}{\cosh \pi(x-t)/a + \cos \pi y/a},$$

and further simplification is not possible except for some particular forms of f.

In the above example the inversion of the factor multiplying $F(w)$ was obtained from Table 7.3, but the explicit inversion can be carried out directly by complex variable methods similar to those used in Example 7.8. We have from equation (7.20) that the inverse $g(x, y)$ of $\sinh wy \operatorname{cosech} wa$ is given by

$$g(x, y) = \frac{1}{(2\pi)^{\frac{1}{2}}} \int_{-\infty}^{\infty} \frac{\sinh wy}{\sinh wa} e^{-iwx} \, dw.$$

The integral is an analytic function of the complex variable w except at the points $w = \pm in\pi/a$, $n = 1, 2, \ldots$, where it has simple poles. For x positive the integral is, by Cauchy's theorem and Jordan's lemma, equal to $-2\pi i$ times the sum of the residues at the poles of the integrand in the lower half plane. The residue of the integrand at $w = -in\pi/a$ is

$$\frac{1}{(2\pi)^{\frac{1}{2}}} e^{-n\pi x/a} \frac{\sinh(-in\pi y/a)}{\{d/dw \sin wa\}_{w=-in\pi/a}}$$

and hence

$$g(x, y) = -\frac{(2\pi)^{\frac{1}{2}}}{a} \sum_{n=1}^{\infty} (-1)^n \sin \frac{n\pi y}{a} e^{-n\pi x/a}$$

$$= -\frac{(2\pi)^{\frac{1}{2}}}{a} \operatorname{Im} \sum_{n=1}^{\infty} (-1)^n e^{-n\pi x/a} e^{in\pi y/a}.$$

Summing the geometric series and taking the imaginary part gives

$$g(x, y) = \left(\frac{\pi}{2}\right)^{\frac{1}{2}} \frac{\sin \pi y/a}{[\cosh \pi x/a + \cos \pi y/a]},$$

in agreement with Table 7.3(8).

Associated with the exponential Fourier transforms are two transforms which are particularly useful for solving boundary-value problems where one of the independent variables ranges from 0 to ∞. These are the Fourier cosine and sine transforms defined, respectively, by

$$F^c(w) = \left(\frac{2}{\pi}\right)^{\frac{1}{2}} \int_0^\infty f(x)\cos wx \; dw, \qquad (7.28)$$

$$F^s(w) = \left(\frac{2}{\pi}\right)^{\frac{1}{2}} \int_0^\infty f(x)\sin wx \; dx. \qquad (7.29)$$

(Capital letters with the appropriate affix will be used to denote cosine and sine transforms.) It can be verified by direct substitution into equation (7.19) that the cosine transform of f is the exponential Fourier transform of the even function equal to $f(|x|)$. By definition, $F(w)$ is clearly an even function of w, and representing f by the right-hand side of equation (7.20) gives the inversion formula

$$f(x) = \left(\frac{2}{\pi}\right)^{\frac{1}{2}} \int_0^\infty F^c(w)\cos wx \; dw. \qquad (7.30)$$

Similarly it can be verified that the sine transform of f is the exponential Fourier transform of the odd function defined by $f(|x|)\operatorname{sgn} x$, and the appropriate inversion formula is

$$f(x) = \left(\frac{2}{\pi}\right)^{\frac{1}{2}} \int_0^\infty F^s(w)\sin wx \; dw. \qquad (7.31)$$

(It should be noted that the right-hand side of equation (7.31) is zero when $x = 0$.)

Integration by parts shows that the cosine transform of f_x is equal to

$$\left[\left(\frac{2}{\pi}\right)^{\frac{1}{2}} f(x)\cos wx \right]_0^\infty + w\left(\frac{2}{\pi}\right)^{\frac{1}{2}} \int_0^\infty f(x)\sin wx \; dx,$$

showing that no simple relationship exists between the cosine transforms of f and f_x. Repeated integration by parts, however, gives that

$$\left(\frac{2}{\pi}\right)^{\frac{1}{2}} \int_0^\infty f_{xx} \cos wx \; dx = -w^2 F^c - \left(\frac{2}{\pi}\right)^{\frac{1}{2}} f_x(0);$$

thus, when $f_x(0)$ is known, there is a simple relationship between the cosine transforms of f and f_{xx}. Similarly

$$\left(\frac{2}{\pi}\right)^{\frac{1}{2}} \int_0^\infty f_{xx} \sin wx \, dx = -w^2 F^s + \left(\frac{2}{\pi}\right)^{\frac{1}{2}} wf(0),$$

exhibiting a relationship between the sine transforms of f and f_{xx}. Hence for a function $u(x, y)$ we see that $U^c(w, y)$ and $U^s(w, y)$, its cosine and sine transforms with respect to x, are such that

$$\left(\frac{2}{\pi}\right)^{\frac{1}{2}} \int_0^\infty u_{xx} \cos wx \, dx = -w^2 U^c - \left(\frac{2}{\pi}\right)^{\frac{1}{2}} u_x(0, y), \quad (7.32)$$

$$\left(\frac{2}{\pi}\right)^{\frac{1}{2}} \int_0^\infty u_{xx} \sin wx \, dx = -w^2 U^s + \left(\frac{2}{\pi}\right)^{\frac{1}{2}} wu(0, y). \quad (7.33)$$

Equation (7.32) shows that taking the cosine transform with respect to x of

$$a_1 u_{xx} + a_3 u + b_1 u_{yy} + b_2 u_y + b_3 u = g(x, y), \quad (7.34)$$

where a_1 and a_3 are constants, gives

$$a_1 \left[-w^2 U^c - \left(\frac{2}{\pi}\right)^{\frac{1}{2}} u_x(0, y) \right] + a_3 U^c + b_1 U_{yy}^c + b_2 U_y^c + b_3 U^c = G^c.$$

This, when $u_x(0, y)$ is known, is an ordinary differential equation for U^c and hence the cosine transform can be used for problems for equation (7.34) when x ranges from 0 to ∞ and where $u \to 0$ as $x \to \infty$ with $u_x(0, y)$ known. Similarly the sine transform may be used when u, rather than u_x, is known on $x = 0$.

It is again possible to obtain, from the cosine or sine transform, some information about the asymptotic behaviour as $x \to \infty$ of a function. In order to do this we first rewrite equations (7.30) and (7.31) in the form

$$f(x) = \frac{1}{(2\pi)^{\frac{1}{2}}} \int_{-\infty}^\infty F^c(|w|) e^{-iwx} \, dw, \quad (7.35)$$

and

$$f(x) = \frac{1}{(2\pi)^{\frac{1}{2}}} \int_{-\infty}^\infty \text{sgn} \, w F^s(|w|) e^{-iwx} \, dw, \quad (7.36)$$

and note that the integrals can then be evaluated asymptotically using Theorem 7.6. The precise asymptotic behaviour will be determined by the singularities (in the sense defined immediately prior to Theorem 7.6) of the integrand. It should be noted that in both cases there will always be singularities near $w = 0$ where

$$F^c(|w|) = F^c(0) + |w| F_w^c(0) + \cdots,$$

$$\text{sgn } wF^s(|w|) = \text{sgn } w(F^s(0) + |w| F_w^s(0) + \cdots).$$

In the case when the only singularity is at $w = 0$ we deduce, from Theorem 7.6 and the above expansions, that

$$\left(\frac{2}{\pi}\right)^{\frac{1}{2}} \int_0^\infty F^c(w)\cos wx \, dw \sim -\left(\frac{2}{\pi}\right)^{\frac{1}{2}}\left[\frac{F_w^c(0)}{x^2} - \frac{F_{www}^c(0)}{x^4} + \cdots\right], \quad (7.37)$$

$$\left(\frac{2}{\pi}\right)^{\frac{1}{2}} \int_0^\infty F^s(w)\sin wx \, dw \sim \left(\frac{2}{\pi}\right)^{\frac{1}{2}}\left[\frac{F^s(0)}{x} - \frac{F_{ww}^s(0)}{x^3} + \cdots\right],$$

$$\text{as} \quad x \to \infty. \quad (7.38)$$

Example 7.13

Solve Laplace's equation in the quadrant $x > 0$, $y > 0$, with $\lim_{y \to \infty} u = 0$, $\lim_{x \to \infty} u = 0$, $u_x(0, y) = 0$, and $u(x, 0) = e^{-x}$.

Laplace's equation is of the general form of equation (7.34) and the condition on u_x on $x = 0$ suggests use of the cosine transform with respect to x. We have that

$$U_{yy}^c - w^2 U^c = 0$$

and hence, in order to satisfy the condition as $y \to \infty$,

$$U^c = A e^{-wy}.$$

The condition on u on $y = 0$ gives

$$A = \left(\frac{2}{\pi}\right)^{\frac{1}{2}} \int_0^\infty e^{-x} \cos wx \, dx$$

$$= \left(\frac{2}{\pi}\right)^{\frac{1}{2}} \frac{1}{1 + w^2}.$$

Hence

$$u = \frac{2}{\pi} \int_0^\infty \frac{e^{-wy} \cos wx}{1 + w^2} \, dw;$$

this integral cannot be evaluated in a simple closed form but it can of course be evaluated numerically. Its behaviour as $x \to \infty$ can be deduced from equation (7.37) and we have that, for fixed y,

$$u \sim \frac{2y}{\pi x^2} \quad \text{as} \quad x \to \infty$$

Example 7.14

Solve Laplace's equation in the quadrant $x > 0$, $y > 0$ with $\lim_{y \to \infty} u = 0$, $\lim_{x \to \infty} u = 0$, $u(0, y) = 0$ and $u_y(x, 0) = 1$ for $0 \le x \le 1$ and zero otherwise.

This problem is very similar to the previous one but the condition on the line $x = 0$ now suggests use of the sine transform. Hence

$$U_{yy}^s - w^2 U^s = 0$$

giving, as the solution satisfying the condition as $y \to \infty$,

$$U^s = A e^{-wy}.$$

The condition on u_y implies that

$$-wA = \left(\frac{2}{\pi}\right)^{\frac{1}{2}} \int_0^1 \sin wx \, dx$$

$$= \left(\frac{2}{\pi}\right)^{\frac{1}{2}} \frac{(1 - \cos w)}{w}.$$

Hence

$$u = \frac{2}{\pi} \int_0^\infty \frac{(\cos w - 1)}{w^2} e^{-wy} \sin wx \, dw,$$

which again cannot be simplified to a closed form, though its behaviour for large x can be determined from equation (7.38) which gives

$$u \sim -\frac{1}{\pi x} \quad \text{as} \quad x \to \infty.$$

Exercises 7.3

1. Obtain, in the form of an infinite integral, the solution of

$$u_{xx} + u_{yy} = 0, \qquad 0 < y < a, \quad \text{all } x,$$

such that $u(x, 0) = 0$, $u(x, a) = e^{-|x|}$, $\lim_{|x| \to \infty} u = 0$.

2. $u(x, t)$ satisfies the equation

$$u_{xx} - 10u = u_t, \qquad x, t > 0,$$

and the conditions

$$u(x, 0) = 0, \qquad u_x(0, t) = e^{-t}, \quad \text{and} \quad \lim_{x \to \infty} u = 0.$$

Show that, for fixed x,

$$u \sim -\tfrac{1}{3}e^{-t}e^{-3x} \quad \text{as} \quad t \to \infty.$$

3. Find the solution of

$$u_{xx} = u_{yy} + 2u_y + u, \qquad x > 0, \qquad y > 0$$

with $u(0, y) = 1$, $u(x, 0) = u_y(x, 0) = 0$.

4. Find, in the form of an infinite integral, the solution of

$$u_{xx} + u_{yy} - u = 0, \qquad x > 0, \qquad y > 0$$

such that

$$u(x, 0) = e^{-2x}, \qquad u(0, y) = 0; \qquad \lim_{(x^2+y^2)^{\frac{1}{2}} \to \infty} u = 0.$$

Determine also the behaviour of u as $x \to \infty$ for fixed y.

5. Obtain, in the form of an infinite integral, that solution of Laplace's equation in $x > 0$, $y > 0$, which vanishes when $y = 0$ and whose normal derivative is equal to e^{-y} when $x = 0$ and tends to zero as $x^2 + y^2 \to \infty$. Determine the behaviour of the solution for fixed x as $y \to \infty$.

7.4. General integral transforms

In the previous sections the Laplace transform and the various forms of Fourier transforms were introduced as purely mathematical entities in their own right, and were then shown to have properties which are helpful in solving particular boundary-value problems. An alternative and possibly more motivated approach would have been to start with a boundary-value problem for a particular partial differential equation and look for an integral transform which reduced it to an ordinary differential equation. We shall now briefly pursue this alternative approach and show how it leads to more general integral transforms than those encountered hitherto. It should be noted that for an integral transform to be useful in solving partial differential equations it must, in addition to reducing a partial differential equation to an ordinary one, be such that a simple inversion formula exists.

A general integral transform with respect to x of the function $u(x, y)$, defined for $a \le x \le b$, will be of the form

$$U^*(w, y) = \int_a^b K(w, x)u(x, y)\,dx, \qquad (7.39)$$

where $K(w, x)$ is some known function (often referred to as the kernel of the transform) of w and x. The Laplace and Fourier transforms are clearly particular examples of the general transform defined by equation (7.39). The equations for the Fourier and Laplace transforms were obtained by formally taking the transforms of the governing partial differential equations, i.e. by multiplying the equation by the appropriate kernel and integrating over the range of x. Carrying out this procedure on equation (7.7) with the general kernel $K(w, x)$ gives

$$\int_a^b [a_1u_{xx} + a_2u_x + a_3u + b_1u_{yy} + b_2u_y + b_3u]K(w, x)\,dx$$

$$= \int_a^b g(x, y)K(w, x)\,dx,$$

i.e.

$$\int_a^b [a_1u_{xx} + a_2u_x + a_3u]K(w, x)\,dx + b_1U_{yy}^* + b_2U_y^* + b_3U^* = G^*, \qquad (7.40)$$

assuming the interchangeability of differentiation and integration. The occurrence of derivatives with respect to x in equation (7.40) prevents it from reducing automatically to an equation for U^*.

The terms involving these derivatives can however be transformed by integration by parts and we have that

$$\int_a^b u_{xx} a_1 K \, dx = [u_x a_1 K]_{x=a}^{x=b} - \int_a^b u_x (a_1 K)_x \, dx$$

$$= [u_x a_1 K]_{x=a}^{x=b} - [u(a_1 K)_x]_{x=a}^{x=b} + \int_a^b u(a_1 K)_{xx} \, dx,$$

$$\int_a^b u_x a_2 K \, dx = [ua_2 K]_{x=a}^{x=b} - \int_a^b u(a_2 K)_x \, dx.$$

Thus equation (7.40) can be rewritten as

$$\int_a^b u[a_3 K + (a_1 K)_{xx} - (a_2 K)_x] \, dx + [u_x a_1 K - u(a_1 K)_x + ua_2 K]_{x=a}^{x=b}$$

$$+ b_1 U_{yy}^* + b_2 U_y^* + b_3 U^* = G^*. \quad (7.41)$$

Equation (7.41) reduces to an ordinary differential equation for U^* if K can be determined such that

$$a_3 K + (a_1 K)_{xx} - (a_2 K)_x = f(w)K, \quad (7.42)$$

for some function $f(w)$. In order that equation (7.41) can be solved for U^* it is necessary to choose K so that the integrated terms are known, and for the particular case when K can be chosen so that these terms vanish we obtain

$$b_1 U_{yy}^* + b_2 U_y^* + b_3 U^* + f(w)U^* = G. \quad (7.43)$$

We are now in a position to investigate what type of transform is the most appropriate for a given partial differential equation and we first consider the case when a_1 to a_3 are constant. In this case we deduce from equation (7.42) that K has to be of the form $e^{\lambda x}$, where λ is some function of w, and taking λ to be μw gives f to be a quadratic in w. In a problem where x ranges from $-\infty$ to ∞, and u and its derivatives vanish as $|x| \to \infty$ we see that the integrated terms in equation (7.41) can be made to vanish by taking μ to be purely imaginary. Thus, an appropriate kernel is $\exp(iwx)$, which yields the exponential Fourier transform which was introduced in §7.3 as being suited to problems of the type being considered. It is possible to deduce in a similar relatively systematic fashion that the Fourier sine (cosine) transform will be suited for problems with $0 \le x < \infty$, a_1, a_3 constant and $a_2 \equiv 0$

when $u(0, y)(u_x(0, y)]$ is known and u and its derivatives are required to vanish as $x \to \infty$. Similarly the Laplace transform can be derived as being suited for problems in $0 \le x < \infty$, with a_1 to a_3 constant and u and u_x known for $x = 0$.

We now look at the new and slightly different problem of finding the transform most suited for solving equation (7.7) in $0 \le x \le a$ with $a_3 \equiv 1$, $a_2 \equiv a_3 \equiv 0$, and u vanishing for $x = 0$ and $x = a$. It follows from equation (7.42) that a possible simple form for K is $A\, e^{\mu wx} + B\, e^{-\mu wx}$ for some constant μ. (f will then be proportional to w^2.) The second and third integrated terms in equation (7.41) will vanish at $x = 0$ and $x = a$ by virtue of the boundary conditions on u but u_x is unknown at $x = 0$ and $x = a$ and hence in order to get a useful relation for U^* we must have $K = 0$ when $x = 0$ and $x = a$. This means that $A = -B$ and that $\sinh \mu wa = 0$.

Choosing μ to be i gives K to be proportional to $\sin n\pi x/a$ which is effectively a transform only defined for discrete values of w. We deduce, therefore, that

$$U^*(n, y) = \int_0^a u(x, y)\sin \frac{n\pi x}{a}\, dx \qquad (7.44)$$

satisfies the equation

$$b_1 U_{yy}^* + b_2 U_y^* + b_3 U^* - \frac{n^2\pi^2}{a^2} U^* = G^*(n, y).$$

If values of u or u_y are prescribed for two values of y, then values of U^* are known for two values of y and hence U^* is completely determined. Comparison of equations (7.44) and (1.27) shows that $U^*(n, y)$ is proportional to the coefficient of $\sin n\pi x/a$ in the sine series for u and hence

$$u = \frac{2}{a} \sum_{n=1}^{\infty} U^*(n, y)\sin \frac{n\pi x}{a}. \qquad (7.45)$$

Equation (7.44) defines the transform referred to as the finite sine transform of a function and, as demonstrated by equation (7.45), leads to the solution in the form of a Fourier series. It can be verified by direct substitution that the equation obtained for $U^*(n, y)$ is precisely that obtained by substituting equation (7.45) into the appropriate form of equation (7.7) and equating corresponding coefficients of $\sin n\pi x/a$. Thus the finite sine transform is a somewhat elaborate way of arriving at a Fourier series solution and it is generally preferable to obtain such a representation directly using the methods of §7.1. The only situation where a finite transform has some slight advantage over the direct approach is when u is known at $x = 0$ and $x = a$, but is not zero. In this case the last two integrated terms in equation (7.41) will not be zero, though they will be known, and hence an equation for U^* can be found. It is possible to define finite transforms corresponding to all the Fourier series previously encountered but we shall not pursue this, or the concept of finite transforms, any further. Finite transforms are, however, discussed in some detail in Sneddon (1972).

So far we have only rederived previous results and methods and in order to try and break new ground we now try and determine the transform appropriate for solving

$$x^2 u_{xx} + x u_x + u_{yy} = 0, \qquad x \ge 0, \qquad a < y < b, \qquad (7.46)$$

with u or u_y prescribed on $y = a$ and $y = b$ and $u \to 0$ as $x \to \infty$. (Replacing x and y in equation (7.46) by ρ and θ respectively, where the latter are the usual plane polar coordinates, shows that it reduces to Laplace's equation in the angular sector bounded by the lines $\theta = a$ and $\theta = b$. The problems posed are, therefore, the Dirichlet and Neumann problems for Laplace's equation within this sector.)

Equation (7.42) becomes

$$x^2 K_{xx} + 3x K_x + K = f(w) K, \qquad (7.47)$$

and a simple solution of equation (7.47) is obtained by taking $f = w^2$ so that $K = x^{w-1}$. The integrated terms in equation (7.41) simplify to

$$[u_x x^{w+1}]_0^\infty - [w u x^w]_0^\infty,$$

and evaluation of these terms requires more information about u near $x = 0$ and as $x \to \infty$. The fact that equation (7.46) is to hold when $x = 0$ implies that u and $x u_x$ are bounded near $x = 0$ and hence the contribution to the integrated terms from $x = 0$ will vanish if w is chosen to be such that Re $w > 0$. If it is assumed that

$$u = B x^{-\mu} + o(x^{-\mu}) \quad \text{as} \quad x \to \infty, \qquad (7.48)$$

where B is a constant, then the integrated terms will vanish completely if

$$0 < \text{Re } w < \mu. \tag{7.49}$$

Thus if U^* is defined by

$$U^* = \int_0^\infty x^{w-1} u(x, y) \, dx \tag{7.50}$$

we deduce that, when u satisfies equations (7.46) and (7.48),

$$U^*_{yy} + w^2 U^* = 0 \tag{7.51}$$

provided that the condition (7.49) holds. This condition is necessary for the existence of U^* and it can be shown that the latter is an analytic function in the strip $0 < \text{Re } w < \mu$.

The transform defined by equation (7.50) is known as the Mellin transform and, as has been demonstrated, it enables equation (7.46) (and the corresponding inhomogeneous equation) to be transformed into a simple ordinary differential equation. The simplicity of equation (7.51) would, however, be of little use if it were not possible to determine u from U^*, and it is fortunate that, once again, there exists an inversion theorem. In this case the inversion theorem is slightly more complicated than those previously encountered and we have (*Sneddon* 1972)

Theorem 7.7

If $f^*(w)$ is an analytic function of w in the strip $a < \text{Re } w < b$, and is absolutely integrable along any line parallel to the imaginary axis and lying in the strip, then $f(x)$ defined by

$$f(x) = \frac{1}{2\pi i} \int_{c-i\infty}^{c+i\infty} x^{-w} f^*(w) \, dw, \tag{7.52}$$

where $a < c < b$, is such that

$$f^* = \int_0^\infty x^{w-1} f(x) \, dx,$$

for $a < \text{Re } w < b$.

In order to illustrate the complete procedure we consider the particular problem of solving equation (7.46) in $0 < y < 1$, $x \geq 0$, with $u(x, 0) = 0$ and $u(x, 1) = 1$ for $0 \le x \le 1$ and zero otherwise. In this case

$$U^*(w, 0) = 0 \quad \text{and} \quad U^*(w, 1) = \int_0^1 x^{w-1} \, dw = \frac{1}{w}.$$

The appropriate solution of equation (7.51) is thus

$$U^* = \frac{1}{w} \frac{\sin wy}{\sin w}.$$

U^* is analytic in the strip $0 < \operatorname{Re} w < \pi$ and hence the inversion theorem gives

$$u(x, y) = \frac{1}{2\pi i} \int_{c-i\infty}^{c+i\infty} \frac{x^{-w}}{w} \frac{\sin wy}{\sin w} \, dw,$$

where $0 < c < \pi$.

The above integral can be evaluated using Cauchy's residue theorem and the steps in this calculation will be sketched for the case $x > 1$. The contour of integration can be completed by a semi-circle in the right-hand half plane and for $x > 1$ the integral over this contour can be shown to vanish. The only singularities of the integrand are simple poles and hence u can be obtained as $2\pi i$ times the sum of the residues at the poles in the right-hand half plane. The latter are at $w = n\pi, n = 1, \ldots$, and it follows that

$$u = \frac{1}{\pi} \sum_{n=1}^{\infty} \frac{(-1)^n x^{-n\pi}}{n} \sin n\pi y.$$

This series can in fact be simplified further and written in a closed form, but we shall not pursue this.

As a final example of determining the most suitable integral transform for solving a particular problem, we consider the problem of solving

$$u_{xx} + x^{-1} u_x + u_{yy} = 0, \qquad x \ge 0, \tag{7.53}$$

with $u \to 0$ as $x \to \infty$ and either u or u_y prescribed for two values of y.

(Replacing x and y in equation (7.53) by ρ and z, where these are the radial and axial coordinates, respectively, of a cylindrical polar coordinate system, shows that the equation is the axisymmetric form of the three-dimensional Laplace equation.)

Equation (7.42) becomes

$$K_{xx} - (x^{-1}K)_x = f(w)K,$$

and this simplifies, on writing K as xK', to

$$K'_{xx} + K'_x = f(w)K'.$$

Taking $f(w) = -w^2$ shows that this equation is Bessel's equation of order zero in the variable wx and thus the general form for K' is $AJ_0(wx) + BY_0(wx)$, in the usual notation for solutions of Bessel's equation. The integrated terms in equation (7.41) now take the form

$$[u_x xK']_0^\infty - [u(xK'_x - K'_x)]_0^\infty;$$

all that is known about u near $x = 0$ is that it is bounded and hence, in view of the singular behaviour of Y_0, the contribution at $x = 0$ to the integrated term will be unknown, and possibly unbounded, unless $B = 0$. $J_0(wx)$ is proportional to $x^{-\frac{1}{2}}$ as $x \to \infty$ and hence, with $B = 0$, the integrated term is zero for functions u such that $\lim_{x \to \infty} x^{\frac{1}{2}} u = 0$.

Thus U^* defined by

$$U^* = \int_0^\infty xJ_0(wx)u(x, y)\, dx, \qquad (7.54)$$

satisfies

$$U^*_{yy} = s^2 U^*,$$

which is again a relatively simple equation. Equation (7.54) defines the Hankel transform of order zero and in this case the inversion theorem is (Sneddon 1972)

$$u = \int_0^\infty wJ_0(wx)U^*(w, y)\, dw. \qquad (7.55)$$

Tables of Hankel transforms are available in Vol. 2 of Erdélyi *et al.* (1954); this will be referred to as IT2.

The value of the Hankel transform can be seen by attempting to solve equation (7.54) in $y > 0$ with $\lim_{y \to \infty} u = 0$ and $u(x, 0) = 1$ for $0 \le x \le 1$ and zero otherwise. The condition on $y = 0$ gives

$$U^*(w, 0) = \int_0^1 xJ_0(wx)\, dx,$$

and the integral can be evaluated using results in Watson (1966) to give $U^*(w, 0) = w^{-1}J_0(w)$. The function U^* satisfying all the prescribed conditions is thus

$$U^* = w^{-1}J_0(w)e^{-wy},$$

and hence

$$u = \int_0^\infty e^{-wy}J_0(w)J_0(xw)\,dw.$$

This integral cannot be reduced to a simple closed form but by using results in IT2 it can be expressed as a complete elliptic integral.

Bibliography

Carslaw, H. S. (1930). *Introduction to the theory of Fourier's series and integrals*, Chapter 10. Dover, New York.

Copson, E. T. (1965). *Asymptotic expansions.* Cambridge University Press, Cambridge.

Erdélyi, A., Magnus, W. and Oberhettinger, F. (1954). *Integral transforms*, Vols. 1 and 2. McGraw-Hill, New York.

Lighthill, M. J. (1962). *Fourier analysis and generalised functions.* Cambridge University Press, Cambridge.

Murray, J. D. (1974). *Asymptotic analysis.* Clarendon Press, Oxford.

Ritt, R. K. (1970). *Fourier series.* McGraw-Hill, New York.

Sneddon, I. N. (1972). *The use of integral transforms.* McGraw-Hill, New York.

Watson, G. N. (1966). *Theory of Bessel functions.* Cambridge University Press, Cambridge.

Williams, W. E. (1973). *Fourier series and boundary-value problems.* Allen and Unwin, London.

8 Some equations involving more than two independent variables

THE theory of partial differential equations involving more than two independent variables becomes rather complicated and the discussion will be confined to the particular cases of Laplace's and Helmholtz's equation in three space variables and the wave equation in two and three space variables. It is impossible, in a relatively small space, to give a complete account of the properties of solutions of even these equations and we therefore concentrate on those properties which are generalizations of those for the corresponding equation in two independent variables and, in particular, on the way in which Green's functions can be defined for these equations and used to obtain suitable integral representations. Laplace's and Helmholtz's equations are dicussed in §8.1, whilst the wave equation and the heat-conduction equation are considered in §8.2 and §8.4 respectively. Some of the particular problems associated with obtaining time-harmonic solutions of the wave equation are examined briefly in §8.3.

8.1. Laplace's and related equations

Laplace's and Poisson's equations in three independent variables x, y, and z are, respectively,

$$\nabla^2 u = u_{xx} + u_{yy} + u_{zz} = 0, \tag{8.1}$$

and

$$\nabla^2 u = u_{xx} + u_{yy} + u_{zz} = g(x, y, z). \tag{8.2}$$

The basic uniqueness theorems proved in §4.2 for solutions of the two-dimensional Laplace and Poisson equations can be generalized in a comparatively obvious way to the corresponding three-dimensional problems and a mean-value theorem and a maximum–minimum principle can also be proved for the three-dimensional Laplace's equation. The other concepts introduced in §§4.2 and 4.3, namely fundamental solutions, Green's functions, and integral representations also prove equally useful for solving equations (8.1) and (8.2). The appropriate generalizations of the results for the two-dimensional problems can be obtained fairly easily once it is noted that the main difference in the proofs and calculations is the need to replace the Green's identities of equations (4.10) and (4.11) by their three-dimensional generalizations, namely

$$\int_V (v \, \nabla^2 u - u \, \nabla^2 v) \, \mathrm{d}V = \int_S \left(v \frac{\partial u}{\partial n} - u \frac{\partial v}{\partial n} \right) \mathrm{d}S, \qquad (8.3)$$

and

$$\int_V (u \, \nabla^2 u + \mathrm{grad}^2 u) \, \mathrm{d}V = \int_S u \frac{\partial u}{\partial n} \, \mathrm{d}S. \qquad (8.4)$$

In equations (8.3) and (8.4) V is the region bounded by a closed surface S; for multiply-connected bounded regions the surface integrals have to be replaced by a sum of integrals over the various bounding surfaces.

We therefore summarize the precise manner in which the various results and concepts of §§4.2 and 4.3 generalize to the case of the three-dimensional equations (8.1) and (8.2) and conclude the section by sketching further generalizations of the results to the more general equation when a negative constant multiple of u is included on the left-hand side of equations (8.1) and (8.2). These latter generalizations are included at this stage as they provide the most effective manner of carrying out some of the detailed calculations for the wave equation in §8.2.

Uniqueness theorems

Uniqueness theorems for the Dirichlet and Neumann problems for equations (8.1) and (8.2) in a bounded region, with conditions imposed on the boundary surface S, may be proved almost exactly as in §4.2, the only difference being that, in the various proofs, D has to be replaced by V, and C by S. The theorems can also be shown to hold in the unbounded region outside a closed surface S (or several surfaces) if u is required to be $O(r^{-1})$ at infinity, where r is the distance from some fixed point. (For such unbounded regions there will be, on the right-hand side of equation (8.4), an integral over the surface of the sphere at infinity and it can be shown, by arguments similar to those of §4.1, that the assumption about the behaviour of u at infinity is sufficient to ensure that there is no contribution from the integral over the sphere at infinity.)

Mean-value theorem and maximum–minimum principle for Laplace's equation

The mean-value theorem states that the mean of the value of the solution of Laplace's equation over a sphere surrounding a given point P is equal to the value of the solution at P. The existence of a third variable means that the technical details of the proof are slightly different from those in the proof of Theorem 4.3 and it therefore seems worthwhile to sketch an appropriate proof.

The surface element on a sphere of radius r is $r^2 \sin \theta \, d\theta \, d\phi$, where θ and ϕ are the usual spherical polar angles, and the mean value of $u(x, y, z)$ over a sphere of radius r and centre O is defined by

$$w(r) = \frac{1}{4\pi r^2} \int_0^\pi \int_0^{2\pi} u(r \sin \theta \cos \phi, r \sin \theta \sin \phi, r \cos \theta) r^2 \times$$
$$[sin \, \theta \, d\phi \, d\theta]$$
$$= \frac{1}{4\pi} \int_0^\pi \int_0^{2\pi} u \sin \theta \, d\phi \, d\theta.$$

On the sphere

$$w_r = \frac{1}{4\pi} \int_0^\pi \int_0^{2\pi} u_r \sin \theta \, d\phi \, d\theta$$

$$= \frac{1}{4\pi r^2} \int_0^\pi \int_0^{2\pi} u_r r^2 \sin \theta \, d\phi \, d\theta$$

$$= \frac{1}{4\pi r^2} \int_S \frac{\partial u}{\partial n} \, dS,$$

where S denotes the surface of the sphere and $\partial/\partial n$ is the derivative in the direction of the outward normal to S. It now follows from equation (8.3) (with $v \equiv 1$) that $w_r = 0$ and hence that w is constant; the mean-value theorem now follows immediately as in §4.2. The mean-value theorem can be used, by an obvious rewording of the arguments used in the two-dimensional case, to show that solutions of equation (8.1) can attain neither their maximum or minimum values in the interior of a bounded region. This maximum–minimum principle can then be used to establish the 'well-posedness' of the Dirichlet problem for Poisson's equation in three independent variables. The arguments are virtually identical with those used for two variables.

Fundamental solutions and Green's functions

Fundamental solutions associated with equation (8.1) are most easily defined by introducing the three-dimensional delta function $\delta_3(\mathbf{r}' - \mathbf{r})$ such that

$$\int_V \delta_3(\mathbf{r}' - \mathbf{r})g(x', y', z')\,\mathrm{d}x'\,\mathrm{d}y'\,\mathrm{d}z' = \begin{cases} g(x, y, z), (x, y, z) & \text{in } V, \\ 0 & \text{otherwise;} \end{cases}$$

(8.5)

$\delta_3(\mathbf{r}' - \mathbf{r})$ can also be interpreted as a product $\delta(x'-x)\delta(y'-y)$ $\delta(z'-z)$ of three one-dimensional delta functions. Fundamental solutions associated with equation (8.1) are then defined to be solutions $F(x, y, z, x', y', z')$ of

$$\nabla'^2 F = \delta_3(\mathbf{r}' - \mathbf{r}).$$

(8.6)

Setting v in equation (8.3) to be a solution F of equation (8.6), and using the delta function in a purely formal manner, shows that any solution of equation (8.2) can be written as

$$u(x, y, z) = \int_V Fg(x', y', z)\,\mathrm{d}V' - \int_S \left(F\frac{\partial u}{\partial n'} - u\frac{\partial F}{\partial n'}\right)\mathrm{d}S',$$

(8.7)

where the integration is over the dashed variables.

The above definition of a fundamental solution is the obvious generalization of that for the two-dimensional situation but it gives no information about the nature of the singularity of a fundamental solution near $\mathbf{r}' = \mathbf{r}$. This information proved useful in determining the Green's functions in Examples 4.2 and 4.5 and, in general, explicit knowledge of the singular behaviour of a fundamental solution often proves helpful in the determination of Green's functions. In view of this we now establish the precise singular behaviour, near $\mathbf{r}' = \mathbf{r}$, of solutions of equation (8.6).

Direct differentiation shows that

$$|\mathbf{r}' - \mathbf{r}|^{-1} = \{(x' - x)^2 + (y' - y)^2 + (z' - z)^2\}^{-\frac{1}{2}}$$

satisfies Laplace's equation except when $\mathbf{r}' = \mathbf{r}$ and setting $v \equiv 1$

and $u \equiv |\mathbf{r}' - \mathbf{r}|^{-1}$ in equation (8.3) gives

$$\int_V \nabla'^2 |\mathbf{r}' - \mathbf{r}|^{-1} \, \mathrm{d}V' = \int_S \frac{\partial}{\partial n'} |\mathbf{r}' - \mathbf{r}|^{-1} \, \mathrm{d}S';$$

this integral is therefore zero when the point $\mathbf{r}$ is outside S. Taking V to be the spherical region with centre $\mathbf{r}$ and radius R so that, on S, $|\mathbf{r}' - \mathbf{r}| = R$ and $\partial/\partial n' = \partial/\partial R$, gives

$$\int_V \nabla'^2 |\mathbf{r}' - \mathbf{r}|^{-1} \, \mathrm{d}V' = \int_S \frac{\partial}{\partial R} \frac{1}{R} \, \mathrm{d}S = -\frac{S}{R^2} = -4\pi.$$

Hence $\nabla'^2 |\mathbf{r}' - \mathbf{r}|^{-1}$ has properties similar to those of the delta function and the above analysis therefore suggests that, near $\mathbf{r}' = \mathbf{r}$,

$$F \sim -\frac{1}{4\pi} |\mathbf{r}' - \mathbf{r}|^{-1}. \tag{8.8}$$

It is possible, by applying equation (8.3) to the region between S and a small sphere round $\mathbf{r}' = \mathbf{r}$, v being a solution of Laplace's equation with singular behaviour of the type specified by equation (8.8), to rederive equation (8.7) using classical analysis.

Green's functions can again be defined as fundamental solutions satisfying particular boundary conditions and inspection of equation (8.7) shows that choosing F to be the Green's function vanishing (having zero normal derivative) on a closed bounded surface S yields an integral representation for the solution of the Dirichlet (Neumann) problem within S.

Fundamental solutions associated with Laplace's equation can be interpreted physically in terms of solutions of electrostatic problems. This interpretation is of some historical interest and may also be helpful to readers with some knowledge of electrostatics. The association with electrostatics stems from the fact that the electrostatic potential satisfies equation (8.2), g being proportional to the volume charge density. A point charge is essentially charge located at a point and the charge density associated with it must therefore be such that its integral is zero over any volume not containing the point and equal to the total charge over any volume including the point. The charge density due to a point charge at $\mathbf{r}'$ is therefore proportional to $\delta_3(\mathbf{r}' - \mathbf{r})$, and fundamental solutions may be regarded as potentials generated by point charges. In particular, the Green's function vanishing on a boundary is the potential produced by a point charge placed near an earthed conductor whose shape is that of the given boundary. The interpretation of fundamental solutions as being related to the potential due to a point charge is, by equation (8.8), consistent with the fact that the potential due to a point charge at $\mathbf{r}'$ is proportional to $|\mathbf{r}' - \mathbf{r}|^{-1}$.

Fourier series and integral transform methods can again be used to determine Green's functions in particular cases; the calculations can however become rather involved and we shall not attempt to illustrate them. The elementary method of images can again be used to find Green's functions for particular geometries (cf. Examples 4.2 and 4.3). The basic result necessary is that

$$\{(x' - x_0)^2 + (y' - y_0)^2 + (z' - z_0)^2\}^{-\frac{1}{2}}$$

satisfies Laplace's equation (in the dashed variables) whenever $(x' - x_0)^2 + (y' - y_0)^2 + (z' - z_0)^2 \neq 0$. It therefore follows, as in Example 4.2, that the solution of equation (8.6) in $y' \geq 0$, vanishing when $y' = 0$, and tending to zero as $x'^2 + y'^2 + z'^2 \to \infty$ is

$$G_1 = -\frac{1}{4\pi} \{(x' - x)^2 + (y' - y)^2 + (z' - z)^2\}^{-\frac{1}{2}}$$

$$+ \frac{1}{4\pi} \{(x' - x)^2 + (y' + y)^2 + (z' - z)^2\}^{-\frac{1}{2}}.$$

The first term provides the correct singular behaviour near $\mathbf{r}' = \mathbf{r}$, the second term is non-singular for $y' > 0$ and setting $y' = 0$ shows that G_1 is zero when $y' = 0$.

The geometrical arguments of Example 4.3 can be adapted to show that if P and P' are inverse points with respect to the sphere of centre O and radius a (i.e. $OP\,OP' = a^2$) then, for any point Q on the sphere, $P'Q = (a/OP)PQ$ so that, for Q on the sphere,

$$\frac{1}{PQ} - \frac{a}{rP'Q} = 0,$$

where $r = OP$. Hence taking Q to be the point (x', y', z') and P to be the point (x, y, z) gives the solution of equation (8.6) which vanishes on the sphere to be

$$-\frac{1}{4\pi} \left(\frac{1}{|\mathbf{r}' - \mathbf{r}|} - \frac{a}{r\,|\mathbf{R} - \mathbf{r}'|} \right),$$

where $\mathbf{OP}' = \mathbf{R}$.

We have, exactly as in Example 4.3, that

$$|\mathbf{r}' - \mathbf{r}|^2 = r^2 - 2r\rho \cos \theta + \rho^2,$$

$$|\mathbf{R} - \mathbf{r}'|^2 = \rho^2 - \frac{2\rho a^2}{r} \cos \theta + \frac{a^4}{r^2},$$

where $\rho = OQ$ and θ is the angle between OP and OQ. On the sphere $\partial/\partial n' = (\partial/\partial \rho)$ and $|\mathbf{R} - \mathbf{r}'| = a \, |\mathbf{r} - \mathbf{r}'|/r$ so that on the sphere $\rho = a$,

$$\frac{\partial}{\partial n'}\left(\frac{1}{|\mathbf{r}' - \mathbf{r}|} - \frac{a}{r \, |\mathbf{R} - \mathbf{r}'|}\right) = \frac{r^2 - a^2}{a \, |\mathbf{r}' - \mathbf{r}|^3}.$$

Therefore, from equation (8.7), we have that the solution of Laplace's equation within the sphere of radius a and equal to $f(x', y', z')$ on its surface is given by

$$\frac{a^2 - r^2}{4\pi a} \int \frac{f(x', y', z') \, \mathrm{d}S'}{|\mathbf{r}' - \mathbf{r}|^3},$$

where the integration is over the surface of the sphere.

Further generalizations

An equation whose properties are very similar to those of Laplace's equation is

$$\nabla^2 u + \lambda u = 0, \tag{8.9}$$

where λ is a real constant. Equation (8.9) is generally referred to as either Helmholtz' equation or, for $\lambda > 0$, the reduced wave equation (its precise connection with the wave equation is amplified in §8.3). The behaviour of solutions of equation (8.9) depends crucially on the sign of λ and, at this stage, we only look at the relatively straightforward case when $\lambda < 0$. The additional complications arising for the case $\lambda > 0$ are intimately connected with the relationship between equation (8.9) and the wave equation and will be discussed separately in §8.3. The case $\lambda < 0$ corresponds to that in §4.4 where the uniqueness theorems for the general elliptic operator are only proved when the coefficient of u is not positive.

When $\lambda < 0$ equation (8.4) can be used to prove uniqueness theorems for the Dirichlet and Neumann problems in a bounded region, and similar theorems also hold in unbounded regions provided that u is required to tend to zero at infinity. The necessary arguments are obvious generalizations of those of §4.4.

Fundamental solutions associated with equation (8.9) are defined to be solutions of

$$\nabla'^2 F + \lambda F = \delta_3(\mathbf{r}' - \mathbf{r}) \qquad (8.10)$$

and equation (8.3), with v taken to be a solution of equation (8.10), can again be used to show that solutions of equation (8.9) can be represented in the form of the right-hand side of equation (8.7) (with $g \equiv 0$), where F now has to be taken to satisfy equation (8.10).

The solution of equation (8.10) which tends to zero as $x'^2 + y'^2 + z'^2 \to \infty$ is of particular significance in obtaining integral representations in unbounded regions and it may be determined, as in Example 4.5, by seeking solutions which depend only on the distance $R[= \{(x - x')^2 + (y - y')^2 + (z - z')^2\}^{\frac{1}{2}}]$ from (x, y, z). The equation then becomes

$$\frac{1}{R^2} \frac{d}{dR}\left(R^2 \frac{dF}{dR}\right) + \lambda F = \delta_3(\mathbf{r}' - \mathbf{r}) \qquad (8.11)$$

and, for $R \neq 0$, solutions are $R^{-1} \exp \pm (-\lambda)^{\frac{1}{2}} R$. The negative branch of the square root has to be chosen in order to satisfy the condition at infinity and it follows from equation (8.8) that G, defined by

$$G = -\frac{1}{4\pi |\mathbf{r}' - \mathbf{r}|} e^{-(-\lambda)^{\frac{1}{2}}|\mathbf{r}' - \mathbf{r}|}, \qquad (8.12)$$

has the correct singular behaviour and is therefore the required function. The result of equation (8.12) is used in §8.2 to determine the Green's function for the wave equation in three space dimensions.

Integral representations of solutions of equation (8.9) in the infinite region bounded internally by a surface S and which tend to zero at infinity can be obtained by applying equation (8.3), with v replaced by G, in the region between S and a sphere of large radius R. The behaviour of u at infinity together with the exponentially damped behaviour of G, are sufficient to ensure that the integral over the large sphere vanishes as $R \to \infty$. Thus a representation of the solution of equation (8.7) is obtained with F replaced by G, though it should be remembered that now the normal derivative $\partial/\partial n'$ is in the direction out of the unbounded region, i.e. into S. A similar representation will be obtained if G is replaced by any solution of equation (8.10) which vanishes at infinity.

Example 8.1

Obtain a solution of equation (8.9), $\lambda < 0$, in the region $z \geq 0$ such that $u_z(x, y, 0) = f(x, y)$ and u tends to zero at infinity.

The above remarks concerning integral representations in unbounded regions show that the solution can be obtained immediately from equation (8.7) as

$$u(x, y, z) = \int\limits_{-\infty}^{\infty}\!\!\int G_2 f(x', y') \, dx' \, dy',$$

where G_2 is a solution of equation (8.10) which has zero normal derivative on $z' = 0$ and vanishes as $x'^2 + y'^2 + z'^2 \to \infty$ (the normal derivative $\partial u/\partial n'$ in equation (8.7) in the direction out of the region $z' > 0$ is $-u_{z'}$).

The function

$$-\frac{1}{4\pi}\left[\frac{e^{-(-\lambda)^{\frac{1}{2}}\{(x-x')^2+(y-y')^2+(z-z')^2\}^{\frac{1}{2}}}}{\{(x-x')^2+(y-y')^2+(z-z')^2\}^{\frac{1}{2}}}\right.$$
$$\left. + \frac{e^{-(-\lambda)^{\frac{1}{2}}\{(x-x')^2+(y-y')^2+(z+z')^2\}^{\frac{1}{2}}}}{\{(x-x')^2+(y-y')^2+(z+z')^2\}^{\frac{1}{2}}}\right]$$

has the correct singular behaviour, vanishes at infinity, and its derivative with respect to z' is zero for $z' = 0$ and hence this is the appropriate Green's function. The final form for u is therefore

$$-\frac{1}{2\pi}\int\limits_{-\infty}^{\infty}\!\!\int \frac{f(x', y')e^{-(-\lambda)^{\frac{1}{2}}\{(x-x')^2+(y-y')^2+z^2\}^{\frac{1}{2}}}}{\{(x-x')^2+(y-y')^2+z^2\}^{\frac{1}{2}}} \, dx' \, dy'$$

and for a given function f this can be evaluated by analytical or numerical methods as appropriate.

Exercises 8.1

1. Prove that if $\nabla^2 u \geq 0$ within a closed region V then u cannot attain its maximum in the interior of V.

2. Prove that if two harmonic functions u and v are such that $u \geq v$ on the boundary of a closed region V then $u \geq v$ everywhere within V.

3. Show that the Green's function G vanishing on the boundary of a closed region V must, within V, satisfy

$$-\frac{1}{4\pi}|\mathbf{r}'-\mathbf{r}|^{-1} \leq G \leq 0, \qquad \mathbf{r}' \neq \mathbf{r}.$$

4. u is harmonic in $a^2 \leq x^2+y^2+z^2 \leq b^2$ and $M(r)$ denotes the maximum value of u on $x^2+y^2+z^2=r^2$. Show that

$$\left(\frac{1}{a}-\frac{1}{b}\right)M(r) \leq \left(\frac{1}{a}-\frac{1}{r}\right)M(b)+\left(\frac{1}{r}-\frac{1}{b}\right)M(a).$$

8.2. The wave equation in two and three space dimensions

The wave equation in two and three space dimensions takes the respective forms

$$c^2(u_{xx}+u_{yy})-u_{tt}=0, \tag{8.13}$$

$$c^2(u_{xx}+u_{yy}+u_{zz})-u_{tt}=0, \tag{8.14}$$

where c is a constant, t is the time, and x, y, z are space variables. Equations (8.13) and (8.14) are very similar in form to equation (3.17) which suggests that they may describe phenomena with 'wave-like' properties but it is not as simple as in §3.2 to establish these properties, particularly for equation (8.13). It is however possible to get some useful information for equation (8.14) if only spherically symmetrical solutions are sought. Setting $u=v(r,t)/r$, where $r[=(x^2+y^2+z^2)^{\frac{1}{2}}]$ is the distance from the origin, shows that v must satisfy

$$c^2 v_{rr}-v_{tt}=0,$$

and the general spherically symmetric solution of equation (8.14) is therefore

$$u=\frac{f(ct-r)+g(ct+r)}{r}, \tag{8.15}$$

where f and g are arbitrary functions. Equation (8.15) shows that solutions of equation (8.14) have some of the characteristics associated with solutions of the one-dimensional wave equation in that they represent phenomena propagated with speed c. It should be observed however that the factor $1/r$ produces a damping effect not present in the one-dimensional equation.

We first establish the general 'wave-like' nature of solutions of equations (8.13) and (8.14) and show the way in which solutions of initial-value problems for the two equations differ considerably in nature. We also derive a generalization of the uniqueness theorem proved in §3.3 for the one-dimensional wave equation and show that the concepts of fundamental solutions and Green's functions can be extended to equations (8.13) and (8.14) and used to derive integral representations of solutions of initial-value problems for these equations. Knowledge of these integral representations is, in fact, necessary in order to determine the general nature of the solutions of equations (8.13) and (8.14). The derivation of these representations is somewhat complicated and it is therefore preferable to use them without proof in order to describe the general nature of the solutions.

'Wave-like' nature of solutions of initial-value problems

The solution at (x, y, t) of equation (8.13), with

$$u(x, y, 0) = f_1(x, y), \qquad u_t(x, y, 0) = f_2(x, y),$$

is given by

$$u(x, y, t) = \frac{1}{2\pi} \int_{D_t} \frac{f_2(x', y') \, dx' \, dy'}{\{c^2 t^2 - (x - x')^2 - (y - y')^2\}^{\frac{1}{2}}}$$

$$+ \frac{\partial}{\partial t} \frac{1}{2\pi} \int_{D_t} \frac{f_1(x', y') \, dx' \, dy'}{\{c^2 t^2 - (x - x')^2 - (y - y')^2\}^{\frac{1}{2}}}, \quad (8.16)$$

where D_t is the disc with centre at (x, y) and of radius ct. The solution at (x, y, z, t) of equation (8.14) such that

$$u(x, y, z, 0) = f_1(x, y, z), \qquad u_t(x, y, z, 0) = f_2(x, y, z),$$

is

$$u = \frac{1}{4\pi t} \int_{S_t} f_2(x', y', z') \, dS' + \frac{\partial}{\partial t} \frac{1}{4\pi t} \int_{S_t} f_1(x', y', z') \, dS',$$

$$(8.17)$$

where S_t is the surface of the sphere of centre (x, y, z) and of radius ct.

The right-hand sides of equations (8.16) and (8.17) are both dependent only on the initial conditions within a finite region of space, namely the disc D_t for equation (8.16) and the sphere S_t for equation (8.17)), thus exhibiting, as for the one-dimensional equation, the existence of a finite domain of dependence. Further information can be obtained about the function represented by the right-hand side of equation (8.16) by taking f_1 and f_2 in the integral to be zero except within a small region R of the (x', y')-plane. The right-hand side of equation (8.16) will therefore be zero when the disc D_t does not contain R (Fig. 8.1) and will first cease to be zero when $t = d/c$, where d is the minimum distance of (x, y) from R. Thus there is effectively a disturbance propagated outwards from the region R with a speed c.

By assuming f_1 and f_2 in equation (8.17) to be only non-zero within a finite region R_1 of three-dimensional space, a similar situation can be seen to prevail for solutions of equation (8.14); u will be zero for points (x, y, z) outside R_1 until the sphere S_t first intersects the boundary of R_1. Thereafter u will generally be non-zero whilst S_t intersects R_1, but immediately t has increased so that R_1 is completely within S_t, u will again become zero as the domain of integration in equation (8.17) is only the surface of S_t. Thus we once again have a disturbance propagated with finite speed, but with the additional feature that the disturbance at a given point will only be non-zero for a finite time. This is an

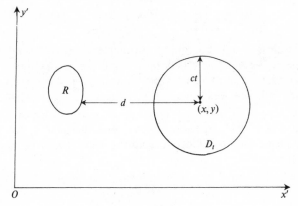

FIG 8.1

example of Huygen's principle which states that an initial state which is sharply localized will be observed later at another point as an equally sharply localized disturbance. In the limiting case when the initial disturbance is localized at a point then Huygen's principle asserts that the disturbance will only be noticed at any other point at one instant.

Huygen's principle does not hold in two dimensions since, even when D_t completely contains R, there will generally be a non-zero contribution to the integrals in equation (8.16). Thus in two dimensions there is a sharp leading edge to a disturbance, but once that leading edge has passed a point there will always be a residual effect at that point. The general form of equation (8.16) shows that, as t increases, this residual effect gradually decreases so the waves are said to diffuse.

The differing nature of wave propagation in two and three dimensions can be seen by considering the effect of the tapping of the skin of a drum (the amplitude, if assumed small, satisfies equation (8.13)) and of making a sharp noise in the atmosphere (sound propagation at small amplitudes is governed by equation (8.14)). In the former case there will eventually be, at all times, some kind of disturbance everywhere on the drum whereas in the latter case the noise at any given point will be of very short duration.

The solution of the initial-value problem for the one-dimensional wave equation with $u(x, 0) = f(x)$, $u_t(x, 0) = g(x)$ is, from equation (3.25),

$$u = \tfrac{1}{2}[f(x-ct)+f(x+ct)]+\frac{1}{2c}\int_{x-ct}^{x+ct} g(w)\,\mathrm{d}w.$$

Thus, unless $g \equiv 0$, Huyghen's principle does not hold in one dimension because, as t becomes large, the second term will in general yield a non-zero contribution. The wave disturbance however will not tend to zero as t becomes large. This can be seen by taking $g(x)$ to be identically zero except for $|x - x_0| < \varepsilon$; substituting in the expression for u then gives that, for t sufficiently large, the contribution from the second term at any point is equal to

$$\int_{x_0-\varepsilon}^{x_0+\varepsilon} g(w)\,\mathrm{d}w$$

and is therefore constant.

Uniqueness theorems

The analogues of the initial-value–boundary-value problem of §3.3 are the problems of solving equations (8.13) and (8.14) with u and u_t known for $t = 0$, and u or its normal derivative prescribed on some surface or surfaces (for equation (8.14)) and on some curve or curves (for equation (8.13)). Existence and uniqueness theorems may again be proved for such problems; the proofs of existence are complicated and will not be presented but the uniqueness proofs are very similar to that for the one-dimensional problem. We therefore derive the uniqueness theorem for solutions of equation (8.13) in a region D bounded by a closed curve C, with u and u_t known at $t = 0$ and u or its normal derivative prescribed on C. (The corresponding proof for equation (8.14) is virtually the same except that integrals over plane areas will have to be replaced by volume integrals and line integrals by surface integrals.) If u_1 and u_2 are two solutions of the problem then their difference u will, together with u_t, vanish for $t = 0$, and u or its normal derivative will vanish on C. A function $E(t)$ is now defined by

$$E(t) = \tfrac{1}{2} \int_D \left[c^{-2} u_t^2 + u_x^2 + u_y^2 \right] \mathrm{d}S,$$

and therefore

$$\frac{\mathrm{d}E}{\mathrm{d}t} = \int_D \left[u_t u_{tt} + u_x u_{xt} + u_y u_{yt} \right] \mathrm{d}S.$$

Also

$$u_x u_{xt} + u_y u_{yt} = (u_t u_x)_x + (u_t u_y)_y - u_t \nabla^2 u$$
$$= \operatorname{div}(u_t \operatorname{grad} u) - u_t \nabla^2 u.$$

Hence

$$\frac{\mathrm{d}E}{\mathrm{d}t} = \int_D \left[u_t (c^{-2} u_{tt} - \nabla^2 u) + \operatorname{div}(u_t \operatorname{grad} u) \right] \mathrm{d}S,$$

and, as u satisfies equation (8.13), the divergence theorem gives

$$\frac{\mathrm{d}E}{\mathrm{d}t} = \int_C u_t \frac{\partial u}{\partial n} \mathrm{d}s.$$

Finally as u (or $\partial u/\partial n$) vanishes on C we deduce that $\mathrm{d}E/\mathrm{d}t = 0$.

Initially u and u_t are zero and hence $E(0) = 0$ so that $E \equiv 0$, it then follows that u has to be a constant and, as u is zero at $t = 0$, the constant must be zero, thus proving uniqueness. The above proof can be modified to prove uniqueness for the initial-value problem in the absence of boundaries (cf. Exercises 8.2, 2).

Fundamental solutions, Green's functions, and integral representations

Integral representations of the solutions of equations (8.13) and (8.14) can be obtained by introducing fundamental solutions and Green's functions. The details of the analysis are very similar to that of §5.3 though the added number of space dimensions does complicate the situation. The theory will be developed for the three-dimensional equation (8.14), the modifications necessary in considering equation (8.13) consisting of replacing surface integrals by line integrals and volume integrals by integrals over plane areas. The starting point is, as in all the other cases considered, the defining relation for the adjoint operator.

The vector identity

$$v \, \nabla^2 u - u \, \nabla^2 v = \mathrm{div}(v \, \mathrm{grad} \, u - u \, \mathrm{grad} \, v)$$

shows that

$$v\left(\nabla^2 u - \frac{1}{c^2} u_{tt}\right) - u\left(\nabla^2 v - \frac{1}{c^2} v_{tt}\right) = \mathrm{div}(v \, \mathrm{grad} \, u - u \, \mathrm{grad} \, v)$$

$$-\frac{1}{c^2}(vu_t - uv_t)_t. \qquad (8.18)$$

The right-hand side of equation (8.18) can be regarded as a divergence in the four-dimensional (x, y, z, t)-space and with this interpretation we deduce, by analogy with the previous cases considered, that equation (8.14) is self-adjoint.

We first consider the problem of obtaining integral representations at time t of solutions of equation (8.14) within the region V bounded by the closed surface S. Integrating equation (8.18) over the volume V and with respect to time from $t = 0$ to T gives, on using the divergence theorem, that

$$\int_0^T \int_V \left[v\left(\nabla'^2 u - \frac{1}{c^2} u_{t't'}\right) - u\left(\nabla'^2 v - \frac{1}{c^2} v_{t't'}\right) \right] dV' \, dt'$$

$$= \int_0^T \int_S \left(v \frac{\partial u}{\partial n'} - u \frac{\partial v}{\partial n'} \right) dS' \, dt' - \frac{1}{c^2} \int_V (vu_{t'} - uv_{t'})_{t'=T} \, dV'$$

$$+ \frac{1}{c^2} \int_V (vu_{t'} - uv_{t'})_{t'=0} \, dV', \quad (8.19)$$

where dashes are used to denote variables of integration. Thus, when T is taken to be such that $T > t$, we see that a representation of $u(x, y, z, t)$ which does not involve values of u for times greater than t is obtained by taking v to be a function $F(x, y, z, x', y', z', t', t)$ such that

$$\nabla'^2 F - \frac{1}{c^2} F_{t't'} = \delta_3(\mathbf{r}' - \mathbf{r})\delta(t' - t), \quad (8.20)$$

with

$$F \equiv 0, \quad t' > t. \quad (8.21)$$

F is therefore, by analogy with previous results, a fundamental solution of equation (8.14) and we have that

$$u(x, y, z, t) = -\int_0^t \int_S \left(F \frac{\partial u}{\partial n'} - u \frac{\partial F}{\partial n'} \right) dS' \, dt'$$

$$- \frac{1}{c^2} \int_V (Fu_{t'} - uF_{t'})_{t'=0} \, dV'. \quad (8.22)$$

Equation (8.22) is still valid when V is the region between two surfaces S and S_1 provided that an additional surface integral over S_1 is included on the right-hand side. If S_1 is taken to be the sphere at infinity and F is chosen to be the function G_1 which satisfies equations (8.20) and (8.21) and which vanishes at infinity, then it is possible to show, assuming u and u_t are bounded at infinity, that the integral over the sphere at infinity will be zero. Thus in these circumstances equation (8.22) gives an integral representation of u in the region outside a closed surface $S_1(\partial/\partial n'$ now denoting the derivative along the inward drawn normal to S). Finally, in the absence of any boundaries, we obtain the representation

$$u(x, y, z, t) = -\frac{1}{c^2} \int (G_1 u_{t'} - (G_1)_{t'})_{t'=0} \, dV', \quad (8.23)$$

the integration now being all over all space. The corresponding representation for solutions of equation (8.13) is

$$u(x, y, t) = -\frac{1}{c^2} \int (G_2 u_{t'} - (G_2)_{t'})_{t'=0} \, dS', \qquad (8.24)$$

where the integration is over all x', y'; G_2 satisfies equation (8.21) and

$$\nabla'^2 G_2 - \frac{1}{c^2}(G_2)_{t't'} = \delta_2(\mathbf{r}' - \mathbf{r})\delta(t' - t). \qquad (8.25)$$

For obvious reasons G_1 and G_2 are referred to as the causal free-space Green's functions and, once they have been found, the solution of any initial-value problem is given by equation (8.23) or (8.24), as appropriate.

It is again possible to define Green's functions appropriate to solving boundary-value problems with u or its normal derivative prescribed on a surface, but the calculations can become fairly complicated and we restrict ourselves to calculating G_1 and G_2. The easiest practical method of calculating these functions is by introducing the adjoint Green's functions G_i^* again satisfying equation (8.20) but with

$$G_i^* \equiv 0, \qquad t' < t. \qquad (8.26)$$

It can be proved as in the previous cases that

$$G_i(x, y, z, x', y', z', t, t') = G_i^*(x', y', z', x, y, z, t', t). \quad (8.27)$$

G_1^* can be obtained fairly directly by introducing its Laplace transform $\bar{G}_1^*$ with respect to t'; taking the Laplace transform of equation (8.20) gives

$$\nabla'^2 \bar{G}_1^* - \frac{s^2}{c^2} \bar{G}_1^* = \delta_3(\mathbf{r}' - \mathbf{r})e^{-st}.$$

It now follows from equation (8.12) that, for Re $s > 0$,

$$\bar{G}_1^* = -\frac{1}{4\pi |\mathbf{r}' - \mathbf{r}|} e^{-|\mathbf{r}' - \mathbf{r}|/c} e^{-st}.$$

The Laplace transform with respect to t' of $\delta(t' - a)$ is e^{-sa}, and hence

$$G_1^* = -\frac{1}{4\pi |\mathbf{r}' - \mathbf{r}|} \delta\left(t' - t - \frac{1}{c}|\mathbf{r}' - \mathbf{r}|\right),$$

and

$$G_1 = -\frac{1}{4\pi |\mathbf{r}' - \mathbf{r}|} \delta\left(t - t' - \frac{1}{c}|\mathbf{r}' - \mathbf{r}|\right). \tag{8.28}$$

Equation (8.28) represents a rather unusual result in that the Green's function is no longer a normal function but a delta function. However this does not pose any serious problems as G_1 only occurs within the integrand of equation (8.23) and the final expression obtained is non-singular. The result that G_1 is not a normal function is a generalization of our previous result that the Green's function for the one-dimensional wave equation is discontinuous and therefore only represents a weak solution of the equation.

Substituting G_1 into equation (8.23) gives

$$u = \frac{1}{4\pi}\left[\int \frac{\delta\left(t - \frac{1}{c}|\mathbf{r}' - \mathbf{r}|\right)u_t(x', y', z', 0)\, \mathrm{d}V'}{|\mathbf{r}' - \mathbf{r}|}\right.$$

$$\left. + \frac{\partial}{\partial t}\int \frac{\delta\left(t - \frac{1}{c}|\mathbf{r}' - \mathbf{r}|\right)u(x', y', z', 0)\, \mathrm{d}V'}{|\mathbf{r}' - \mathbf{r}|}\right]. \tag{8.29}$$

Note that t and t' only occur in the combination $t - t'$, so that $G_{t'} = -G_t$ and the differential operator with respect to t may be taken outside the integral. The volume integral may be evaluated by introducing spherical polar co-ordinates (R, θ, ϕ) with origin at (x, y, z) so that, for example, the first integral becomes

$$\frac{1}{4\pi}\int_0^\infty \int_0^\pi \int_0^{2\pi} \delta\left(t - \frac{1}{c}R\right)$$

$$\times \frac{f_2(x + R\sin\theta\cos\phi, y + R\sin\theta\sin\phi, z + R\cos\theta)R^2}{R}$$

$$\times \sin\theta\, \mathrm{d}\theta\, \mathrm{d}\phi\, \mathrm{d}R,$$

where $f_2(x', y', z') = u_t(x', y', z', 0)$. The integral with respect to R may be evaluated by setting $R = cv$ and, on using the fundamental property of the delta function, the integral simplifies to

$$\frac{c^2 t}{4\pi}\int_0^\pi \int_0^{2\pi} f_2(x + ct\sin\theta\cos\phi, y + ct\sin\theta\sin\phi, z + ct\cos\theta)$$

$$\times \sin\theta\, \mathrm{d}\theta\, \mathrm{d}\phi.$$

Since $dS = c^2 t^2 \sin\theta\, d\theta\, d\phi$ this integral is the first term on the right-hand side of equation (8.17) and evaluating, in a similar fashion, the term involving $u(x', y', z', 0)$ in equation (8.29) gives the second term on the right-hand side of equation (8.17).

If the right member of equation (8.14) is not zero, but a known function $g(x, y, z, t)$, the analysis leading to equation (8.22) is still generally valid but an additional term

$$\frac{1}{c^2}\int_0^t\int_V Gg\,dV'\,dt'$$

has to be included in the right member of equation (8.22). The corresponding term to be included in the right member of equation (8.23) is therefore

$$\frac{1}{c^2}\int\int_0^t g(x', y', z', t')G_1\,dV'\,dt'$$

$$= \frac{1}{4\pi c^2}\int\int_0^t \frac{g(x', y', z', t')\delta\left(t - t' - \frac{1}{c}|\mathbf{r}' - \mathbf{r}|\right)}{|\mathbf{r}' - \mathbf{r}|}\,dt'\,dV'$$

$$= -\frac{1}{4\pi c^2}\int \frac{g\left(x', y', z', t - \frac{1}{c}|\mathbf{r}' - \mathbf{r}|\right)}{|\mathbf{r} - \mathbf{r}'|}\,dV'H\left(t - \frac{1}{c}|\mathbf{r}' - \mathbf{r}|\right),$$

where the volume integral is over all space. In a physical context g represents a source term and therefore the effect at (x, y, z) at time t of a source at (x', y', z') involves the actual source strength at time $t - |\mathbf{r} - \mathbf{r}'|/c$, again illustrating that disturbances are propagated with speed c. The contribution to u due to g is, for obvious reasons, generally referred to as a retarded potential.

The Green's function G_2 can be obtained in a similar fashion by first determining the Laplace transform, $\bar{G}_2^*$, of the adjoint Green's function G_2^*. We have that

$$(\bar{G}_2^*)_{x'x'} + (\bar{G}_2^*)_{y'y'} - \frac{s^2}{c^2}\bar{G}^* = e^{-st}\delta_2(\mathbf{r}' - \mathbf{r})$$

and, for Re $s > 0$, it follows from Example (4.5) that

$$\bar{G}_2^* = -e^{-st}K_0(s\,|\mathbf{r}' - \mathbf{r}|/c).$$

The inverse G_2^* can now be obtained from result (8) on p. 277 of ITI and is equal to

$$-\frac{1}{2\pi} H\left(t'-t-\frac{1}{c}|\mathbf{r}'-\mathbf{r}|\right)\left\{(t'-t)^2-\frac{1}{c^2}|\mathbf{r}'-\mathbf{r}|^2\right\}^{-\frac{1}{2}},$$

where $H(z)$ is the Heaviside unit function. Hence

$$G_2 = -\frac{1}{2\pi} H\left(t-t'-\frac{1}{c}|\mathbf{r}'-\mathbf{r}|\right)\left\{(t'-t)^2-\frac{1}{c^2}|\mathbf{r}'-\mathbf{r}|^2\right\}^{-\frac{1}{2}},$$

and substituting this result into equation (8.24) immediately gives the form of u quoted in equation (8.16). Once again we note that the Green's function is a weak solution in that it is discontinuous across $|\mathbf{r}'-\mathbf{r}| = c(t-t')$.

Exercises 8.2

1. u satisfies equation (8.13) and, at $t=0$, $u=u_t=0$ outside the disc $x^2+y^2=c^2$. Up to what value of t is it certain that u will be zero at $(4c, 0)$, $(0, 5c)$, $(3c, 4c)$?

2. E is defined by

$$E = \int_D [c^{-2}u_t^2 u_x^2 + u_y^2]\,\mathrm{d}S,$$

where u satisfies equation (8.13) and D is the disc of radius $c(t_0-t)$ and centre (x_0, y_0) where (x_0, y_0, t_0) are constants with $t<t_0$. Show that

$$\frac{\mathrm{d}E}{\mathrm{d}t} \leq 0$$

and hence deduce that u at (x_0, y_0, t_0) is determined uniquely by Cauchy conditions at $t=0$ within the disc $(x-x_0)^2+(y-y_0)^2 \leq c^2 t_0^2$.

3. Find the function $G(x, y, t, x', y', t')$ vanishing for $t'>t$ and satisfying the equation

$$u_{x'x'}+u_{y'y'}-u_{t't'}-k^2u = \delta(x'-x)\delta(y'-y)\delta(t'-t),$$

where k is a real constant. (The Laplace transform with respect to t of $[\cos a(t^2-b^2)^{\frac{1}{2}}/(t^2-b^2)^{\frac{1}{2}}]H(t-b)$ is $K_0[(s^2+a^2)^{\frac{1}{2}}b]$.)

4. Find the function $G(x, y, z, t, x', y', z't')$ vanishing for $t' > t$ and satisfying the equation

$$u_{x'x'} + u_{y'y'} + u_{z'z'} - u_{t't'} + k^2 u = \delta(x'-x)\delta(y'-y)\delta(z'-z)\delta(t'-t).$$

(The Laplace transform of $[ab/(t^2-b^2)^{\frac{1}{2}}]J_1[a(t^2-b^2)^{\frac{1}{2}}]H(t-b)$ is $e^{-bs} - e^{-b(s^2+a^2)^{\frac{1}{2}}}$.)

8.3. The reduced wave equation

The particular problems associated with finding solutions of equation (8.14) which can be represented in the form $\phi(x, y, z)e^{i\omega t}$, where ω is real, are of considerable importance in the theory of acoustic and electromagnetic scattering. The detailed calculations associated with these problems are somewhat complicated and therefore most of the details will be omitted and only the simple basic results quoted, without proof.

Writing $u = \phi(x, y, z)e^{i\omega t}$ in equation (8.14) shows that

$$(\nabla^2 + k^2)\phi = 0, \tag{8.30}$$

where $k = \omega/c$. It has already been pointed out that the problems associated with the reduced wave equation (8.30) are more complicated when the factor multiplying ϕ is positive than when it is negative. The main difference occurs in unbounded regions and stems from the necessity, in the case when the time dependence is harmonic, for a condition which is effectively equivalent to the causal condition (i.e. the past not depending on the future) imposed previously. The difficulty is most easily illustrated by looking at the radially symmetric solutions of equation (8.30), which are given by

$$\phi_1 = \frac{e^{ikr}}{r} \quad \text{and} \quad \phi_2 = \frac{e^{-ikr}}{r},$$

where r is the distance from some fixed origin. The corresponding solutions of equation (8.14) are

$$u_1 = \frac{e^{i\omega(t+r/c)}}{r} \quad \text{and} \quad u_2 = \frac{e^{i\omega(t-r/c)}}{r},$$

and clearly u_1 represents a disturbance propagating in the direction of decreasing r (i.e. from infinity) whilst u_2 represents a disturbance propagating outwards. Our previous causal arguments imply that u_1 has to be rejected and the basic difficulty is to translate this criterion into a condition on ϕ. It is insufficient to require that ϕ tends to zero at infinity as both ϕ_1 and ϕ_2 satisfy this condition though there is some slight difference in their behaviour at infinity in that

$$\frac{\partial \phi_1}{\partial r} - ik\phi_1 = O\left(\frac{1}{r}\right) \quad \text{and} \quad \frac{\partial \phi_2}{\partial r} + ik\phi_2 = O\left(\frac{1}{r^2}\right).$$

Thus possible conditions at infinity which would exclude ϕ_1 are

$$\lim_{r \to \infty} r\phi \text{ bounded}, \qquad \lim_{r \to \infty} r(\phi_r + ik\phi) = 0, \qquad (8.31)$$

and these conditions are known as Sommerfeld's radiation conditions. It can be shown that they are sufficient to eliminate disturbances which are coming in from infinity, and it is also possible to prove uniqueness theorems for Dirichlet and Neumann problems in regions bounded internally provided the condition (8.31) is imposed.

It is now possible to define uniquely the free-space Green's function associated with equation (8.30) by requiring it to be the solution of

$$(\nabla^2 + k^2)G = \delta(\mathbf{r}' - \mathbf{r})$$

which also satisfies condition (8.31). It may therefore be deduced from equation (8.12) that the Green's function is

$$-\frac{1}{4\pi} \frac{e^{-ik|\mathbf{r}'-\mathbf{r}|}}{|\mathbf{r}'-\mathbf{r}|}.$$

The methods of §8.1 can be used to deduce that a solution of equation (8.30) in the region exterior to a closed surface S and satisfying the radiation condition at infinity is given by

$$\phi = -\frac{1}{4\pi} \int_S \left\{ \frac{e^{-ik|\mathbf{r}-\mathbf{r}'|}}{|\mathbf{r}-\mathbf{r}'|} \frac{\partial \phi}{\partial n'} - \phi \frac{\partial}{\partial n'} \left(\frac{e^{-ik|\mathbf{r}-\mathbf{r}'|}}{|\mathbf{r}-\mathbf{r}'|} \right) \right\} dS', \quad (8.32)$$

and this representation is of considerable importance in scattering theory. The derivation of equation (8.32) is not entirely straight-forward as it is not possible to infer immediately that the integral over the sphere at infinity is zero, but the fact that both G and ϕ satisfy the radiation conditions turns out to be sufficient to make this integral zero.

Another difficulty associated with equation (8.30) is that there are no uniqueness theorems in bounded regions for the Dirichlet and Neumann problems and, in fact, there can be non-trivial solutions of the homogeneous Neumann and Dirichlet problems for particular values of k.

It is a matter of taste as to whether the harmonic time variation is taken to be proportional to $e^{i\omega t}$ or to $e^{-i\omega t}$ but it should be noted that in the latter case the appropriate radiation condition is obtained by replacing i by $-i$ in equation (8.31).

It turns out also that radiation conditions have to be imposed at infinity for the two-dimensional form of equation (8.30) and the appropriate conditions can be shown to be

$$\lim_{r \to \infty} r^{\frac{1}{2}}\phi \text{ is bounded,} \qquad \lim_{r \to \infty} r^{\frac{1}{2}}(\phi_r + ik\phi) = 0,$$

where r is now $(x^2 + y^2)^{\frac{1}{2}}$.

8.4. The heat-conduction equation

In three-dimensional space the temperature u in a conductor of heat satisfies

$$u_{xx} + u_{yy} + u_{zz} - u_t = \nabla^2 u - u_t = 0, \tag{8.33}$$

the corresponding equation in two-dimensional space being

$$u_{xx} + u_{yy} - u_t = 0. \tag{8.34}$$

An uniqueness theorem and a maximum–minimum principle for the one-dimensional heat equation are established in §6.2 and it is shown in §6.3 that integral representations of solutions of this equation can be obtained by introducing suitable Green's functions. These results can all be generalized to solutions of equations (8.33) and (8.34), the methods of establishing these generalizations being obvious extensions of those of §6.2 and §8.1 and therefore, to avoid undue repetition, the relevant arguments will only be sketched briefly.

The analysis will also be almost entirely confined to equation (8.33), there is no fundamental difference between the case of two and three space dimensions and results for the former can be deduced from those for the latter by making obvious notational changes (mainly replacing volumes by plane areas and surfaces by curves).

Uniqueness theorem

For all twice-differentiable functions u and v, we have that

$$v(\nabla^2 u - u_t) - u(\nabla^2 v + v_t) = \text{div}(v \text{ grad } u - u \text{ grad } v) - (uv)_t,$$
(8.35)

and interpreting the right-hand side of equation (8.35) as a four-dimensional divergence shows that the operator adjoint to that of equation (8.33) is $\nabla^2 + \partial/\partial t$. Setting $v \equiv 1$ and replacing u by u^2 in equation (8.35) gives

$$u(\nabla^2 u - u_t) + \tfrac{1}{2}(u^2)_t + \text{grad}^2 u = \text{div}(u \text{ grad } u), \qquad (8.36)$$

and equation (8.36) is a generalization of equation (6.8) to the three-dimensional case.

The problem which corresponds in three dimensions to the initial-value–boundary-value problem of §6.2 is the solution of equation (8.33) in the region V bounded externally by a closed surface S with u prescribed on S. It can again be shown that there exists a unique solution to this problem, and uniqueness can be proved fairly directly by integrating equation (8.36), with u equal to the difference between two possible solutions, over V and with respect to time from $t = 0$ to $t = T$, giving

$$\int_V \int_0^T \{\tfrac{1}{2}(u^2)_t + \text{grad}^2 u\} \, dV = \int_V \text{div}(u \text{ grad } u) \, dV.$$

The right-hand side of this identity can, by using the divergence theorem, be converted into a surface integral which is zero since u vanishes on S. Hence

$$\int_V \int_0^T \{\tfrac{1}{2}(u^2)_t + \text{grad}^2 u\} \, dV = 0,$$

and following previous arguments gives $u \equiv 0$. This proof is effectively the three-dimensional analogue of the first proof given

of Theorem 6.1. Uniqueness can also be proved by the second
method used in proving Theorem 6.1 if the function I is now
defined by

$$I = \tfrac{1}{2} \int_V u^2 \, dV,$$

where u is again the difference between two possible solutions, so
that

$$\frac{dI}{dt} = \int_V u u_t \, dV = \int_V u \, \nabla^2 u \, dV$$

$$= \int_V \mathrm{div}(u \, \mathrm{grad} \, u) \, dV - \int_V \mathrm{grad}^2 u \, dV.$$

It has already been shown above that the first volume integral is
zero and hence $dI/dt \le 0$ and uniqueness again follows as in
Theorem 6.1. This second proof also confirms the dissipative
nature of the heat-conduction process in three dimensions and, as
before, suggests that difficulties may occur with the 'backward-
time' problem. It can be shown that, in general, there does not
exist a solution of this latter problem.

It can also be proved that the maximum (minimum) value of u
in V for $0 \le t \le T$ is not greater (less) than the values of u on S or
at $t = 0$. The proof, with obvious changes of notation, is that of
Theorem 6.2 with v now defined by

$$v = u + \frac{\varepsilon}{4a^2} [(x - x_0)^2 + (y - y_0)^2 + (z - z_0)^2].$$

Here (x_0, y_0, z_0) is the point at which it is assumed that u takes its
maximum value M, and the maximum value of u on S or at $t = 0$
is assumed to be $M - \varepsilon$. Uniqueness and well-posedness are again
a direct consequence of this maximum–minimum principle.

Similar results may be shown to hold in unbounded regions
provided that u is required to be bounded at infinity. (Proof in
this case may be found in Stakgold (1968).)

Fundamental solutions and Green's functions can be associated
with equation (8.34) and it can be deduced from equation (8.35),
by the methods used to derive previous integral representations,
that a causal representation of solutions of equation (8.34) is

given by

$$u(x, y, z, t) = -\int_0^t \int_V \left(F \frac{\partial u}{\partial n'} - u \frac{\partial F}{\partial n'} \right) dV' dt'$$

$$-\int_V (Fu)_{t=0} \, dV', \quad (8.37)$$

where $F(x, y, z, t, x', y', z', t')$ satisfies

$$\nabla'^2 u + u_{t'} = \delta_3(\mathbf{r'} - \mathbf{r})\delta(t' - t), \quad (8.38)$$

and

$$F \equiv 0, \quad t' > t. \quad (8.39)$$

The representation of equation (8.37) can be extended to hold in unbounded regions provided that F is taken to be the function G (the causal free space Green's function) satisfying equations (8.38) and (8.39) and tending to zero at infinity. This latter condition is necessary to ensure, as in previous cases, the vanishing of the integral over the sphere at infinity.

The free-space Green's function is most easily calculated by introducing the adjoint function G^* satisfying

$$\nabla'^2 G^* - G_{t'}^* = \delta_3(\mathbf{r'} + \mathbf{r})\delta(t' - t),$$
$$G^* \equiv 0, \quad t' < t,$$

and $G^* \to 0$ as $x^2 + y^2 + z^2 \to \infty$. The Laplace transform $\bar{G}^*$ of G^* with respect to t' satisfies

$$\nabla'^2 \bar{G}^* - s\bar{G}^* = \delta_3(\mathbf{r'} - \mathbf{r})e^{-st},$$

and hence, by equation (8.12) we have that

$$\bar{G}^* = -\frac{1}{4\pi} e^{-st} \frac{e^{-s^{\frac{1}{2}} |\mathbf{r} - \mathbf{r'}|}}{|\mathbf{r'} - \mathbf{r}|},$$

where that branch of the square root is taken which has positive real part. The inverse of $\exp\{-s^{\frac{1}{2}} |\mathbf{r} - \mathbf{r'}|\}$ is found, from result on (1) on p. 245 of IT1, to be $\frac{1}{2}\pi^{-\frac{1}{2}}(t')^{-\frac{3}{2}} |\mathbf{r} - \mathbf{r'}| \exp\{-\frac{1}{4} |\mathbf{r} - \mathbf{r'}|^2/t'\}$ and hence, on using the shift theorem, we have that

$$G^* = -\frac{1}{8\pi^{\frac{3}{2}}(t' - t)^{\frac{3}{2}}} \exp\left\{ -\frac{1}{4} \frac{|\mathbf{r} - \mathbf{r'}|^2}{(t' - t)} \right\} H(t' - t).$$

It can again be proved that

$$G^*(x, y, z, t, x', y', z', t') = G(x', y', z', t', x, y, z, t)$$

and therefore

$$G = -\frac{1}{8\pi^{\frac{3}{2}}(t-t')^{\frac{3}{2}}} \exp\left\{-\frac{1}{4}\frac{|\mathbf{r}-\mathbf{r}'|^2}{(t-t')}\right\} H(t-t'). \qquad (8.40)$$

The volume integral on the right-hand side of equation (8.37), with F set equal to G defined by equation (8.40), gives an integral representation of the solution of the initial-value problem for equation (8.33) in an unbounded region when there are no internal boundaries present.

In a similar way the result of Example (4.5) can be used to show that the Laplace transform of the adjoint Green's function associated with equation (8.34) is

$$-\frac{1}{2\pi} e^{-st} K_0(s^{\frac{1}{2}} |\mathbf{r}'-\mathbf{r}|),$$

This can be inverted using result no. (35) on p. 283 of IT1 to give

$$G = -\frac{1}{4\pi(t-t')} \exp\left\{-\frac{1}{4}\frac{|\mathbf{r}'-\mathbf{r}|^2}{(t-t')}\right\} H(t-t').$$

Exercises 8.3

1. Show that, if $u_{xx} + u_{yy} - u_t \geq 0$, for $t \geq 0$, at all points within a bounded region D of the (x, y)-plane, then the maximum value of u is either attained when $t = 0$ or on the boundary of D.

2. $u(x, y, t)$ satisfies

$$u_{xx} + u_{yy} - u_t \geq -4 \quad \text{for} \quad t > 0, \qquad 0 < x^2 + y^2 = r^2 < 1,$$

and $u = 0$ for $r = 1$ and $u(x, y, 0) = 1 - r^2$.

Show, that for all $t > 0$ and $r \leq 1$,

$$u(x, y, t) \leq 1 - r^2.$$

3. Find, in $z' \geq 0$, the function $G(x, y, z, t, x', y', z', t')$ satisfying equations (8.38) and (8.39) and the condition

$$G(x, y, z, t, x', y', 0, t') = 0.$$

Bibliography

Stakgold, I (1968). *Boundary value problems in mathematical physics*, Vol. 2, p. 226. Macmillan, New York.

9 Systems of first-order partial differential equations

THE basic definitions associated with systems of first-order partial differential equations in two independent variables are given in §9.1, where examples of such systems are given together with an account of their relationship to second-order equations and of the reduction of such equations to a system of first-order ones. A method of classification, akin to that for second-order equations, is described in §9.2 and in §9.3 the particular case of totally hyperbolic linear systems is considered in some detail. The chapter concludes in §9.4 with a brief discussion of some of the problems associated with solving quasi-linear hyperbolic systems. Such systems are of importance in many practical contexts and occur in, for example, the theory of gas flow and in water-wave theory.

9.1. Basic definitions and examples

A system of n first-order partial differential equations in n dependent variables $u^{(1)}, \ldots, u^{(n)}$ and two independent variables x and y is, as in the case of a single first-order equation, said to be *quasi-linear* if it can be written in the form

$$\sum_{k=1}^{n} (a_{ik} u_x^{(k)} + b_{ik} u_y^{(k)}) = c_i, \qquad i = 1, \ldots, n, \qquad (9.1)$$

where the a_{ik}, b_{ik}, and c_i are independent of the derivatives of $u^{(1)}, \ldots, u^{(n)}$. The system is said to be *semi-linear* when the a_{ik}, b_{ik} depend only on x and y, and to be *linear* if, in addition, the c_i are linear in $u^{(1)}, \ldots, u^{(n)}$, i.e.

$$c_i = \sum_{k=1}^{n} d_{ik} u^{(k)},$$

where the d_{ik} depend only on x and y. Equation (9.1) can also be rewritten in matrix form as

$$A\mathbf{u}_x + B\mathbf{u}_y = \mathbf{c}, \qquad (9.2)$$

where A and B are the matrices defined by

$$A = \begin{pmatrix} a_{11} & \cdots & a_{1n} \\ \cdot & & \cdot \\ \cdot & & \cdot \\ \cdot & & \cdot \\ a_{n1} & \cdots & a_{nn} \end{pmatrix}, \qquad B = \begin{pmatrix} b_{11} & \cdots & b_{1n} \\ \cdot & & \cdot \\ \cdot & & \cdot \\ \cdot & & \cdot \\ b_{n1} & \cdots & b_{nn} \end{pmatrix},$$

and **c** and **u** are the column vectors

$$\mathbf{c} = (c_1, \ldots, c_n)^T, \qquad \mathbf{u} = (u^{(1)}, \ldots, u^{(n)})^T$$

where the affix T denotes the transpose.

We shall confine our discussion almost entirely to the case of linear systems and in particular to those properties of linear systems that are obvious generalizations of corresponding properties of first-and second-order linear equations. A relationship between first-order equations and systems of such equations is clearly to be expected on account of the similarity between the forms of equations (2.1) and (9.1), but the relationship between first-order systems and second-order equations is possibly more unexpected. The existence of such a relationship stems from the fact that second-and higher-order equations can be reduced to systems of first-order ones. For example, the equation

$$au_{xx} + 2bu_{xy} + cu_{yy} = -F, \tag{9.3}$$

where F is a function of x and y, is equivalent to the system

$$\begin{align} au_x^{(1)} + 2bu_y^{(1)} + cu_y^{(2)} &= -F, \\ u_y^{(1)} - u_x^{(2)} &= 0, \end{align} \tag{9.4}$$

where $u^{(1)} = u_x$, $u^{(2)} = u_y$. There is no unique method of reducing a given second-order equation to a first-order system and an alternative reduction of equation (9.3) is provided by setting $u^{(1)} = u_x$, $u^{(2)} = u_x + u_y$ so that

$$\begin{align} au_x^{(1)} + (2b - c)u_y^{(1)} + cu_y^{(2)} &= -F, \\ u_y^{(1)} - u_x^{(2)} + u_x^{(1)} &= 0. \end{align}$$

It is important, when reducing a given equation to a system of first-order equations, to make certain that the system and the equation are exactly equivalent in the sense that a solution of one is always a solution of the other.

Failure to do this may result in a system being unnecessarily large and possessing features not present in the original equation. An example of this type of situation is provided by the equation

$$u_{yy} - u_{xx} + u = 0 \qquad (9.5)$$

which reduces both to

$$\left.\begin{array}{l} u_y^{(2)} - u_x^{(2)} + u^{(1)} = 0, \\ u_y^{(1)} + u_x^{(1)} - u^{(2)} = 0, \end{array}\right\} \qquad (9.6)$$

where

$$u^{(1)} = u \quad \text{and} \quad u^{(2)} = u_y + u_x,$$

and to

$$\left.\begin{array}{r} u_y^{(1)} + u_x^{(1)} - u_x^{(2)} + u^{(3)} = 0, \\ u_y^{(2)} - u_x^{(2)} + u^{(3)} = 0, \\ u_y^{(3)} + \tfrac{1}{2} u_x^{(3)} - \tfrac{1}{2} u^{(1)} - \tfrac{1}{2} u^{(2)} = 0, \end{array}\right\} \qquad (9.7)$$

where $u^{(1)} = u_y$, $u^{(2)} = u_y + u_x$, $u^{(3)} = u$. Eliminating $u^{(2)}$ from equation (9.6) yields equation (9.5), but eliminating $u^{(1)}$ and $u^{(2)}$ from equation (9.7) gives

$$\left(2\frac{\partial}{\partial y} + \frac{\partial}{\partial x}\right)(u_{yy}^{(3)} - u_{xx}^{(3)} + u^{(3)}) = 0,$$

so that the system (9.7) is not exactly equivalent to equation (9.5).

It is also necessary to avoid reducing an equation to a system such that all linear combinations of the matrices A and B are singular (the difficulties associated with such systems are elaborated upon in §9.2). A system with this property is

$$\left.\begin{array}{r} u_x^{(1)} - u^{(2)} = 0, \\ u_y^{(1)} - u^{(3)} = 0, \\ u_y^{(3)} - u_x^{(2)} + u^{(1)} = 0, \end{array}\right\} \qquad (9.8)$$

the system (9.8) is equivalent, when $u^{(1)}$ is set equal to u, to equation (9.5). The fact that all linear combinations of A and B are singular normally indicates that the system is unnecessarily large.

It is also worth noting that the general non-linear equation discussed in §2.5, viz.

$$F(x, y, u, p, q) = 0, \qquad (9.9)$$

where $p = u_x$, and $q = u_y$, can be reduced to a system of first-order quasi-linear equations. Differentiating equation (9.9) with respect to x gives

$$F_x + pF_u + p_x F_p + q_x F_q = 0, \qquad (9.10)$$

whilst differentiation with respect to y gives

$$F_y + qF_u + p_y F_p + q_y F_q = 0. \qquad (9.11)$$

Equations (9.10) and (9.11) can be rewritten, on using the identity $p_y = q_x$, as

$$p_x F_p + p_y F_q = -F_x - pF_u, \qquad (9.12)$$

$$q_x F_p + q_y F_q = -F_y - qF_u. \qquad (9.13)$$

Equations (9.12) and (9.13) form, together with the identity

$$F_p u_x + F_q u_y = pF_p + qF_q, \qquad (9.14)$$

a system of three quasi-linear equations for u, p, and q which is equivalent to equation (9.9). It also follows from this analysis that a system of n non-linear first-order equations is equivalent to a system of $3n$ quasi-linear first-order equations.

Apart from being the consequence of the reduction of higher-order equations and non-linear equations, systems of first-order equations can arise directly in the practical formulation of a problem. An example of this occurs in one-dimensional isentropic gas flow down a tube; the gas velocity u and the density ρ satisfy

$$\left.\begin{array}{r} u_t + uu_x + c^2\rho^{-1}\rho_x = 0, \\ \rho_t + u\rho_x + \rho u_x = 0, \end{array}\right\} \qquad (9.15)$$

where x denotes the displacement along the tube, t denotes time, and c (the local speed of sound) is a known function of ρ.

9.2. Classification of systems of equations

Analogy with first-order equations suggests that there should exist a solution of the Cauchy problem for the system of equation (9.2) with **u** prescribed on some curve C. This same analogy also suggests that there may be certain families of curves such that prescribing **u** on them does not determine **u** uniquely off them, and we now show that seeking these latter curves leads, as in the case of second-order equations, to a broad classification of systems of first-order equations.

If, on some curve C,

$$x = x(s), \qquad y = y(s), \qquad \mathbf{u}(x(s), y(s)) = \mathbf{u}(s),$$

where s is a parameter characterizing C, then differentiation with respect to s gives

$$\mathbf{u}_s(s) = x_s \mathbf{u}_x + y_s \mathbf{u}_y.$$

Hence, on using equation (9.2), we have that

$$(A y_s - B x_s)\mathbf{u}_x = \mathbf{c} y_s - B \mathbf{u}_s, \qquad (9.16)\text{a}$$

$$(B x_s - A y_s)\mathbf{u}_y = \mathbf{c} x_s - A \mathbf{u}_s. \qquad (9.16)\text{b}$$

Equations (9.16) (cf. equations (2.13)) show that $\mathbf{u}_x$ and $\mathbf{u}_y$ (and hence **u**) cannot usually be found when C is a member of the family of curves defined by

$$\frac{\mathrm{d}y}{\mathrm{d}x} = \mu, \qquad (9.17)$$

where

$$\det(A\mu - B) = 0. \qquad (9.18)$$

The curves given by equations (9.17) and (9.18) with μ real are, as for second-order equations, defined to be the characteristics of the system. For linear systems the characteristics can be found without knowing **u**, but in the quasi-linear case the equations defining the characteristics involve **u** and hence the characteristics cannot be determined explicitly. It should be noted, as mentioned in §3.1, that the characteristics of systems of equations are equivalent to the characteristic traces of first-order equations.

At any point of C the derivative $\mathbf{u}_n$ of **u** normal to C will be a linear combination $\alpha\mathbf{u}_x + \beta\mathbf{u}_y$, for some α, β, and equations (9.16) then give that, on C,

$$(A y_s - B x_s)\mathbf{u}_n = \mathbf{c}(\alpha y_s - \beta x_s) - (\alpha B - \beta A)B \mathbf{u}_s. \qquad (9.19)$$

If it is assumed that C is such that $\mathbf{u}_n$ is discontinuous across it then subtraction of the appropriate forms of equation (9.19) at points on opposite sides of C gives

$$(Ay_s - Bx_s)[\mathbf{u}_n] = 0, \tag{9.20}$$

where $[\mathbf{u}_n]$ is the discontinuity in $\mathbf{u}_n$ across C. Equation (9.20) shows explicitly that the only curves across which the derivatives of $\mathbf{u}$ are discontinuous are the characteristics, and on the characteristics

$$(A\mu - B)[\mathbf{u}_n] = 0. \tag{9.21}$$

This agrees with the results already obtained for first- and second-order equations.

A system for which the roots of equation (9.18) are all real (but not necessarily distinct), and such that there exist n independent solutions of equation (9.21), is said to be hyperbolic. The system is said to be totally hyperbolic when the roots of equation (9.18) are distinct and real and to be parabolic when these roots are real but not distinct and there exist less than n independent solutions of equation (9.21). An elliptic system is defined to be one such that none of the roots of equation (9.18) are real.

The normal derivative of $\mathbf{u}$ can therefore be discontinuous across the characteristics of a hyperbolic system and hence the characteristics can, as for second-order hyperbolic equations, separate a region where $\mathbf{u}$ is uniform from one where $\mathbf{u}$ is non-uniform. Thus we would expect (cf. §5.1) that problems associated with 'wave-like' phenomena would lead to hyperbolic systems.

If no restrictions are placed on the matrices A and B it is possible to construct systems in which all directions at a point are characteristic and hence no Cauchy problem is soluble. This kind of situation normally arises when the system is unecessarily large and the possibility of all curves being characteristic can be excluded by imposing a simple restriction on the matrices A and B. If all directions are characteristic then, in particular, the x- and y-directions are characteristic, and equation (9.18) shows that the matrices A and B must be singular. If the axes are rotated through any angle the matrices multiplying the derivatives of $\mathbf{u}$ in the new system are linear combinations of A and B and repeating the above argument shows that these new matrices are also singular. We can therefore exclude systems in which all directions are characteristic by requiring that there must exist some λ, μ such that

$$\det(\lambda A + \mu B) \neq 0, \qquad (9.22)$$

and we assume henceforth that this condition can be satisfied.

It turns out to be possible, by using independent variables directed along the characteristics, to define a system using a somewhat simpler set of relations than those of equation (9.1). The necessary transformations are carried out most effectively in general by using elementary results concerning the diagonalization of matrices and for totally hyperbolic systems the procedure is described in §9.3. It is however helpful, before looking at this general reduction, to see how a comparison of the structures of equation (9.1) and (2.1) can lead to a method of obtaining a simpler set of equations, and also to another method of establishing the existence of characteristics. All methods of solving equation (2.1) were based on the fact that it reduced to an ordinary differential equation on the characteristics and it seems worth exploring whether the characteristics of linear systems can play a similar part in developing methods of solution. The left-hand side of equation (9.1) is somewhat more complicated than that of equation (2.1) in that it represents a linear combination of the derivatives of $u^{(k)}$ along the curves defined by

$$\frac{dy}{dx} = \frac{b_{ik}}{a_{ik}}.$$

The derivatives of $u^{(k)}$ will all be in the same direction if the above ratio is independent of k; it is unrealistic to expect that this condition will automatically hold for a general system but the linearity of the system does suggest that it might be possible to rearrange the equations of the system so as to produce one in which all the differentiations in a given equation are in the same direction.

Multiplying the ith equation by some scalar ξ_i and summing over all i gives

$$\sum_{i=1}^{n} \sum_{k=1}^{n} [\xi_i a_{ik} u_x^{(k)} + \xi_i b_{ik} u_y^{(k)}] = \sum_{i=1}^{n} \xi_i c_i, \qquad (9.23)$$

and in equation (9.23) $u^{(k)}$ is differentiated in the direction defined by

$$\frac{dy}{dx} = \frac{\sum_{i=1}^{n} \xi_i b_{ik}}{\sum_{i=1}^{n} \xi_i a_{ik}}.$$

All the $u^{(k)}$ will therefore be differentiated in the same direction if the ξ_i can be chosen so that the above ratio is some constant μ so that

$$\sum_{i=1}^{n} \xi_i(\mu a_{ik} - b_{ik}) = 0, \qquad k = 1, \ldots, n. \qquad (9.24)$$

For hyperbolic systems there will be n independent sets of solutions ξ_i of equation (9.24) (which is equivalent to equation (9.21) with μ satisfying equation (9.18)). Thus the original system of equations can be rearranged to give n equations of the form

$$\sum_{k=1}^{n} \alpha_{ik} Du^{(k)} = c_i, \qquad i = 1, \ldots, n, \qquad (9.25)$$

where the α_{ik}, c_i are known and D denotes the derivative along one of the characteristics of the system. Any equation in the form of equation (9.25) is said to be in *characteristic form*. Thus for hyperbolic systems there will exist n independent real characteristic forms whilst for parabolic systems all the characteristics will be real but there will be less than n independent characteristic forms. Equation (9.25) can be further simplified by defining a new set of dependent variables $v^{(i)}$ by

$$v^{(i)} = \sum_{k=1}^{n} \alpha_{ik} u^{(k)},$$

so that

$$Dv^{(i)} = f_i, \qquad i = 1, \ldots, n \qquad (9.26)$$

where, for linear equations, the f_i are known functions of x, y, and $v^{(1)}, \ldots, v^{(n)}$. The set of equations (9.26) is defined to be the *canonical form* of the system and is a generalization of equation (2.8). A more direct method of deriving the canonical form for general values of n is described in §9.3 but for the case $n = 2$ the method used above is as convenient as any other.

We conclude this section with some examples of systems of different types and start by relating the present system of classification to that of classifying second-order equations, taking as our example the system defined by equation (9.4), which is equivalent to the second-order equation (9.3). We also illustrate the above method of deriving the characteristic form. Equation (9.18) gives

$$a\mu^2 - 2b\mu + c = 0,$$

which is equation (3.5) with dy/dx replaced by μ; thus the present method of classification corresponds to that of §3.1 whenever the roots of the above quadratic are not equal. In the case of equal roots (i.e. $b^2 = ac$) we have that $\mu = b/a$ and making the relevant substitutions in the appropriate form of equation (9.24) yields two equations, one of which is directly proportional to the other. Thus the present classification and nomenclature is entirely consistent with that used previously for second-order equations.

Multiplying the first of equations (9.4) by ξ_1, and the second by ξ_2 gives

$$a\xi_1 u_x^{(1)} + (2b\xi_1 + \xi_2)u_y^{(1)} - \xi_2 u_x^{(2)} + c\xi_1 u_y^{(2)} = -F\xi_1.$$

The derivatives of both $u^{(1)}$ and $u^{(2)}$ will be along the curves defined by equation (9.17) provided that

$$\frac{2b\xi_1 + \xi_2}{a\xi_1} = \mu = -\frac{c\xi_1}{\xi_2},$$

i.e.

$$\xi_1(2b - a\mu) + \xi_2 = 0,$$
$$c\xi_1 + \mu\xi_2 = 0,$$

which are equations (9.24). The values of μ are the roots of the above quadratic and appropriate values of ξ_1 and ξ_2 are $-\mu$ and c respectively. Substituting these values for ξ_1 and ξ_2 gives the characteristic form to be

$$a\mu u_x^{(1)} + (2b\mu - c)u_y^{(1)} + cu_x^{(2)} + c\mu u_y^{(2)} = -F\mu,$$

i.e.

$$a\mu(u_x^{(1)} + \mu u_y^{(1)}) + c(u_x^{(2)} + \mu u_y^{(2)}) = -F\mu$$

which simplifies, on using equation (9.17), to

$$a\mu \frac{du^{(1)}}{dx} + c \frac{du^{(2)}}{dx} + \mu F = 0.$$

This is an ordinary differential equation on the characteristics and it is also derived in §10.2 (as equation (10.21)) directly from the definition of characteristics for second-order equations. For the special case of the one-dimensional wave equation, the differential equations along the characteristics can be integrated directly and we recover the result of §3.2 that $cp \pm q$ are constant on the characteristics.

A simple example of a hyperbolic system which is not totally hyperbolic is provided by

$$u_y^{(1)} + u_x^{(1)} = 0,$$
$$u_y^{(2)} + u_x^{(2)} = u^{(1)}.$$

In this case equation (9.18) gives $(\mu - 1)^2 = 0$, so that there exists only one family of characteristics though there are two independent characteristic forms.

It should be noted that equations (9.12) to (9.14) define a hyperbolic system which is already in characteristic form with the characteristics being given, in parametric form, by

$$\frac{dx}{dt} = F_p, \qquad \frac{dy}{dt} = F_q. \tag{9.27}$$

Substituting for F_p and F_q from equation (9.24) into equations (9.12) to (9.14) gives

$$\frac{dp}{dt} = -F_x - pF_u, \qquad \frac{dq}{dt} = -F_y - qF_u, \qquad \frac{du}{dt} = pF_p + qF_q, \tag{9.28}$$

and equations (9.27) and (9.28) are seen to be the equations defining a characteristic strip which were derived in a different way in §2.5.

Example 9.1

Determine the region in which the system

$$u_y + (x-1)v_x + (y+1)v_y = 0,$$
$$(x+1)u_x + (y-1)u_y + 2v_x = 0,$$

has real characteristics.

Multiplying the first equation by ξ_1 and the second by ξ_2 gives

$$\xi_1[u_y + (x-1)v_x + (y+1)v_y] + \xi_2[(x+1)u_x + (y-1)u_y + 2v_x] = 0.$$

The condition that the derivatives of both v and u are in the direction defined by equation (9.17) gives, as above,

$$\xi_1 + \xi_2[y - 1 - \mu(x+1)] = 0,$$
$$\xi_1[y + 1 - \mu(x-1)] - 2\mu\xi_2 = 0.$$

Therefore, in order that these equations can be solved for ξ_1 and ξ_2,

$$\mu^2(x^2 - 1) - 2\mu xy + (y^2 - 1) = 0,$$

i.e.

$$(x^2 - 1)\mu = xy \pm (x^2 + y^2 - 1)^{\frac{1}{2}}.$$

The system therefore has real characteristics for $x^2 + y^2 > 1$. The characteristic form can again be obtained, as above, by substituting for the appropriate values of ξ_1 and ξ_2 in terms of μ.

Exercises 9.1

1. In what region are the characteristics of the system

$$3u_y + (x - 3)v_x + (y + 3)v_y = 0,$$
$$(x + 3)u_x + (y - 3)u_y + 6v_x = 0,$$

real?

2. Show that outside a certain circle in the (x, y)-plane the characteristics of the system

$$(xy + 1)u_x + (x + y)(u_y + v_x) + 2v_y = 0,$$
$$u_y - v_x + u^2 = 0,$$

are real and distinct.

3. Reduce the system

$$u_x - v_y = 0,$$
$$-c^2 v_x + u_y = 0,$$

to characteristic form, where c is a function of v. Show that, when $c \equiv v$, $u - \frac{1}{2}v^2$ is constant on the curves $dx/dy = v$.

4. Determine the differential equations defining the characteristics of the following system and reduce the system to characteristic form:

$$u_x + (\cos 2v)v_x + (\sin 2v)v_y = 0,$$
$$u_y + (\sin 2v)v_x - (\cos 2v)v_y = 0.$$

Show that $v + u$ is constant on one family of characteristics and $v - u$ is constant on the other.

9.3. Totally hyperbolic linear systems

In order to simplify the algebra we only consider systems defined by

$$\mathbf{u}_y + A\mathbf{u}_x = \mathbf{c}. \tag{9.29}$$

This restriction involves no loss of generality as it is assumed that any system is such that equation (9.22) holds, so that it is always possible to rotate the axes so that at least one of the matrices in equation (9.2) is non-singular. Pre-multiplication of equation (9.2) by the inverse of that matrix then yields the form shown in equation (9.29).

In many situations which give rise to systems of the above form the variable y can be identified with time and, as in the case of second-order hyperbolic equations, it is then only sensible to seek solutions of Cauchy problems on only one side of an initial curve. The slopes μ of the characteristics of the above system are the roots of $\det(\mu A - I) = 0$ and hence, as $\mu = 0$ is not a root of this equation, the y-coordinate has a special role in that it is not possible to have a characteristic parallel to lines $y = $ constant.

Limiting the discussion to totally hyperbolic systems means that the matrix A in equation (9.29) has n distinct real eigenvalues and can be reduced to diagonal form. This algebraic property of A can be used to give a simple and direct method of reducing the system of equation (9.29) to a canonical form equivalent to that shown in equation (9.26). We have that if $\lambda_1, \ldots, \lambda_n$ are the eigenvalues of the matrix A and $\mathbf{p}_1, \ldots, \mathbf{p}_n$ are corresponding eigenvectors then

$$P^{-1}AP = \begin{pmatrix} \lambda_1 & 0 & 0 & \cdots \\ 0 & \lambda_2 & \cdots & \\ & & & \lambda_n \end{pmatrix} = \Lambda,$$

where

$$P = (\mathbf{p}_1, \mathbf{p}_2, \ldots, \mathbf{p}_n).$$

Setting $\mathbf{u} = P\mathbf{v}$ in equation (9.29) gives

$$P\mathbf{v}_y + AP\mathbf{v}_x + P_y\mathbf{v} + AP_x\mathbf{v} = \mathbf{c},$$

and therefore

$$\mathbf{v}_y + \Lambda \mathbf{v}_x = \mathbf{e}, \qquad (9.30)$$

where

$$\mathbf{e} = P^{-1}\mathbf{c} - P^{-1}P_y\mathbf{v} - P^{-1}AP_x\mathbf{v}.$$

In component form equation (9.30) becomes

$$v_y^{(i)} + \lambda_i v_x^{(i)} = e_i, \qquad i = 1, \dots, n, \qquad (9.31)$$

and the left-hand side of this equation is directly proportional to the derivative of $v^{(i)}$ along the characteristics C_i defined by $x = x_i(y)$, where

$$\frac{dy}{dx_i} = \frac{1}{\lambda_i}, \qquad (9.32)$$

so that equations (9.31) and (9.26) are equivalent canonical forms.

The above reduction is formal in the sense that it has been assumed that P is non-singular in some neighbourhood of a point Q at which the system is totally hyperbolic and that the system is totally hyperbolic within some such neighbourhood. It can be shown rigorously that this will be the case provided that the elements of A are continuously differentiable within some neighbourhood of Q.

Example 9.2

Reduce the system

$$\mathbf{u}_y + \begin{pmatrix} 4x - y & 2y - 2x \\ 2x - 2y & 4y - x \end{pmatrix} \mathbf{u}_x = \mathbf{0}$$

to canonical form in that region of the (x, y)-plane where it is totally hyperbolic.

The eigenvalues of A are easily found to be $3x$ and $3y$ with corresponding eigenvectors $(2, 1)^T$ and $(1, 2)^T$, so that

$$P = \begin{pmatrix} 2 & 1 \\ 1 & 2 \end{pmatrix}.$$

The system is therefore totally hyperbolic for $x \neq y$ and can be transformed into the canonical form

$$\mathbf{v}_y + \begin{pmatrix} 3y & 0 \\ 0 & 3x \end{pmatrix}\mathbf{v}_x = 0,$$

where

$$\mathbf{v} = \tfrac{1}{3}\begin{pmatrix} 2 & -1 \\ -1 & 2 \end{pmatrix}\mathbf{u}.$$

Example 9.3

Reduce the system

$$u_y^{(1)} + 4u_x^{(2)} = 0,$$
$$u_y^{(2)} + 9u_x^{(1)} = 0,$$

to canonical form and obtain its general solution and the particular solution such that, on $y = 0$, $\mathbf{u} = (2x, 3x)^T$.

The matrix A is

$$\begin{pmatrix} 0 & 4 \\ 9 & 0 \end{pmatrix},$$

the eigenvalues are ± 6 with corresponding eigenvectors $(2, 3)^T$ and $(2, -3)^T$, so that

$$\mathbf{u} = \begin{pmatrix} 2 & 2 \\ 3 & -3 \end{pmatrix}\mathbf{v}$$

and

$$\mathbf{v}_y + \begin{pmatrix} 6 & 0 \\ 0 & -6 \end{pmatrix}\mathbf{v}_x = \mathbf{0}.$$

The canonical equations become, in component form,

$$v_y^{(1)} + 6v_x^{(1)} = 0,$$
$$v_y^{(2)} - 6v_x^{(2)} = 0.$$

These equations are both of the type that can be solved by the method of §2.2, though in this particular case it is obvious by inspection that the general solutions are

$$v^{(1)} = f(6y - x), \qquad v^{(2)} = g(6y + x),$$

where f and g are arbitrary functions. Hence

$$u^{(1)} = 2f(6y-x) + 2g(6y+x), \qquad u^{(2)} = 3f(6y-x) - 3g(6y+x).$$

The conditions on $y=0$ give

$$2x = 2f(-x) + 2g(x), \qquad 3x = 3f(-x) - 3g(x)$$

and hence $g = 0$ and $f(x) = -x$ so that

$$u^{(1)} = 2x - 12y, \qquad u^{(2)} = 3x - 18y.$$

Example 9.4

Reduce the system

$$\mathbf{u}_y + \begin{pmatrix} 4 & 2 & -2 \\ -5 & 3 & 2 \\ -2 & 4 & 1 \end{pmatrix} \mathbf{u}_x = \mathbf{0}$$

to canonical form and hence find $\mathbf{u}$ such that $\mathbf{u} = (x, 2x, 3x)^T$ when $y = -x$.

The eigenvalues of the matrix are 1, 2, 5 with corresponding eigenvectors $(2, 1, 4)^T$, $(1, 1, 2)^T$, $(0, 1, 1)^T$ and therefore the canonical form is

$$\mathbf{v}_y + \begin{pmatrix} 1 & 0 & 0 \\ 0 & 2 & 0 \\ 0 & 0 & 5 \end{pmatrix} \mathbf{v}_x = \mathbf{0},$$

where

$$\mathbf{u} = \begin{pmatrix} 2 & 1 & 0 \\ 1 & 1 & 1 \\ 4 & 2 & 1 \end{pmatrix} \mathbf{v}$$

and hence

$$\mathbf{v} = \begin{pmatrix} -1 & -1 & 1 \\ 3 & 2 & -2 \\ -2 & 0 & 1 \end{pmatrix} \mathbf{u}.$$

In this case there are three families of characteristics, namely the lines $y - x = \text{constant}, 2y - x = \text{constant}$ and $5y - x = \text{constant}$. The general solutions of the equations in the canonical form can again be found by inspection and we have that

$X = [f(x - y), \; g(2y - x), \; h(5y - x)]^T$ where f, g, and h are arbitrary functions. When $y = -x$ we have that $\mathbf{v} = (0, x, x)^T$ and this gives

$$f = 0, \qquad g(x) = -\tfrac{1}{3}x, \qquad h(x) = -\tfrac{1}{6}\mathbf{v}$$

and finally,

$$u = (\tfrac{1}{3}(x - 2y), \tfrac{1}{2}x - \tfrac{3}{2}y, \tfrac{5}{6}x - \tfrac{13}{6}y)^T.$$

One of the advantages of reducing a system to canonical form is that the simple nature of this latter form enables one to see fairly easily if there exists, as in Examples 9.2 and 9.3, a simple analytic method of solution. In general however the e_i in equation (9.31) will vary with the components of $\mathbf{v}$ and the likelihood of obtaining an exact solution will be small. Even in these circumstances canonical forms prove to be valuable as their existence forms the basis of one powerful method of solving linear hyperbolic equations numerically (the method of characteristics §10.2), and of a formal method of proving the existence of the solution of the Cauchy problem for a linear hyperbolic system. The value of canonical forms in these latter contexts is that they are effectively ordinary differential equations on the characteristics. If $v^{(i)}$ on the ith characteristic is denoted by $V^{(i)}$, i.e.

$$V^{(i)}(y) = v^{(i)}(x_i(y), y),$$

then

$$\frac{dV^{(i)}}{dy} = \frac{dx_i}{dy} v_x^{(i)} + v_y^{(i)}.$$

Therefore we deduce from equations (9.31) and (9.32) that

$$\frac{dV^{(i)}}{dy} = e_i(x_i(y), y, \mathbf{v}), \qquad i = 1, \ldots, n, \qquad (9.33)$$

which is a system of ordinary differential equations for the $V^{(i)}$.

One immediate conclusion that can be drawn from equation (9.33) is that the Cauchy problem on C_i, with

$$\mathbf{v}(x_i(y), y) = \mathbf{w}(y),$$

will not have a solution unless $\mathbf{w}$ is such that

$$\frac{\mathrm{d}w_i}{\mathrm{d}y} = e_i(\mathbf{w}, y), \tag{9.34}$$

which is of course consistent with our definition of characteristics in §9.2. It can also be shown that, even if equation (9.34) is satisfied, there is no unique solution to a Cauchy problem on a characteristic. An example of this non-uniqueness is provided by attempting to solve the Cauchy problem on the characteristic $x = 6y$ for the system of Example 9.3 with $\mathbf{w} = (1, 1)^T$ [this satisfies equation (9.34)]. Thus $f(0) = 1$ and $g(12y) = 1$ and f cannot be found uniquely.

The set of ordinary differential equations (9.34) can be used to develop a method of proving the existence of the solution of the Cauchy problem for linear systems and the initial steps of this proof will be sketched for the case when conditions are imposed on $y = 0$. Problems of this type normally occur when y is a time variable and therefore we shall only determine $\mathbf{v}$ at points $P(x', y')$ near the line $y = 0$ with $y' > 0$, though a similar approach can be used for the case $y' < 0$. For P sufficiently near $y = 0$ it is possible to draw the characteristics back from P to intersect the line $y = 0$ (none of the characteristics are parallel to this line) and it is assumed that the characteristic C_i intersects $y = 0$ at $(x_i^{(0)}, 0)$, the $x_i^{(0)}$ therefore depending on (x', y'). Integrating equation (9.33) from $y = 0$ to $y = y'$ gives

$$V^{(i)}(y') - V^{(i)}(0) = \int_0^{y'} e_i(x_i(y), y, \mathbf{v}(x_i(y))) \, \mathrm{d}y,$$

and, as $V^{(i)}(0) = w^{(i)}(x_i^{(0)})$, this simplifies to

$$v^{(i)}(x_i(y'), y') = w^{(i)}(x_i^{(0)}) + \int_0^{y'} e_i(x_i(y), \mathbf{v}(x_i(y), y)) \, \mathrm{d}y,$$

$$i = 1, \ldots, n \quad (9.35)$$

It is possible to prove that the system of integral equations defined in equation (9.35) has a unique solution for sufficiently small y', but the proof is technically complicated and will not be considered. Some brief details are given, together with further references, in Jeffrey (1976).

Equation (9.35) also provides some additional information concerning the solution of the Cauchy problem as the form of the right-side shows that the solution at a point P depends only on e_i and $w^{(i)}$ in the region bounded by the line $y = 0$ and the extreme characteristics C_1 and C_n (see Fig. 9.1) drawn backwards from P to $y = 0$. Thus there exists for linear hyperbolic systems a finite domain of dependence, and domains of determinancy and influence can also be defined as for second-order equations.

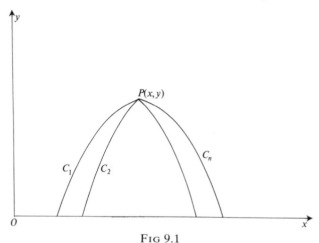

FIG 9.1

It is possible to generalize equations (9.35) so as to obtain a set of integral equations which define the Cauchy problem on an arbitrary non-characteristic curve, and thus prove the existence and uniqueness for Cauchy problems on such curves. If y is a time variable it is however necessary, as for second-order equations, to distinguish between 'space-like' and 'time-like' curves. A curve Γ is said to be 'space-like' at a point if all the characteristics drawn through the point in the direction of increasing time are all on the same side of the curve (Fig. 9.2); if these characteristics are not on the same side then the curve is said to be 'time-like' (Fig. 9.3). It follows, therefore, that in general in order to preserve causality it is only sensible to solve Cauchy problems on 'space-like' curves.

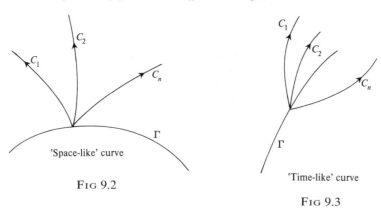

FIG 9.2

'Space-like' curve

'Time-like' curve

FIG 9.3

The above results concerning the solution of Cauchy problems posed on space-like curves can be used to determine the type of conditions that need to be imposed on a composite curve such as *ABC* in Fig. 9.4, where the arc *AB* is time-like.

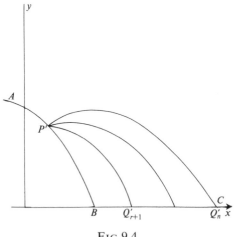

FIG 9.4

We assume that Cauchy conditions are imposed on *BC* and that, for points P' on *AB* and sufficiently near $y = 0$, $n - r$ characteristics drawn backwards from P' intersect *BC* at $Q'_{r+1}, \ldots, Q'_n$. Our arguments above show that the Cauchy conditions on *BC* will have some effect on $\mathbf{u}$ on *AB* and it is therefore certainly not possible to prescribe $\mathbf{u}$ independently on *AB* and

BC. Since $n-r$ of the characteristics from P' intersect BC, the Cauchy conditions on BC will impose $n-r$ relations between the $u^{(i)}$ at P'. Therefore, if r relations between the $u^{(i)}$ are prescribed on AB, it would seem feasible, by using these relations and integrating along the $n-r$ characteristics through $Q'_{r+1}, \ldots, Q'_n$ on BC to determine all the $u^{(i)}$ on AB. Once all the $u^{(i)}$ are known on AB and BC it is then possible, by the earlier arguments, to find $\mathbf{u}$ at the general point P shown in Fig. 9.5. The above heuristic arguments can be made formal and it can be proved rigorously that imposition of Cauchy conditions on BC and of r relations between the $u^{(i)}$ on AB leads to a unique solution in a neighbourhood of ABC.

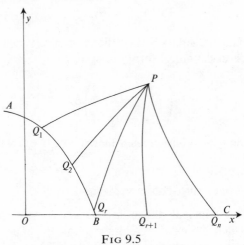

FIG 9.5

The above arguments can also be extended to determine the appropriate conditions to impose on U-shaped curves and more general curves that are partly 'time-like' and partly 'space-like'. Problems posed on such curves are effectively generalizations of the initial-value–boundary-value problem for the one-dimensional wave equation

$$u_{xx} - u_{tt} = 0.$$

This equation is equivalent to the system

$$u_x^{(1)} - u_t^{(2)} = 0,$$
$$u_t^{(1)} - u_x^{(2)} = 0,$$

where $u^{(1)} = u_x$, $u^{(2)} = u_t$, and the characteristics are the lines $x \pm t = \text{constant}$. Only one characteristic drawn backwards from a given point of either of the lines $x = 0$ and $x = a$ intersects the line $t = 0$ in the region $0 \le x \le a$. Thus the above argument shows that a unique solution can be obtained if $u^{(1)}$, $u^{(2)}$ are prescribed for $t = 0$, $0 \le x \le a$, and one of $u^{(1)}$, $u^{(2)}$ prescribed on lines $x = 0$ and $x = a$; this agrees with the results of §3.3.

Exercises 9.2

1. Reduce to canonical form the system

$$\mathbf{u}_y + \begin{pmatrix} 5 & 2 \\ 3 & 4 \end{pmatrix} \mathbf{u}_x = \mathbf{0}$$

and hence find the solution such that $\mathbf{u} = (x, 6x)^T$ when $y = -x$.

2. Reduce to canonical form the system

$$\mathbf{u}_y + \begin{pmatrix} 3 & 2 & 2 \\ 1 & 4 & 1 \\ -2 & -4 & -1 \end{pmatrix} \mathbf{u}_x = \mathbf{0}$$

and hence find the solution such that $\mathbf{u} = (x, 0, 0)^T$ when $y = 2x$.

9.4. Quasi-linear hyperbolic systems

One of the principal difficulties in dealing with quasi-linear equations is that the characteristics are now dependent on $\mathbf{u}$ and this considerably complicates the situation. The basic steps which lead to equation (9.30) can be carried out for quasi-linear systems, but the matrix P will vary with the components of $\mathbf{u}$ and it becomes difficult to obtain explicit relationships between $\mathbf{u}$ and $\mathbf{v}$. Even if such relationships can be found, further difficulties would arise as the e_i in equation (9.31) would depend on the derivatives of $\mathbf{v}$ and the essentially simple nature of equation (9.31) is then lost. For quasi-linear equations therefore there is no advantage in carrying out a complete reduction to canonical form, though it is still useful to reduce the system to the form exhibited in equation (9.25) where each equation involves only one directional derivative. The formal derivation of the characteristic forms shown in equation (9.25) is most easily carried out in general by introducing the eigenvectors $\mathbf{q}_i$ of the transpose A^T of A. These vectors satisfy

$$A^T\mathbf{q}_i = \lambda_i\mathbf{q}_i$$

(the ξ_i in equation (9.24) are seen to be proportional to the components of $\mathbf{q}_i$) and pre-multiplying equation (9.29) by $\mathbf{q}_i^T$ gives

$$\mathbf{q}_i^T\mathbf{u}_y + \lambda_i\mathbf{q}_i^T\mathbf{u}_x = \mathbf{q}_i^T\mathbf{c}, \tag{9.36}$$

which is equivalent to equation (9.25).

Equation (9.15) governing one-dimensional isentropic gas flow is a particular case of equation (9.29) with y identified with t, and $\mathbf{u}$ and A being defined by

$$\mathbf{u} = \begin{pmatrix} u \\ \rho \end{pmatrix}, \qquad A = \begin{pmatrix} u & c^2\rho^{-1} \\ \rho & u \end{pmatrix}.$$

The eigenvalues of A are $u \pm c$ and the corresponding eigenvectors $\mathbf{q}_1$ and $\mathbf{q}_2$ of the transposed matrix satisfy

$$\begin{pmatrix} -c & \rho \\ c^2\rho^{-1} & -c \end{pmatrix}\mathbf{q}_1 = \mathbf{0} \quad \text{and} \quad \begin{pmatrix} c & \rho \\ c^2\rho^{-1} & c \end{pmatrix}\mathbf{q}_2 = \mathbf{0}.$$

Hence

$$\mathbf{q}_1 = \begin{pmatrix} 1 \\ c\rho^{-1} \end{pmatrix} \quad \text{and} \quad \mathbf{q}_2 = \begin{pmatrix} 1 \\ -c\rho^{-1} \end{pmatrix},$$

and equation (9.36) now gives

$$\left.\begin{array}{r} \dfrac{c}{\rho}[\rho_t + (u+c)\rho_x] + [u_t + (u+c)u_x] = 0, \\[2mm] -\dfrac{c}{\rho}[\rho_t + (u-c)\rho_x] + [u_t + (u-c)u_x] = 0. \end{array}\right\} \tag{9.37}$$

In the first equation both ρ and u are differentiated along the characteristics defined by

$$\frac{\mathrm{d}x}{\mathrm{d}t} = u + c,$$

whilst in the second equation the differentiation is along the characteristics satisfying

$$\frac{\mathrm{d}x}{\mathrm{d}t} = u - c.$$

In this case the results could have been obtained as directly by using the approach described in §9.2 and the above formal approach is most useful for $n > 2$.

The general theory associated with systems of quasi-linear equations is technically complicated but it does turn out to be possible to generalize the concepts of weak solutions and of shock waves and these aspects are treated in some detail in Jeffrey (1976) and Whitham (1974).

However when the matrix A is not explicitly dependent on x and y and there are only two dependent variables, it is possible to make some progress towards obtaining a solution. Problems of this type do often occur in practical situations (cf. equation (9.15)). However the details can become complicated even in comparatively simple cases, and therefore our discussion will be limited to sketching the basic approach. Particular problems of this type are however discussed in some detail in Jeffrey (1976), Jeffrey and Taniuti (1964), and Whitham (1974).

Each of the equations of the system of equation (9.36) is of the form

$$l_1(u_y^{(1)} + \lambda u_x^{(1)}) + l_2(u_y^{(2)} + \lambda u_x^{(2)}) = f_1,$$

where l_1 and l_2 are known functions of $u^{(1)}$ and $u^{(2)}$. It is possible, in principle, to find an integrating factor m such that

$$l_1 \, \mathrm{d}u^{(1)} + l_2 \, \mathrm{d}u^{(2)} = m \, \mathrm{d}r,$$

where r is a known function.

We must have

$$m \frac{\partial r}{\partial u^{(1)}} = l_1, \qquad m \frac{\partial r}{\partial u^{(2)}} = l_2,$$

so that m could be any solution of

$$\frac{\partial}{\partial u^{(2)}} \left(\frac{l_1}{m} \right) = \frac{\partial}{\partial u^{(1)}} \left(\frac{l_2}{m} \right).$$

Thus by an appropriate choice of integrating factors it is possible to put the system (9.36) into the form

$$r_y^{(i)} + \lambda_i r_x^{(i)} = e_i, \qquad i = 1, 2, \tag{9.37}$$

where the $r^{(i)}$ are known functions of $u^{(i)}$ (at least in principle), and e_i may be functions of $r^{(i)}$, x, and y. The system (9.37) is identical in form with that of equation (9.31) and hence reduces to a system of ordinary differential equations on the characteristics. It is therefore possible to solve them by the method of characteristics (§10.2).

Variables analogous to the $r^{(i)}$ were introduced by Riemann in connection with problems of one-dimensional isentropic gas flow. In that particular case the e_i are zero so that the $r^{(i)}$ are constant on the characteristics and are referred to in these circumstances as the Riemann invariants (usually denoted by r and s). It therefore seems reasonable, when $e_i \neq 0$, to follow Whitham (1974) and refer to the $r^{(i)}$ as Riemann variables.

In the case when the e_i are zero so that the $r^{(i)}$ are invariant it often proves useful to use the Riemann invariants as independent variables. Clearly $r^{(2)}$ can be taken as a parameter on C_1 and $r^{(1)}$ as a parameter on C_2 so that the equations defining the characteristics can (cf. eqn. (9.32)) be written as

$$\lambda_1 \frac{\partial y}{\partial r^{(2)}} = \frac{\partial x}{\partial r^{(2)}}, \qquad \lambda_2 \frac{\partial y}{\partial r^{(1)}} = \frac{\partial x}{\partial r^{(1)}}.$$

These latter equations are linear in x and y and in some circumstances can be solved in a closed form.

The Riemann invariants can be written down immediately for the case of isentropic gas flow as equations (9.37) can be rewritten as

$$[u + L(\rho)]_t + (u + c)[u + L(\rho)]_x = 0,$$
$$[u - L(\rho)]_t + (u - c)[u - L(\rho)]_x = 0,$$

where

$$L(\rho) = \int^{\rho} \frac{c(\rho')}{\rho'} \, d\rho'.$$

It now follows immediately that the Riemann invariants are $u \pm L$.

Bibliography

Jeffrey, A. (1976). *Quasi-linear hyperbolic systems and waves.* Pitman, London.

Whitham, G. B. (1974). *Linear and non-linear waves.* Wiley, New York.

Jeffrey, A. and Taniuti, T. (1964). *Non-linear wave propagation with applications to physics and magnetohydrodynamics.* Academic Press, New York.

10 Approximate methods of solution

THE emphasis throughout the previous chapters has been on the determination of analytical (i.e. exact) solutions of partial differential equations but it is an unfortunate fact of life that for many equations (particularly non-linear ones) such solutions cannot be found and approximate methods have had to be devised to deal with problems that cannot be solved analytically. In this chapter, therefore, some of the principal approximate approaches to the solution of partial differential equations are summarized. Each of the topics discussed is itself the subject of several texts and cannot really be adequately summarized in a short space but, nevertheless, it does seem worthwhile to attempt to give some idea of the principles involved in the various approaches.

One of the principal methods of solving complicated problems for partial differential equations is by numerical means and the basis of the finite-difference method of solution is described in §10.1 and a brief account given in §10.2 of a numerical method, known as the method of characteristics, which is well suited for solving problems for hyperbolic equations. §10.3 deals with the use of variational methods to obtain approximate solutions and the use of integral equation formulations is sketched in §10.4. Finally a description of a class of methods referred to as perturbation methods is described in §10.5.

10.1. Finite-difference methods

The essence of all finite-difference methods of solving partial differential equations is the replacement, at each of a discrete number of given points, of the partial derivatives by approximations involving the dependent variable evaluated at each given point and at appropriate neighbouring ones. This transforms the problem of solving the partial differential equation to one of solving a set of linear algebraic equations. Difference methods are therefore essentially simple in nature though, as with any approximate method, their actual application in a particular case can often present considerable difficulties. The main difficulty is in ensuring that the numerical values obtained are good approximations to the exact solution. The techniques that have been devised to attain accuracy in varying circumstances are numerous and it is well beyond the scope of this text even to try to summarize such techniques. We shall therefore merely sketch the numerical methods suitable for solving simple problems for the archetypal second-order linear equations, and attempt to relate the various methods used to the theoretical properties associated with the various types of equations. Detailed accounts of numerical methods of solving partial differential equations are available in many specialist text books such as Forsythe and Wasow (1960), Mitchell (1969), and Smith (1969).

The points at which the derivatives are replaced by approximations can have an arbitrary distribution but the problem is considerably simplified by choosing a completely regular distribution of points. Such a distribution is most easily defined by choosing the points to be at the nodes of a suitable net or mesh. We restrict ourselves to problems which can be solved by using a mesh of equal rectangles with the vertices of the rectangles at the points (ih, jk) where i, j are integers positive, negative, or zero (see Fig. 10.1). The point (ih, jk) will be denoted by $P_{i,j}$, and $u_{i,j}$ will be used to denote $u(ih, jk)$.

The basic formulae used in making finite difference approximations can be derived from Taylor's theorem. For the x-direction this theorem gives

$$u(x \pm h, y) = u(x, y) \pm h u_x(x, y) + \tfrac{1}{2} h^2 u_{xx}(x, y) \pm \tfrac{1}{6} h^3 u_{xxx}(x, y) \\ + O(h^4). \quad (10.1)$$

Equation (10.1) can be used to obtain the following expressions for u_x, u_{xx} at P_{ij} in terms of $u_{i,j}, u_{i+1,j}$, and $u_{i-1,j}$:

$$u_x = \frac{1}{h}(u_{i+1,j} - u_{i,j}) + O(h), \quad (10.2)$$

$$u_x = \frac{1}{h}(u_{i,j} - u_{i-1,j}) + O(h), \quad (10.3)$$

$$u_x = \frac{1}{2h}(u_{i+1,j} - u_{i-1,j}) + O(h^2), \quad (10.4)$$

$$u_{xx} = \frac{1}{h^2}(u_{i+1,j} - 2u_{i,j} + u_{i-1,j}) + O(h^2). \quad (10.5)$$

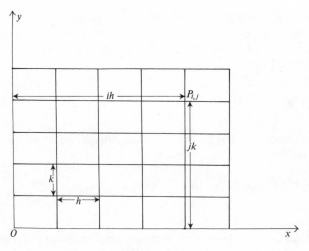

FIG 10.1

Expressions (10.2) and (10.3), neglecting the order terms, are referred to as forward-difference and backward-difference formulae respectively. A similar set of expressions can be obtained for the y-derivatives of u at P_i, viz

$$u_y = \frac{1}{k}(u_{i,j+1} - u_{i,j}) + O(k), \tag{10.6}$$

$$u_y = \frac{1}{k}(u_{i,j} - u_{i,j-1}) + O(k), \tag{10.7}$$

$$u_y = \frac{1}{2k}(u_{i,j+1} - u_{i,j-1}) + O(k^2), \tag{10.8}$$

$$u_{yy} = \frac{1}{k^2}(u_{i,j+1} - 2u_{i,j} + u_{i,j-1}) + O(k^2). \tag{10.9}$$

The above formulae are sufficient to permit numerical solution of simple problems for the archetypal second-order equations. We consider first the Dirichlet problem for Laplace's equation in the rectangle $0 \le x \le a$, $0 \le y \le b$. The most convenient kind of mesh to choose in this case is one where the vertices of the rectangle are nodal points of the mesh. This can be achieved by choosing $h = a/m$ and $k = b/n$, where m and n are positive integers. The mesh then subdivides the rectangle into mn equal rectangular regions. Equations (10.5) and (10.9) show that, excluding terms of orders h^2 and k^2, Laplace's equation at $P_{i,j}$ may be replaced by

$$\frac{1}{h^2}(u_{i+1,j} - 2u_{i,j} + u_{i-1,j}) + \frac{1}{k^2}(u_{i,j+1} - 2u_{i,j} + u_{i,j-1}) = 0. \quad (10.10)$$

For equation (10.10) to hold at each interior point of the mesh we must have

$$2(h^2 + k^2)u_{i,j} = k^2(u_{i+1,j} + u_{i-1,j}) + h^2(u_{i,j+1} + u_{i,j-1}),$$
$$1 \le i \le m-1, \qquad 1 \le j \le n-1. \quad (10.11)$$

Equation (10.11) shows that u at any interior nodal point can be expressed in terms of the values of u at the four immediately adjacent nodal points (Fig. 10.2), and it also follows directly from this equation that the value of $|u|$ at any interior mesh point does not exceed its values at any of the four nodal points immediately adjacent to it.

Applying this argument successively at all interior mesh points leads to the conclusion that $|u|$ at the interior mesh points cannot exceed the maximum value of $|u|$ on the boundary. This is the

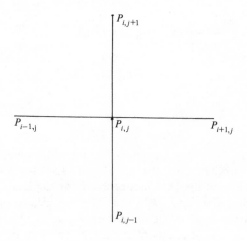

Fig 10.2

finite-difference analogue of the maximum modulus principle of §4.2 and is in fact a consequence of this latter principle.

In equation (10.11) the values of $u_{0,j}$, $u_{m,j}$, $u_{i,0}$, $u_{i,n}$ are determined by the given conditions on the boundary of the rectangle. Hence, as i and j vary through their permitted range, we obtain a set of $(m-1)(n-1)$ inhomogeneous linear algebraic equations for the $(m-1)(n-1)$ unknown values of u at the interior mesh points. There is of course a prior possibility that, for particular choices of h and k, the matrix of the system could be singular but fortunately it can be shown that this is not the case and also that, as h and k tend to zero, the values of u satisfying (10.11) at the mesh points tend to the exact solution of the problem (Greenspan 1965).

It is worth noting that the proof of the existence of the solution of the equations (10.11) and of its convergence to the exact solution as $h, k \rightarrow 0$ is based on the existence of the maximum modulus principle. Thus the success of the numerical method of solution is directly related to the existence of the maximum modulus principle which, as indicated in §4.4, is associated with a class of elliptic equations.

In order to achieve reasonable numerical accuracy it is necessary for h and k to be small and this means that the number of equations becomes large. The computational problems associated with the numerical solution of large sets of equations are considerable and are described in detail in the texts referred to above.

The Dirichlet problem for Laplace's equation within a rectangle can be solved analytically and it would therefore be, in many cases, pointless to try and solve it numerically. However one of the advantages of numerical methods is that they are not unduly dependent on particular geometries and the steps outlined above can equally well be applied to the Dirichlet problem within an L-shaped region. The only change from the numerical point of view is the number of points at which equation (10.11) holds and, in terms of programming, it requires very few instructions to change a program suitable for solving a Dirichlet problem for a rectangle into one which is appropriate for solving a Dirichlet problem within an L-shaped region.

The numerical solution of boundary-value problems for elliptic equations will not be pursued any further. We next examine how the above numerical approach has to be modified to solve the

Cauchy problem on $y = 0$ for the archetypal hyperbolic equation, viz.

$$u_{xx} - u_{yy} = 0, \qquad y > 0. \tag{10.12}$$

Equation (10.12) can, by using equations (10.5) and (10.9), be approximated at $P_{i,j}$, $j \geq 1$ and all i by

$$\frac{1}{h^2}(u_{i+1,j} - 2u_{i,j} + u_{i-1,j}) - \frac{1}{k^2}(u_{i,j+1} - 2u_{i,j} + u_{i,j-1}) = 0,$$

$$j \geq 1 \quad \text{and all } i. \tag{10.13}$$

Equation (10.13) is similar in form to equation (10.10), but it would not be reasonable to solve it in the same way since j and i now vary over an infinite range and reduction to a finite system of algebraic equations is impossible. Any method of solving equation (10.13) has to reflect the fact that, regarding y as a time variable, equation (10.12) is an equation of evolution. This means that $u_{i,j+1}$ should be obtainable directly from values of $u_{i,j}$, $u_{i,j-1}$, etc. and it therefore seems reasonable to exhibit this relationship by rearranging equation (10.13) to give

$$u_{i,j+1} = \frac{k^2}{h^2} u_{i+1,j} + 2\left(1 - \frac{k^2}{h^2}\right) u_{i,j} + \frac{k^2}{h^2} u_{i-1,j} - u_{i,j-1},$$

$$j \geq 1 \quad \text{and all } i. \tag{10.14}$$

An equation of the form (10.14), which expresses $u_{i,j+1}$ directly in terms of the values $u_{i,j}$, $u_{i,j-1}$, etc., is said to be an explicit formula. Equation (10.14) shows that $u_{i,j}$ can be found for all i and j provided that the values of $u_{i,j}$ are known for two consecutive values of j. The known values of u on $y = 0$ enable $u_{i,0}$ to be found, whilst the known values of u_y on $y = 0$ can be substituted into equation (10.6) for $j = 0$ to give $u_{i,1}$. These two sets of values can then be used, in conjunction with equation (10.14), to generate all the $u_{i,j}$ for successive j.

The value of u at $P_{i,j+1}$ is obtained in terms of its previously calculated values at $P_{i\pm1,j}$, $P_{i,j}$, and $P_{i,j-1}$ (Fig. 10.2), and these latter quantities are themselves obtained from previously calculated values on the lines $y = (j-1)k$, $(j-2)k$, and $(j-3)k$. Thus, on following the calculation through from the lines $y = 0$ and $y = k$, we see that u at $P_{i,j}$ will be a function of the values of u

within the region bounded by the lines drawn backwards towards
$y = 0$ from $P_{i,j}$ and whose slopes are $\pm k/h (= p)$ (Fig. 10.3).

The triangular region bounded by these two lines and the line
$y = 0$ therefore is the domain of the dependence at $P_{i,j}$ of the
solution of the set of difference equations defined by equation
(10.14). The domain of dependence at $P_{i,j}$ of the solution of
equation (10.12) is however, the triangular region bounded by
the line $y = 0$ and the lines with slopes ± 1 drawn backwards from
$P_{i,j}$ towards $y = 0$ (i.e. between $y = 0$ and the dotted lines in Fig.
10.3). If the domain of dependence of the difference equations is
contained within that for the differential equation, then it would
be possible to change some of the data for the differential
equation and yet have no effect on the solution of the difference
equations. This could clearly lead to an incorrect numerical
solution and it is possible to prove rigorously (Forsythe and
Wasow 1960) that the $u_{i,j}$ determined by the above scheme only
converge, in the limit as $h, k \to 0$, to the correct solution when the
domain of dependence of the difference equation includes that
for the differential equation, i.e. when $p < 1$.

In progressive computational schemes of the type outlined
above, where each new set of calculated values depends on
previously calculated sets of values, there is a possibility that a
small numerical error (e.g. a rounding-off error) can be magnified
by repeated calculation. A scheme where such a magnification
can exist is said to be unstable and it is therefore desirable to
avoid using such schemes. The existence of instability can be seen
fairly easily by introducing a line of numerical errors of the form

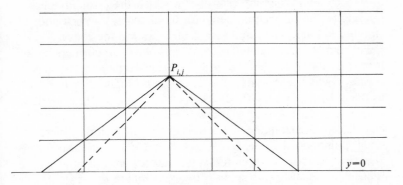

Fɪɢ 10.3

$\varepsilon \exp i\beta x$ (ε small) in the values used on $y = 0$ to start the calculation. (The most general form of error can, by the theory of Fourier integrals, be expressed as a linear superposition of such terms.) As the equations are linear the effect of such errors can be found by looking at the possible forms of solution of the system defined by equation (10.14) with $u_{i,0} = \varepsilon \exp i\beta h$. Assuming that $u_{i,j}$ is of the form $\varepsilon \exp(\alpha jk)\exp i\beta h$ shows, on making this substitution in equation (10.14), that α must be such that

$$z^2 - (2 - 4p^2 \sin^2 \tfrac{1}{2}\beta h)z + 1 = 0, \qquad (10.14)$$

where $z = \exp(\alpha k)$. The solutions of the difference equation will become unbounded as $j \to \infty$ if the modulus of the roots of equation (10.14) exceeds unity. Since the product of the roots is 1 it follows that one of the roots will always have modulus greater than 1 unless both roots have unit modulus. This can only occur when the roots are not real which means that the condition for stability is $(1 - 2p^2 \sin^2 \tfrac{1}{2}\beta h)^2 \leq 1$, that is,

$$p^2 \leq \operatorname{cosec}^2 \tfrac{1}{2}\beta h.$$

Thus, in order to have stability for all values of β, p cannot exceed 1.

In this case the condition for convergence of the difference scheme as $k, h \to 0$ is the same as that for stability. This is a particular example of a general result that, for a properly posed initial-value–boundary-value problem with a consistent finite-difference approximation (i.e. one in which the error in approximating to the differential equation tends to zero with h and k), stability is a necessary and sufficient condition for convergence.

For initial-value–boundary-value problems it is possible, by using a different finite-difference approximation, to devise schemes which are stable for all h and k. The simplest such problem for equation (10.12) arises when u and u_y are prescribed on $y = 0$, $0 \leq x \leq a$, and u is prescribed on $x = 0$ and $x = a$. The calculations are simplified by choosing a mesh such that there are nodal points on the lines $y = 0$, $x = 0$, and $x = a$, and this means choosing h equal to a/m for some integer m. An appropriate finite-difference scheme again is to approximate u_{yy} at (ih, jk) by equation (10.4) and to approximate u_{xx} by the mean of its values at $[ih, (j+1)k]$ and $[ih, (j-1)k]$. At points in $0 < x < a$ this gives

$$u_{i,j+1} - 2u_{i,j} + u_{i,j-1} = \frac{p^2}{2}[u_{i+1,j+1} - 2u_{i,j+1} + u_{i-1,j+1}$$
$$u_{i+1,j-1} - 2u_{i,j-1} + u_{i-1,j-1}],$$

which can be rearranged as

$$-p^2 u_{i+1,j+1} + 2(1+p^2)u_{i,j+1} - p^2 u_{i-1,j+1}$$
$$= 4u_{i,j} + p^2 u_{i+1,j-1} - 2(1+p^2)u_{i,j-1}$$
$$+ p^2 u_{i-1,j-1}, \quad 1 \le i \le m-1. \tag{10.15}$$

If the values are known on the lines $y = jk$ and $y = (j-1)k$ then equation (10.15) provides, as i ranges over its possible values, $m-1$ relations between the values of u on the line $y = (j+1)k$ and hence these values are found by solving a set of linear equations. Such a scheme is termed an implicit scheme. Implicit schemes are not suitable for pure initial-value problems as the range of i in equation (10.15) would then be infinite and it would be necessary to solve an infinite set of algebraic equations. The stability of the above implicit scheme can be investigated by again assuming an error of the form $\varepsilon \exp i\beta x$ on $y = 0$. The error on $y = jk$ will again be of the form $\varepsilon \exp \alpha jk \exp i\beta h$ where α now has to satisfy the equation

$$z^2 - \frac{2z}{(1 + 2p^2 \sin^2 \frac{1}{2}\beta h)} + 1 = 0,$$

where $z = \exp \alpha k$. The roots of this quadratic are always complex and of unit modulus and hence the scheme is always stable. There are several possible different approximations to u_{yy} which lead to stable implicit schemes and some are discussed in Mitchell (1969).

It should be noted that the above numerical approach may also be used to obtain numerical solutions for initial-value–boundary-value problems for the more general non-linear equation

$$u_{xx} - u_{yy} = f(x, y, u),$$

which cannot be solved analytically. The above methods can also be generalized to cover more general second-order hyperbolic equations (Mitchell 1969) though, as a rule, the most effective method of solving such equations numerically is the 'method of characteristics' described in §10.2.

Similar finite difference approaches to those analysed above can also be used to obtain numerical solutions for the Cauchy problem on $y = 0$ for the archetypal parabolic equation

$$u_{xx} = u_y, \qquad y > 0. \qquad (10.16)$$

The approximating difference equation in this case is, from equations (10.5), and (10.6),

$$\frac{1}{h^2}(u_{i+1,j} - 2u_{i,j} + u_{i-1,j}) = \frac{1}{k}(u_{i,j+1} - u_{i,j}), \qquad j = 1, 2, \ldots,$$

which can be rearranged as

$$u_{i,j+1} = ru_{i-1,j} + (1 - 2r)u_{i,j} + ru_{i+1,j}, \qquad j = 1, \qquad (10.17)$$

where $r = k/h^2$. Equation (10.17) provides an explicit formula for values of u on $y = (j+1)k$ in terms of values on $y = jk$, $u_{i,0}$ is known and hence $u_{i,j}$ can be found for all j by successive application of equation (10.17).

Problems of convergence and stability arise exactly as for hyperbolic equations and it can be shown fairly easily (Mitchell 1969) that, as $h, k \to 0$, the solution of the difference equation tends to that of the differential equation provided that $r \le \frac{1}{2}$. The maximum–minimum principle of §6.2 requires that the maximum value of $u_{i,j+1}$ cannot exceed that of $u_{i,j}$ and equation (10.17) shows that this is true when $r \ge \frac{1}{2}$. Stability can again be investigated by considering the particular solution of equation (10.7) with $u_{i,0} = \varepsilon \exp i\beta h$. It follows exactly as for hyperbolic equations that $u_{i,j} = \varepsilon \exp i\beta h \exp \alpha jk$, where

$$e^{\alpha k} = 1 - 4r \sin^2 \tfrac{1}{2}\beta h.$$

In this case $\exp \alpha k$ is automatically less than 1 and the scheme will therefore be stable provided that $\exp \alpha k > -1$, i.e.

$$r \le \tfrac{1}{2} \operatorname{cosec}^2 \tfrac{1}{2}\beta h.$$

and, for this to hold for all β, r must not exceed $\frac{1}{2}$.

The condition $k < \frac{1}{2}h^2$ implies that, for reasonable accuracy, the mesh dimension parallel to Oy has to be very small and this can lead to very lengthy calculations. It is therefore desirable to have methods without such a restrictive stability criterion and it is possible, exactly as for hyperbolic equations, to develop suitable implicit schemes to solve initial-value–boundary-value problems. Explicit schemes do not really model parabolic equations accurately because they imply, exactly as for hyperbolic equations, the existence of a finite domain of dependence. For equation (10.17) the domain of dependence of the point P shown in Fig. 10.4 is the triangular domain bounded by the line $y = 0$ and the lines drawn backwards from P towards $y = 0$ with slopes $\pm k/h$. The domain of dependence at P of solutions of equation (10.17) is, when boundary conditions are imposed on $x = 0$, $x = a$, and $y = 0$, the whole of the rectangle $OABC$. In particular the explicit scheme of equation (10.17) implies that the value of u at a point D is completely independent of the prescribed value of u at E on OC. An implicit scheme would remedy this defect and several such schemes have been proposed, a popular one being the Crank–Nicolson scheme where u_{xx} at (ih, jk) is replaced by the mean of its finite-difference representations at $[ih, jk]$ and $[ih, (j+1)k]$ so that the approximating difference equation is

$$(2/r)(u_{i,j+1} - u_{i,j}) = u_{i+1,j+1} - 2u_{i,j+1} + u_{i-1,j+1}$$
$$+ u_{i+1,j} - 2u_{i,j} + u_{i-1,j}. \qquad (10.18)$$

It can be shown that the implicit scheme based on equation (10.18) is stable for all r.

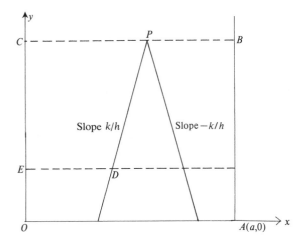

FIG 10.4

10.2. The method of characteristics

The method of characteristics is a numerical method based on the fact that, on a characteristic, a second-order quasi-linear partial differential equation becomes an ordinary differential equation. This property has already been illustrated in §3.2 for the particular case of the one-dimensional wave equation and established in §9.2 by reducing a second-order equation to a system of first-order ones and it will now be established, independently of the theory of systems of equations, for

$$au_{xx} + 2bu_{xy} + cu_{yy} + F = 0, \qquad (10.19)$$

where a, b, c, and F may be functions of x, y, u, u_x, and u_y.

If p and q denote u_x and u_y respectively then

$$dp = p_x \, dx + p_y \, dy = u_{xx} \, dx + u_{xy} \, dy,$$

$$dq = q_x \, dx + q_y \, dy = u_{xy} \, dx + u_{yy} \, dy,$$

and solving these equations for u_{xx} and u_{yy} gives

$$u_{xx} = \frac{dp}{dx} - u_{xy} \frac{dy}{dx},$$

$$u_{yy} = \frac{dq}{dy} - u_{xy} \frac{dx}{dy}.$$

Substituting for u_{xx} and u_{yy} from these identities into equation (10.19) gives, after some slight rearrangement,

$$u_{xy}\left[a\left(\frac{dy}{dx}\right)^2 - 2b\frac{dy}{dx} + c\right] - \left[a\frac{dp}{dx}\frac{dy}{dx} + c\frac{dq}{dx} + F\frac{dy}{dx}\right] = 0. \quad (10.20)$$

There are two sets of characteristics defined by

$$\frac{dy}{dx} = f \quad \text{and} \quad \frac{dy}{dx} = g,$$

where f and g are the roots of the quadratic

$$a\lambda^2 - 2b\lambda + c = 0.$$

Thus on the 'f' and 'g' characteristics respectively equation

(10.20) gives

$$a\frac{\mathrm{d}p}{\mathrm{d}x}f + c\frac{\mathrm{d}q}{\mathrm{d}x} + fF = 0, \qquad (10.21)$$

and

$$a\frac{\mathrm{d}p}{\mathrm{d}x}g + c\frac{\mathrm{d}q}{\mathrm{d}x} + gF = 0, \qquad (10.22)$$

which are equivalent to the result derived in §9.2. If p and q are known at one point on each of the characteristics through a point P then integrating equations (10.21) and (10.22) along the respective characteristics will give, in principle, two relations between p and q at P and hence both can be found and u then obtained by a further integration. It proved possible in §3.2 to carry out the complete process analytically for the wave equation, but in many cases it will not be possible to do this. In the general quasi-linear case the characteristics will depend on u and direct integration along them would not be possible without knowing u, but even when the characteristics are known explicitly analytic integration would not generally be possible. It is however possible to develop a numerical approach which can be used when analytical methods prove useless and we sketch the basic procedure for the particular case of the Cauchy problem posed on a non-characteristic curve Γ.

Two points P and Q are chosen close together on Γ and it is assumed that the 'f' characteristic through P and the 'g' characteristic through Q intersect at R (Fig. 10.5). As a first approximation these characteristics may be approximated by straightlines

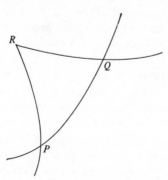

Fig 10.5

whose slopes are the slopes f_P and g_Q of the relevant characteristics at P and Q so that

$$y_R - y_P = f_P(x_R - x_P), \qquad (10.23)$$

$$y_R - y_Q = f_Q(x_R - x_Q), \qquad (10.24)$$

where suffixes are used to denote values evaluated at the corresponding points. Along the characteristic PR equation (10.21) gives

$$a_P f_P(p_R - p_P) + c_P(q_R - q_P) + F_P(y_R - y_P) = 0, \qquad (10.25)$$

and equation (10.22) gives

$$a_Q g_Q(p_R - p_Q) + c_Q(q_R - q_Q) + F_Q(y_R - y_Q) = 0. \qquad (10.26)$$

Equations (10.23) and (10.24) enable approximate values of x_R and y_R to be found and these values can then be used in equations (10.25) and (10.26) to estimate p_R and q_R. Finally an estimate for u_R can be found by replacing p and q in the identity

$$du = p\,dx + q\,dy$$

by the means of their values at P and R so that

$$u_R - u_P = \tfrac{1}{2}(p_P + p_R)(x_R - x_P) + \tfrac{1}{2}(q_P + q_R)(y_R - y_P). \qquad (10.27)$$

The value thus obtained for u_R can now be used to obtain a better approximation for x_R and y_R by replacing f_P and g_Q in equations (10.23) and (10.24) by $\tfrac{1}{2}(f_P + f_R)$ and $\tfrac{1}{2}(f_Q + f_R)$ respectively. (This is effectively the Euler predictor method of solving ordinary differential equations.) More accurate values of p_R and q_R may then be obtained by making the above replacement of f_P and g_Q in equations (10.25) and (10.26) and a new estimate obtained for u_R from equation (10.27). This cycle can be repeated until the difference between two successive values of u_R is smaller than some prescribed quantity. By carrying out the complete process for a large number of closely separated points on Γ it is possible to obtain u at a series of points in a close neighbourhood of Γ. These latter values can then in turn be used to advance the solution further away from Γ and eventually u can be determined at all points. For quasi-linear equations it is possible that shocks may develop and the above numerical approach breaks down in the neighbourhood of such shocks. It is however possible to develop methods which can cope with the occurrence of shocks, and some of these methods are discussed in Forsythe and Wasow (1960) and Richtmeyer and Morton (1967).

The methods sketched above can clearly also be used to obtain numerical solutions for quasi-linear systems of the type discussed in §9.4.

10.3. Variational methods

The variational approach introduced in §1.3 can be used to obtain approximations to the eigenvalues of the problems discussed in §4.5, and also to obtain reasonable approximations for quantities associated with the solutions of the boundary-value problems encountered in §4.4. Variational methods also form the basis of a very powerful numerical method (the finite-element method) of solving the latter problems. We consider first the eigenvalue problem for

$$Eu + \lambda wu = 0, \quad \text{(i.e. equation (4.52))}$$

where the quantities have the same meaning as in §4.5, with $u = 0$ on the boundary C on the region D. The basic result used is that I, defined by

$$I(u) = \frac{\displaystyle\int_D \{au_x^2 + cu_y^2 - fu^2\} \, dx \, dy}{\displaystyle\int_D wu^2 \, dx \, dy}, \qquad (10.28)$$

has a stationary value in the sense described in §1.3 whenever u is one of the eigenfunctions of the above problem, the stationary value being the eigenvalue corresponding to the particular eigenfunction. Also $I \geq \lambda_1$ where λ_1 is the lowest eigenvalue and equality only holds when u is the eigenfunction corresponding to λ_1. A stationary value implies that the difference $I(u + \varepsilon p) - I(u)(p = 0$ on $C)$ is of order ε^2 when u is one of the eigenfunctions and hence substituting into I a form for u which is a first-order approximation produces effectively a second-order approximation to the eigenvalue. In a sense a better estimate is obtained than it would normally be realistic to expect from the approximation.

The fact that I is stationary when u is an eigenfunction of equation (4.52) can be proved by substituting $u + \varepsilon p$ for u in equation (10.28), expanding in powers of ε, and equating the coefficient of ε to zero. This gives

$$\int_D [ap_x u_x + cp_y u_y - fup] \, dx \, dy - I_0 \int_D upw \, dx \, dy = 0, \quad (10.29)$$

where

$$I_0 = \frac{\displaystyle\int_D \{au_x^2 + cu_y^2 - fu^2\} \, dx \, dy}{\displaystyle\int_D wu^2 \, dx \, dy}.$$

Equation (10.29) can be rewritten as

$$\int_D [(apu_x)_x + (cpu_y)_y] \, dx \, dy - \int_D p[Eu + I_0 wu] \, dx \, dy = 0,$$

and the first integral can be transformed using the divergence theorem to give a line integral which vanishes since $p = 0$ on C. The remaining integral has to vanish for all appropriate functions p and this, by arguments used in §1.3, means that

$$Eu + I_0 wu = 0,$$

showing that u is an eigenfunction with I_0 being the corresponding eigenvalue. The proof that the resulting stationary value is a minimum is slightly more complicated and is most easily effected in the following indirect manner. For functions u vanishing on C equation (10.28) simplifies, on using equation (4.43), to

$$I = \frac{-\displaystyle\int_D uEu \, dx \, dy}{\displaystyle\int_D u^2 w \, dx \, dy}, \quad (10.30)$$

which is slightly easier to use. For such functions u we have the result quoted in §4.5 that there exist constants c_r such that

$$u = \sum_{r=1}^{\infty} c_r u_r,$$

where u_r are the eigenfunctions of equation (4.52). Substituting this into equation (10.30) and assuming the validity of interchange of summation and integration gives, after using the orthogonality property of the eigenfunctions,

$$I = \frac{\displaystyle\sum_{r=1}^{\infty} c_r^2 \lambda_r \int_D w u_r^2 \, dx \, dy}{\displaystyle\sum_{r=1}^{\infty} c_r^2 \int_D w u_r^2 \, dx \, dy} \geq \lambda_1,$$

thus verifying that the minimum value of I is λ_1.

In practice various suitable functions u (trial functions) are substituted into the right-hand side of equation (10.30) and an attempt made to estimate the possible minimum value of I. In general the trial functions involve arbitrary constants which are chosen so as to provide a minimum value.

The one-dimensional analogue of the above results concerning the eigenvalues of equation (4.52) is that, for all functions $u(x)$ vanishing at $x = 0$ and $x = b$, $J(u)$, defined by

$$J(u) = \frac{\displaystyle\int_0^b [a(x) u_x^2 - f u^2] \, dx}{\displaystyle\int_0^b w u^2 \, dx},$$

is such that $J(u) \geq \mu_1$ where μ_1 is the smallest eigenvalue of

$$(a u_x)_x + f u + \mu w u = 0$$

with $u = 0$ at $x = 0$ and $x = b$. The proof of this result is almost identical to that given above with the double integral replaced by a single integral. The equation corresponding to equation (10.29) is

$$\int_0^b [a p_x u_x - f u p] \, dx - J_0 \int_0^b u p w \, dx = 0,$$

where

$$J_0 = \frac{\displaystyle\int_0^b [au_x{}^2 - fu^2]\,\mathrm{d}x}{\displaystyle\int_0^b wu^2\,\mathrm{d}x}.$$

Integration by parts gives

$$\int_0^b ap_x u_x\,\mathrm{d}x = [pau]_{x=0}^{x=b} - \int_0^b p(au_x)_x\,\mathrm{d}x,$$

and the result now follows exactly as before.

In the particular case $a \equiv 1, f \equiv 0, w \equiv 1$ we deduce that $J(u) \geq \mu_1$ where μ_1 is the smallest eigenvalue of

$$u_{xx} + \mu u = 0$$

with $u = 0$ for $x = 0$ and $x = b$. The eigenvalues in this case are $n^2\pi^2/b^2$, $n = 1, 2, \ldots$ and hence

$$J(u) \geq \frac{\pi^2}{b^2}.$$

Example 10.1

Determine an upper bound for the lowest eigenvalue of

$$\nabla^2 u + \lambda u = 0,$$

with $u = 0$ on $r^2 = x^2 + y^2 = 1$.

The simplest approach is to assume no angular variation, this gives an equation which is even in r so that the simplest possible suitable trial function is $1 - r^2$, leading to an estimate of 6 for the lowest eigenvalue. A slightly more sophisticated approximation is $u = \cos \frac{1}{2}\pi r$ and the corresponding value of $I(u)$ is 5.83. A further refinement would be to assume u to be of the form $c_1 \cos \frac{1}{2}\pi r + c_2 \cos \frac{3}{2}\pi r$ and choose c_1 and c_2 so as to minimize I, this gives $I = 5.79$. This particular eigenvalue problem is one that can be solved exactly (cf. Example 7.3) and $\lambda_1{}^2$ turns out to be the first zero of the Bessel function of order zero giving $\lambda_1 = 5.78$.

The method of assuming an expansion for the trial function as a finite series of a given set of functions and choosing the coefficients so as to minimize a functional is generally known as the Rayleigh–Ritz method.

If, when $a = c$, the additional term $\mu \int_C wu^2\,\mathrm{d}s$ is included in the numerator of equation (10.28) then it can be shown by generalizing the above argument that I is stationary when u is an eigenfunction of equation (4.52) and $\partial u/\partial n + \mu u = 0$ on C.

The variational method can also be used to give some useful approximate information about solutions of

$$Eu = g \quad \text{(i.e. equation (4.21))}$$

with $u = 0$ on C. It can be shown by an argument very similar to that of §1.3 that

$$J = \int_D \{au_x{}^2 + cu_y{}^2 - fu^2 + 2gu\} \, dx \, dy \qquad (10.31)$$

has, when $u = 0$ on C, a minimum value when u also satisfies equation (4.21). Equation (4.43) shows that

$$J = \int_D (-uEu + 2gu) \, dx \, dy$$

and when u satisfies equation (4.21) this reduces to $\int_D gu \, dx \, dy$. Hence the minimum value gives an accurate estimate for $\int_D gu \, dx \, dy$ and this is often a quantity of interest. In the simplest application of this method it is assumed that u is of the form $c_1 v$, where v is some suitable function, and c_1 is chosen so as to minimize J. Substituting $u = c_1 v$ into equation (10.31) gives

$$c_1 = \frac{\displaystyle\int_D gv \, dx \, dy}{\displaystyle\int_D vEv \, dx \, dy} \quad \text{and} \quad J = \frac{\left(\displaystyle\int_D gv \, dx \, dy\right)^2}{\displaystyle\int_D vEv \, dx \, dy}.$$

More accurate estimates can be obtained by assuming a linear combination of suitable functions and choosing the coefficients so as to minimize J.

The method can be applied to give an estimate of $\int_D u \, dx \, dy$ when

$$u_{xx} + u_{yy} = -2$$

and D is the region $|x| \le 1$, $|y| \le 1$. This problem occurs in elasticity in the theory of torsion of a rectangular cylinder and the integral of u over the square enables the moment required to produce a given twist per unit length to be calculated. The problem is symmetric in x and y and the simplest trial function is $c_1(1-x^2)(1-y^2)$ giving $\int u\, dx\, dy = 1.1104$. A more elaborate trial function is $(1-x^2)(1-y^2)[c_1 + c_2(x^2 + y^2)]$ and, when c_1 and c_2 are chosen so as to minimize $\int u\, dx\, dy$, this gives $\int u\, dx\, dy = 1.1232$. The exact value can be obtained by separation of variables and is 1.1248. It would appear at first sight that these values underestimate the integral, suggesting that the stationary value is a maximum rather than a minimum but it should be noted that $g = -1$ and that the minimum for $-\int u\, dx\, dy$ corresponds to the maximum numerical value of the integral.

The fact that the right-hand side of equation (10.31) is stationary when u satisfies equation (4.1) forms the basis of the finite-element method of solving elliptic partial differential equations. In this method the region D is subdivided into a large number of subareas (or elements) and in each element some form of polynomial approximation is made for u in terms of its values u_i at a few discrete points on the boundary of the element. The most commonly used element is the triangle and u is assumed to be of the form $Ax + By + C$ where A, B, and C can be expressed in terms of the values of u at the vertices of the triangle, this kind of representation ensuring that u is continuous across the boundaries of the element. The relatively simple form assumed for u in each element means that the integration in equation (10.31) can be carried out over each element means that the integration in equation (10.31) can be carried out over each element and hence J can be obtained in terms of the u_i. The stationary property of J requires that $\partial J/\partial u_i = 0$ at all interior points and this provides sufficient equations to determine all the u_i and hence u. The programming effort involved in the finite-element method is considerable, but it has the advantage of being a method which can be adapted to solve problems within rather complicated regions. The finite-element method and its various ramifications are described in detail in Mitchell and Wait (1977) and Strang and Fix (1973).

10.4. Integral equation methods

It often proves useful, when solving boundary-value problems for elliptic equations, to use the integral representations of §§4.3,

4.4, and 8.1 to reduce the problem to the solution of an integral equation. In many cases, particularly for problems in unbounded regions the resulting integral equation is much more amenable to numerical solution than was the original problem, since finite-difference and finite-element methods are not well suited for solving problems where one boundary is at infinity. There is an extensive literature on the numerical solution of integral equations, two recent texts being Baker (1977) and Delves and Walsh (1974). The literature on the reduction of boundary-value problems is itself large and the principal methods are described in Jaswon and Symm (1977), Kanwal (1971), and Sneddon (1966).

In view of the extent of the literature we only sketch one method of using the integral representations of §§4.3, 4.4, and 8.2 to reduce a boundary-value problem to an integral equation, and carry out the detailed calculation for Laplace's equation when the dependent variable is constant on a given lamina. (In electrostatics this is the problem of finding the capacitance of a lamina.)

It follows from equation (4.51) of §4.4 that the solution of

$$(au_x)_x + (cu_y)_y + fu = 0,$$

within the region D bounded by the closed curve C, has the representation

$$u = \int_C F_1 u \, ds' + \int_C F_2 \frac{\partial u}{\partial n'} \, ds', \qquad (10.32)$$

where F_1 and F_2 are known. If u is required to take a given value g on C then equation (10.32) becomes, on C,

$$g = \int_C F_1 g \, ds' + \int_C F_2 \frac{\partial u}{\partial n'} \, ds',$$

which is an integral equation to determine $\partial u/\partial n'$ on C. The value of u at any other point may then be calculated by using equation (10.32).

The detailed calculation associated with the above process can possibly be seen more clearly by looking in slightly more detail at the three-dimensional problem of solving

$$\nabla^2 u = 0,$$

with $u = 1$ on some finite lamina L lying in the plane $z = 0$.

Symmetry shows that u will be an even function of z so that $u = \phi(x, y, z)$ for $z \geq 0$ and $u = \phi(x, y, -z)$ for $z \leq 0$ where ϕ is a solution of Laplace's equation in the region $z \geq 0$. (Fig. 10.6). Both u and $\partial u/\partial z$ are continuous across $z = 0$ except possibly across L and hence $\phi_z = 0$ on $z = 0$ at points off L. Thus ϕ satisfies Laplace's equation in $z \geq 0$ and $\phi(x, y, 0) = 1$ for points (x, y) on L whilst $\phi_z(x, y, 0) = 0$ for points (x, y) off L. It follows from equation (8.7) that

$$\phi = \int [F(x, y, z, x', y', 0)\phi_{z'}(x', y', 0)$$

all x', y'

$$- \phi(x', y', 0)F_{z'}(x, y, z, x', y', 0)] \, dS',$$

where F is some fundamental solution of Laplace's equation (the outward normal is in the direction of $-z'$). The term involving ϕ in the above integral may be eliminated by taking F to be such that $F_{z'} = 0$, $z' = 0$. The method of images then gives

$$F = -\frac{1}{4\pi} \left\{ \frac{1}{\{(x'-x)^2 + (y'-y)^2 + (z'-z)^2\}^{\frac{1}{2}}} \right.$$

$$\left. + \frac{1}{\{(x'-x)^2 + (y'-y)^2 + (z'+z)^2\}^{\frac{1}{2}}} \right\}$$

Thus, on imposing the condition $\phi = 1$ on $z' = 0$ for points of L and using the fact that $\phi_{z'} = 0$ on $z' = 0$ for points off L, we obtain

$$1 = -\frac{1}{2\pi} \int_L \phi_{z'} \frac{dx' \, dy'}{[(x'-x)^2 + (y'-y)^2]^{\frac{1}{2}}}, \qquad (10.33)$$

which is the integral equation sought. One method of solving equation (10.33) is to divide L into a large number of small

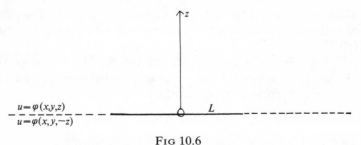

$u = \varphi(x,y,z)$

$u = \varphi(x,y,-z)$

L

FIG 10.6

subareas and assume $\phi_{z'}$ to be constant in each subarea, the
integral equation then reduces to a set of algebraic equations for
the constant values of $\phi_{z'}$ in the various subareas.

10.5. Perturbation methods

In many practical problems involving the solution of a partial
differential equation it turns out that the governing equation
contains some small parameter (ε say) and reduces, when $\varepsilon = 0$,
to a much simpler equation which can often be solved fairly
easily. In more formal terms a solution u is often sought of an
equation of the form

$$L_\varepsilon u = g, \qquad (10.34)$$

where the suffix denotes a dependence on ε, when solutions ϕ of

$$L_0 \phi = g \qquad (10.35)$$

can be found fairly easily. In such circumstances it seems a
reasonable proposition to try and see whether, for small ε, a
solution of a boundary-value problem for equation (10.34) can be
formed from appropriate solutions of equation (10.35). Methods
of constructing such solutions are generally referred to as 'perturbation' or 'asymptotic' methods and in many cases provide the
only reasonable method, other than large-scale computation, of
obtaining a useful approximate solution to a given problem.
Perturbation methods are, however, often difficult to use, primarily because it seldom turns out to be possible to find a solution ϕ
of equation (10.35) such that, over the complete region of
interest, $\lim_{\varepsilon \to 0} u = \phi$, where u is the required solution of equation (10.34). In many cases setting $\varepsilon = 0$ completely changes the
nature of the problem and it becomes impossible to satisfy all the'
boundary conditions, so that it would be unreasonable to expect
$\lim_{\varepsilon \to 0} u$ to be equal to ϕ everywhere. For example, in the limit
as $\varepsilon \to 0$, the equation

$$\varepsilon \nabla^2 u - u = 1, \qquad (10.36)$$

reduces to $u = -1$ and hence solutions of Dirichlet problems for
equation (10.36) will not everywhere be approximately equal to
solutions of the approximating equation unless u is prescribed to
be equal to -1 on the bounding curves. Another example where

we would not expect the solution for $\varepsilon \neq 0$ to tend uniformly to a solution of equation (10.35) is the Dirichlet problem for

$$\frac{\partial^2 \phi}{\partial x^2} + \varepsilon \frac{\partial^2 \phi}{\partial y^2} - \frac{\partial \phi}{\partial y} = 0.$$

The equation changes type when $\varepsilon = 0$ and becomes parabolic and in general the Dirichlet problem is therefore not soluble for the limiting equation.

Problems when $\lim_{\varepsilon \to 0} u \neq \phi$ uniformly within the region of interest are referred to as singular perturbation problems whilst those where $\lim_{\varepsilon \to 0} u = \phi$ are said to be non-singular or regular. Unfortunately singular perturbation problems are the rule rather than the exception and in recent years many techniques have been developed to handle them. Recent texts on singular perturbation theory include Cole (1968), Eckhaus (1973), Nayfeh (1974), and van Dyke (1964).

The methods available are too many and varied to be summarized effectively and we shall confine ourselves to giving the solution to one regular and to one singular perturbation problem. The problems to be considered are both linear, but it should be noted that a particular advantage of singular perturbation methods is that they can be used to solve both linear and non-linear problems.

We consider first the problem of obtaining an approximate solution for small ε of

$$\nabla^2 u - \varepsilon u = 1, \tag{10.37}$$

within the circle C of unit radius, and with $u = 0$ on C. There is in fact no need to use approximate methods in this case as the exact solution is given by

$$u = -\frac{1}{\varepsilon} \left\{ 1 - \frac{I_0(\varepsilon^{\frac{1}{2}} r)}{I_0(\varepsilon^{\frac{1}{2}})} \right\},$$

where I_0 is the modified Bessel function of order zero and r is the distance from the centre O of C. It is however a very useful problem to use for illustrating the perturbation approach as the detailed analysis is sufficiently simple so as not to obscure the basic steps of the perturbation method. The problem for $\varepsilon = 0$ is similar to that for $\varepsilon \neq 0$; this suggests a regular perturbation problem and that the first approximation will be the solution of

$$\nabla^2 u = 1 \tag{10.38}$$

with $u = 0$ on C. The boundary conditions imply radial symmetry so that equation (10.38) simplifies to

$$\frac{1}{r}\frac{d}{dr}\left(r\frac{du}{dr}\right) = 1,$$

and the appropriate solution of this is $u = \frac{1}{4}(r^2 - 1)$. A more accurate estimate of u can be obtained by replacing εu in equation (10.37) by $\frac{1}{4}\varepsilon(r^2 - 1)$ and solving the resulting equation. This process can be repeated with each successive approximation and a power series in ε obtained. This suggests that a more direct approach might be to assume a power series of this type and attempt to determine the successive coefficients directly. We therefore assume an expansion of the form

$$u = \frac{1}{4}(r^2 - 1) + \sum_{n=1}^{\infty} \varepsilon^n u_n,$$

where the u_n are independent of ε, and substitute this into equation (10.37) and equate powers of ε. This gives

$$\nabla^2 u_1 = -\frac{1}{4}(r^2 - 1)$$

and

$$\nabla^2 u_n = u_{n-1}, \qquad n > 1.$$

The condition $u = 0$ on C implies that $u_n = 0$ on C. The boundary-value problems for all the u_n can, in principle, be solved and an infinite series obtained for u in powers of ε. There is no simple method of establishing the convergence of such a series and the most that can be said is that it will provide a good approximation to u for sufficiently small ε. u_1 is easily found to be

$$\tfrac{1}{64}(r^4 - 4r^2) + \tfrac{3}{64},$$

and $u_2, u_3 \ldots$, can also be found in similar forms.

We next consider the apparently similar problem of solving equation (10.36) with $u = 0$ on $r = 1$. It has already been pointed out that this is a singular perturbation problem and the solution obtained for $\varepsilon = 0$, viz. $u = -1$, does not satisfy the condition on $r = 1$. It is the need to satisfy this condition which complicates the issue and it seems worth trying to see whether $u = -1$, except near the boundary where it varies rapidly. We therefore write u as $-1 + v$ where v is to be negligible except near $r = 1$. v therefore satisfies

$$\varepsilon \nabla^2 v - v = 0$$

and, as v varies very rapidly near $r = 1$, the term $\varepsilon \nabla^2 v$ must be of the same order as v near $r = 1$. The behaviour near $r = 1$ can be seen more clearly by using a local 'stretched' coordinate system defined by $1 - r = \varepsilon^\alpha \xi$, where $\alpha > 0$. The equation for v now becomes

$$\varepsilon \left[\varepsilon^{-2\alpha} \frac{\mathrm{d}^2 v}{\mathrm{d}\xi^2} - \frac{\varepsilon^{-\alpha}}{(1 - \varepsilon^\alpha \xi)} \frac{\mathrm{d}v}{\mathrm{d}\xi} \right] - v = 0, \qquad (10.39)$$

and we try and determine α so that v can be found satisfying the conditions imposed on it (it has to be negligible away from $\xi = 0$ and equal to unity when $\varepsilon = 0$). A solution of equation (10.39), correct to $O(\varepsilon^{\frac{1}{2}})$ and with the derivatives of v of $O(1)$, can be found by setting $\alpha = \frac{1}{2}$ and the appropriate solution for v is then $\exp(-\xi)$, and hence

$$u = -1 + e^{-(1-r)\varepsilon^{-\frac{1}{2}}}.$$

The second term in this expression is negligible for small ε except for a small region, (a 'boundary layer'), whose width is of order $\varepsilon^{\frac{1}{2}}$, near $r = 1$. It therefore represents a correction to the solution of the equation for $\varepsilon = 0$ and it should be noted that it cannot be expanded as a power series in ε.

Bibliography

Baker, C. T. H. (1977). *Numerical solution of integral equations.* Clarendon Press, Oxford.

Cole, J. D. (1968). *Perturbation methods in applied mathematics.* Blaisdell, New York.

Delves, L. M. and Walsh, J. (1974). *Numerical solution of integral equations* Clarendon Press, Oxford.

Eckhaus, W. (1973). *Matched asymptotic expansions and singular perturbations.* North-Holland, Amsterdam.

Forsythe, G. E. and Wasow, W. (1960). *Finite difference methods for partial differential equations.* Wiley, New York.

Greenspan, D. (1965). *Introductory numerical analysis of elliptic boundary value problems.* Harper and Row, New York.

Jaswon, M. A. and Symm, G. T. (1977). *Integral equation methods in potential theory.* Academic Press, New York.

Kanwal, R. P. (1971). *Linear integral equations, Chapter 6.* Academic Press, New York.

Mitchell, A. R. (1969). *Computational methods in partial differential equations.* Wiley New York.

—— and Wait, R. (1977). *Finite element method in partial differential equations.* Wiley, New York.

Nayfeh, A. H. (1974). *Perturbation methods.* Academic Press, New York.

Richtmeyer, R. D. and Morton, K. W. (1967). *Difference methods for initial-value problems.* Wiley, New York.

Smith, G. D. (1969). *Numerical solution of partial differential equations.* Clarendon Press, Oxford.

Sneddon, I. N. (1966). *Mixed boundary value problems in potential theory.* North-Holland, Amsterdam.

Strang, G. and Fix, G. (1973). *An analysis of the finite element method.* Prentice-Hall, New York.

van Dyke, M. D. (1964). *Perturbation methods in fluid mechanics.* Academic Press. New York.

344

Solutions to exercises

Exercises 2.1.

1. (i) $x = x_0 + t,$ $y = y_0 - 2t$; $x = t$, $y = s - 2t$, $u = se^t = (y + 2x)e^x$.
   (ii) $x = x_0 e^t$, $y = y_0 e^t$; $x = se^t$, $y = e^t$, $u = s^2 e^{2t} = x^2$.
   (iii) $x = x_0 \cos t + y_0 \sin t$, $\quad y = y_0 \cos t - x_0 \sin t$; $\quad x = s(\cos t + \sin t)$, $y = s(\cos t - \sin t)$, $u = s^2 \exp(s^2 \sin 2t) = \frac{1}{2}(x^2 + y^2)\exp\frac{1}{2}(x^2 - y^2)$.

2. (i) $\phi = x + y$, $u = \exp(x/\phi) g(\phi)$.
   (ii) $\phi = y^2 - x^2$, $u = \operatorname{arsinh}(x/\phi^{\frac{1}{2}}) + g(\phi)$.
   (iii) $\phi = y/x$, $u = x^{-(1-\phi)/(1+\phi)} g(\phi) + (\phi^2 - 1)x$.

3. $u = \dfrac{x^2 + y^2}{x} + x\dfrac{(x^2 + y^2 + a^2 \sin^2\alpha)^{\frac{1}{2}}}{a \cos \alpha} - x\dfrac{(x^2 + y^2)}{a^2 \cos^2\alpha}$, $a \cos \alpha \le |x| \le a$.

4. $u = -4x + 2x^{\frac{1}{2}} + 2x^{\frac{1}{2}}(y^2 + 4x)^{\frac{1}{2}}$, between parabolae $y^2 + 4x = 4$ and $y^2 + 4x = 12$.

5. $x = st + \frac{1}{2}t^2 + s^2$, $\quad y = s + t$, $\quad u = \frac{1}{6}(t^3 + 3st^2 + 6s^2t + 4s^3) = xy - \frac{1}{3}y^3$, $1 < 2x - y^2 < 4$.

6. $x = x_0/(1 - x_0 t)$, $y = y_0(1 - x_0 t)$; $x = s/(1 - st)$, $y = s^2(1 - st)$, $u = \dfrac{s^3}{3}(1 - st)^3 + \frac{5}{3}s^3 = \dfrac{y^2}{3x} + \frac{5}{3}xy$, $1 < xy < 8$, $x > 0$.

7. Differentiate $u(x, 4x^2) = f(x)$ and show inconsistent with equation unless $f = cx$. Two solutions cx, $cy/4x$.

8. $x \times se^t$, $y = se^{-t}$, $u = s^2 e^t = x^{\frac{3}{2}} y^{\frac{1}{2}}$. Between hyperbolae $xy = 1$ and $xy = 4$ in $y > 0$.
   Differentiating $u(s, 1/s) = f$ produces an inconsistency with the differential equation unless $f = cs$. Two solutions cx, c/y.

Exercises 2.2

1. $F(\phi, \psi) = 0$ with
   (i) $\phi = lx + my + nu$, $\psi = x^2 + y^2 + u^2$.
   (ii) $\phi = u$, $\psi = y + u/(x + u)$.
   (iii) $\phi = \dfrac{u}{y}$, $\psi = y(x^2 + y^2 + u^2)$.

2. (i) $5(x + y + u)^2 = 9(x^2 + y^2 + u^2)$.
   (ii) $4(x^2 + y^2 - u)^3 = -u^2(x + y)^2$.
   (iii) $2^{\frac{1}{2}}(x + 3u + 2y) = 3(x^2 + y^2 + u^2)^{\frac{1}{2}}$.

3. (i) $s(1 - st)x = 1$, $(1 - 2st)y = 2s$, $(1 - st)u = s$; $y^2 xu = 4(y - u)^2$.
   (ii) $(1 - st)x = s$, $y = \frac{1}{2}t^2 + (1 - s)$, $u = t$; $(1 + xu)(2 - 2y + u^2) = 2x$.
   (iii) $x = \frac{1}{2}t(t + s) + s$, $y = s + t$, $u = \frac{1}{2}s + t = \dfrac{2x + y(y - 4)}{2(y - 2)}$.

The traces intersect at $(2, 2)$ and the solution is only defined in the region between the parabolae $y^2 = 2x$, $y(y-1) = 2x - 2$, with $y < 2$.

4. $u\left(\dfrac{y^2}{4}, y\right) = \frac{1}{2}y$. $\frac{1}{2}yu_x + u_y = \frac{1}{2} = uu_x + u_y$, which gives a contradiction.

Two solutions are y and $2^{\frac{1}{2}}x^{\frac{1}{2}}$.

5. $x = 2\sin s(1 - e^{-\frac{1}{2}t}) + s$, $y = t$, $u = \sin s e^{-\frac{1}{2}t}$.

Jacobian vanishes when $2(1 - e^{-\frac{1}{2}t})\cos s = -1$ and no solutions for s when $t < 2\log 2$. For $t > 2\log 2$, $x_s = 0$ has, for any given t, a solution $s = s_0$ and multi-valuedness now follows us in Example 2.8.

Exercises 2.3

1. At $t = 0$, $p = \frac{1}{2}s$, $q = 1$; $q \equiv 1$, $p = \frac{1}{2}se^t$, $x = se^t$, $y = e^t$,
   $u = \frac{1}{4}s^2e^{2t} + e^t = \frac{1}{4}x^2 + y$.

2. At $t = 0$, $p = q = 1$; $p \equiv q \equiv 1$, $x = (s-2)e^t + 2$, $y = 1 + e^t$,
   $u = (s-1)e^t + 2 = x + y - 1$.

3. At $t = 0$, $p = q = 1$; $p = q = u = 1/(1-t)$, $x = s - 2\log(1-t)$,
   $u = \exp(y + x - 1)$.

4. At $t = 0$, $p = 0$, $q = 1/s$; $sp = -\tanh(t/s)$, $sq = \operatorname{sech}(t/s)$,
   $x = -2s^2 \operatorname{sech}(t/s)\tanh(t/s)$, $y = -s^2 + 2s^2 \operatorname{sech}^2(t/s)$.
   $u = 2s \operatorname{sech}(t/s)$, $(2y - u^2)^2 = 4(x^2 + y^2)$.

Exercises 2.4

1. $u = -\lambda$ to left of $x + \lambda y = 0$ and $u = \lambda$ to right of $x - 1 = \lambda y$. Between these lines $u = (2x - 1)\lambda/(1 + 2\lambda y)$.

   For $\lambda < 0$ the characteristic traces from $0 \leq x \leq 1$, $y = 0$, all intersect at $x = \frac{1}{2}$, $y = -1/2\lambda$ and traces from $x \leq 0$, $x \geq 1$ also intersect in region $y \geq -1/2\lambda$ and hence no single-valued solution possible. Shock parallel to y-axis from $x = \frac{1}{2}$, $y = -1/2\lambda$ and solution is $(2x - 1)\lambda/(1 + 2\lambda y)$ in triangular region bounded by $x + \lambda y = 0$, $x = 1 + \lambda y$, $y = 0$ and otherwise is $-\lambda$ to left of shock and $+\lambda$ to its right.

2. $\lambda < 0$. Shock along $2x - \lambda y = 0$, $-\lambda$ to left of shock, 2λ to right. $\lambda > 0$. $-\lambda$ to left of $x = -\lambda y$ and 2λ to right of $x = 2\lambda y$. In between these lines $u = x/y$.

3. Shock $y = 2(x - 1)$ from $(1, 0)$ to $(2, 2)$ and shock $x = \sqrt{2}y^{\frac{1}{2}}$ from $(2, 2)$. $u = 0$ for $x < 0$, and, in $x \geq 0$, $u = x/y$ to left of $y = x$ and $x = \sqrt{2}y^{\frac{1}{2}}$, $u = 0$ to right of shocks and $u = 1$ in triangular region bounded by $y = 0$, $y = x$, $y = 2(x - 1)$.

Exercises 3.1

1. (i) $u_{\phi\psi} = 0$; $\phi = 3x - y$, $y = 2x - y$.
   (ii) $u_{\mu\mu} + u_{\lambda\lambda} = 0$; $\lambda = x - y$, $\mu = 2x$.

(iii) $u_{\psi\psi} = 0$; $\phi = y - x$, $\psi = x$.

(iv) $2\phi u_{\phi\psi} - u_{\psi} = 0$; $\phi = xy$, $\psi = y/x$.

(v) $2\lambda\mu(u_{\lambda\lambda} + u_{\mu\mu}) + \mu u_{\lambda} + \lambda u_{\mu} = 0$, $\lambda = y^2$, $\mu = x^2$.

(vi) $(\phi^2 + \psi^2)u_{\psi\psi} - 2\phi u_{\phi} = 0$. $\phi = y \tan \frac{1}{2}x$, $\psi = y$.

(vii) $y > 0$; $\mu(u_{\lambda\lambda} + u_{\mu\mu}) - u_{\mu} = 0$, $\lambda = x$, $\mu = 2y^{\frac{1}{2}}$.

  $y < 0$; $2(\phi - \psi)u_{\phi\psi} + u_{\phi} - u_{\psi} = 0$; $\phi = x + 2(-y)^{\frac{1}{2}}$, $\psi = x - 2(-y)^{\frac{1}{2}}$.

2. (i) $f(x + y - \cos x) + g(x - y + \cos x)$.

  (ii) $y^3 + yf(y^2 + 2x) + g(y^2 + 2x)$.

3. See §10.2.

Exercises 3.2

1. (i) $\phi\left(\dfrac{x + ct}{2c}\right) + \psi\left(\dfrac{ct - x}{2c}\right) - \psi(0)$.

  Determined in parallelogram $0 \le x + ct \le 2ac$, $-2bc \le x - ct \le 0$.

  (ii) $\psi\left(\dfrac{x - ct}{c}\right) - \phi\left(\dfrac{3x - 3ct}{2c}\right) + \phi\left(\dfrac{x + ct}{2c}\right)$.

  Determined in parallelogram $0 < x + ct < 2ac$, $0 < x - ct < \frac{2}{3}ac$.

  (iii) For $x > ct$ solution is given by D'Alembert's formula and for $x < ct$

$$u = \phi_2(ct - x) + \tfrac{1}{2}[\phi_0(ct + x) - \phi_0(3ct - 3x)] + \frac{1}{2c}\int_{3(ct-x)}^{ct+x} \phi_1(w)\, dw$$

$$2\phi_1(0) = c[\phi_2'(0) - \phi_0'(0)],$$

  where the prime denotes a derivative.

2. $\dfrac{8a}{\pi^2} \displaystyle\sum_{n=1}^{\infty} \dfrac{(-1)^n \cos\{(2n-1)\pi ct/2a\}\sin\{(2n-1)\pi x/2a\}}{(2n-1)^2}$.

3. General solution is $F(3x - y) + G(x + y)e^{-\frac{1}{2}(3x-y)}$. G can be determined if $f' - 6f = 4g$; F is then indeterminate and the only condition it has to satisfy is $2F'(0) + F(0) = 0$.

Exercises 3.3

1. (i) $f(3y - x) + g(2y - 3x)$.

  (ii) $e^{-x}f(2x - y) + e^{-2x}g(x + y)$.

  (iii) $f(xy) + g(x/y)$.

2. $4u_{\phi\psi} + (4\psi + 2\phi)u_{\phi} + (2\psi + 4\phi)u_{\psi} + u(2\phi^2 + 2\psi^2 + 5\phi\psi + 2) = 0$,

  $\phi = x - y$, $\psi = x + y$.

  General solution is $e^{-\frac{1}{2}\phi\psi}[e^{-\frac{1}{2}\phi^2}f(\psi) + e^{-\frac{1}{2}\psi^2}g(\phi)]$.

  $u = 2e^{-x^2} \cosh xy$.

  Triangular region in $y > 0$ between lines $y = \pm x$.

Exercises 4.1

1. (i) $\dfrac{4}{\pi} \displaystyle\sum_{n=1}^{\infty}$

$$\frac{\sin\{(2n-1)\pi x/2a\}\sinh\{(2n-1)\pi(b-y)/2a\}\operatorname{cosech}\{(2n-1)\pi b/2a\}}{(2n-1)}.$$

(ii) $\dfrac{4}{\pi} \displaystyle\sum_{n=1}^{\infty}$

$$\frac{\sin\{(2n-1)\pi x/a\}\sinh\{(2n-1)\pi(b-y)/a\}\operatorname{cosech}\{(2n-1)\pi b/a\}}{(2n-1)}.$$

$$+\frac{4}{\pi} \sum_{n=1}^{\infty}$$

$$\frac{\sin\{(2n-1)\pi y/b\}\sinh\{(2n-1)\pi(a-x)/b\}\operatorname{cosech}\{(2n-1)\pi a/b\}}{(2n-1)}.$$

2. Use equation (4.11).
3. If $u \geq v$ on C and $v > u$ at a point in D then the harmonic function $v - u$ is positive in the interior of D and negative on the boundary which contradicts the maximum principle. If v is harmonic so is $b + v$, $a + v$, and inequality follows as above.
4. Proof as in Theorem 4.3, w_r is now positive.
5. On $|x| = 1$, $|y| = 1$, $1 < u < 2$; hence result follows by exercise 3, above.
6. On boundaries $-0.025 < v - u < 0.025$; hence bounds follow by exercise 3, above.
7. $v = [m(b)\log r/a - m(a)\log r/b]/\log b/a$ is harmonic and on $r = a$ and $r = b$, v is equal to $m(a)$ and $m(b)$ respectively and hence on these boundaries $v \geq u$ and therefore $v \geq u$ for all r and the inequality then follows.

Exercises 4.2

1. $\dfrac{1}{4\pi} \log \dfrac{[(x-x')^2+(y-y')^2][(x+x')^2+(y+y')^2]}{[(x-x')^2+(y+y')^2][(x+x')^2+(y-y')^2]}$

$$\frac{y}{\pi} \int_0^{\infty} \left\{ \frac{1}{(x-x')^2+y^2} - \frac{1}{(x+x')^2+y^2} \right\} f(x')\, dx'.$$

2. In the notation of Example 4.3, if the images of P and P' in the line $y = 0$ are S and S' then the required Green's function is

$$\frac{1}{2\pi} \log \frac{PQ}{SQ} \frac{S'Q}{P'Q}.$$

3. (i) $\dfrac{4}{\pi} \sum\limits_{n=1}^{\infty} \dfrac{1}{(2n-1)} \text{sech}(2n-1)\dfrac{\pi b}{2a} \sinh(2n-1)\dfrac{\pi}{2a}(y'-b)$

$\times \cosh(2n-1)\dfrac{\pi y}{2a} \sin(2n-1)\dfrac{\pi x}{2a}\sin(2n-1)\dfrac{\pi x'}{2a}, \; y' > y.$

Interchange y and y' for $y' < y$.

(ii) $-\dfrac{2}{\pi} \sum\limits_{n=1}^{\infty} \dfrac{1}{n} e^{-(n\pi/b)x'} \sinh\dfrac{n\pi x}{b} \sin\dfrac{n\pi y}{b} \sin\dfrac{n\pi y'}{b}, \; x' > x.$

Interchange x' and x for $x' < x$.

4. Green's Function is $-\dfrac{1}{2\pi}[K_0\{[(x-x')^2+(y-y')^2]^{\frac12}\}-$
$K_0\{[(x-x')^2+(y+y')^2]^{\frac12}\}].$

$u = \dfrac{1}{\pi} y \displaystyle\int_{-\infty}^{\infty} \dfrac{K_1\{[(x-x')^2+y^2]^{\frac12}\}}{[(x-x')^2+y^2]^{\frac12}} f(x') \, dx'.$

Exercises 4.3

1. At a maximum, $u_x = u_y = 0$, $u_{xx} \le 0$, $u_{yy} \le 0$ and hence $Eu < 0$. $-u$ will have a positive maximum when u has a negative minimum so, in this case $Eu > 0$.
2. If u is not identically zero it will have either a positive maximum or negative minimum at an interior point so, at such a point, either $Eu < 0$ or $Eu > 0$, both of which lead to a contradiction.

Exercises 4.4

1. Separated solutions are $\cos m\theta J_m(j_{nm}r)$, $m = 0, \ldots,$ $\sin m\theta J_m(j_{nm}r)$, $m = 1, \ldots$, with $\lambda = j_{nm}^2$.
2. $-2 \sum\limits_{n=1}^{\infty} \dfrac{J_2(j_{2n}r)}{j_{2n}^3 J_3(j_{2n})} \cos 2\theta$, where $x = r\cos\theta$, $y = r\sin\theta$.
3. Construct two new integral identities from equation (4.10) by (i) replacing u by $\nabla^2 u$, without changing v, (ii) replacing v by $\nabla^2 v$ without changing u. Subtract these identities to give an identity for

$$\int_D (v\nabla^4 u - u\nabla^4 v) \, dS.$$

This identity can be used as in Theorem 4.8 to prove eigenvalues real and eigenfunctions orthogonal. To prove eigenvalues negative, use equation (4.10), with v replaced by u and u replaced by $\nabla^2 u$, to calculate $\mu \int u^2 \, dS$.

This turns out to be negative unless $\nabla^2 u = 0$ in D and Theorem 4.2 then gives $u \equiv 0$.

4. $(\nabla^4 - \mu_n{}^2)u = (\nabla^2 - \mu_n)(\nabla^2 + \mu_n)u$.

All w_n are therefore eigenfunctions of previous exercise. If there exists an eigenvalue not equal to $-\mu_n{}^2$ then there would exist a function orthogonal to all w_n and this is not possible.

Exercises 5.1

1. $A = (x, x)$, $B = (y, y)$. On $y = x$, $q = 2x$.

$$u = (y^2 - x^2) + x.$$

2. $\partial u/\partial n$ proportional to $2xu_x - u_y$, $u(x, x^2) = x$ then gives

$$p = \frac{1}{1 + 4x^2}, \qquad q = \frac{2x}{1 + 4x^2} \quad \text{when} \quad y = x^2.$$

$$u = \tfrac{1}{2}(x + y^{\frac{1}{2}}) + \tfrac{1}{2}(\tan^{-1} 2x - x) - \tfrac{1}{2}(\tan^{-1} 2y^{\frac{1}{2}} - y^{\frac{1}{2}}).$$

3. (i) u given on characteristics therefore solution possible;
   (ii) given conditions on $y = 1 - x$ would have to be consistent with equation for solution to exist and even then not unique;
   (iii) prescribing u on x, $1 - x$ determines u on third side.

4. (iv) Solution determined in $y \le x$ by conditions on $y = 0$, $x \ge 0$ and value on characteristic $y = x$ together with condition on $x = 0$ determines solution completely.

   General solution is $e^{-2x} \sin y + f(x - y) + g(x + y)$ and

   $$y \le x, \qquad u = e^{-2x} \sin y - \tfrac{1}{2}e^{-2x} \sinh 2y,$$

   $$y \ge x, \qquad u = e^{-2x} \sin y - 2 + 2\cos(x - y) - \tfrac{1}{2} + \frac{e^{-2y}}{2} \cosh 2x.$$

5. Continuity of u at $x = 0$ gives $\phi_0(0) = \phi_2(0)$.

   Also $\phi_2'(x) = u_x(x, 2x/c) + (2/c)u_t(x, 2x/c)$, where the prime denotes a derivative. Setting $x = 0$ gives $c[\phi_2'(0) - \phi_0'(0)] = 2\phi_1(0)$.

6. Cauchy conditions give $u = 0$, $x \ge ct$. Conditions on u and u_t consistent at $t = 0$ so no discontinuities to be propagated $\therefore$ u_t continuous across $x = ct$, set $\phi = x - ct$, $\psi = x + ct$. Only $u_{\phi\phi}$ can be discontinuous across $\phi = 0$, differentiating equation in canonical form with respect to ϕ and subtracting on opposite sides of $\phi = 0$ gives

   $$2[u_{\phi\phi}]_\psi + kc[u_{\phi\phi}] = 0.$$

   Also $[u_{tt}] = c^2[u_{\phi\phi}]$ and $[u_{tt}] = 1$ at $(0, 0)$.

350   *Solutions to exercises*

Exercises 5.2

1. $-\frac{1}{2}H(t'-t-|x'-x|)+\frac{1}{2}H(t'-t-|x'+x|)$.

2. $\dfrac{4}{\pi}\displaystyle\sum_{n=1}^{\infty}\dfrac{1}{(2n-1)}\sin\dfrac{(2n-1)\pi x}{2a}\sin\dfrac{(2n-1)\pi x'}{2a}\sin\dfrac{(2n-1)\pi}{2a}(t'-t)$ for $t'<t$,

and G is zero otherwise. Substituting into equation (5.35) gives

$$u=\frac{2}{a}\sum_{n=1}^{\infty}\sin\frac{(2n-1)\pi x}{2a}\cos\frac{(2n-1)\pi t}{2a}\int_0^a f(x')\sin\frac{(2n-1)\pi x}{2a}\,dx'.$$

3. Argument is as for wave equation.

$$G=-\tfrac{1}{2}I_0(k[(t-t')^2-|x-x'|^2]^{\frac{1}{2}})H(t-t'-|x-x'|).$$

Exercises 5.3

1. $R=\dfrac{x'-y'}{x+y}$, $u=(x-y)(x^2+y^2)$.

2. $R=\dfrac{x'y}{y'x}$, $u=\dfrac{y}{x}(\sin x-\sin y)$.

3. Setting $\phi=xy$, $\psi=y/x$ gives

$$2\phi u_{\phi\psi}-u_{\psi}=0.$$

Riemann function satisfies

$$2\phi'R_{\phi'\psi'}+R_{\psi'}=0,$$

and $R=(\phi/\phi')^{\frac{1}{2}}$. $u=x^2y^2+2x^2/y$.

Exercises 6.1

1. If u is the difference of two solutions then it follows that $I(t)$ defined in §6.2 still satisfies $dI/dt\le0$.
2. $u-B$ is a solution of the heat equation and is negative on $x=0$, $x=a$, and $t=0$. Thus if $u>B$ at a point in $0<x<a$, $t>0$, this would contradict the maximum principle.
3. $u-(1-\frac{1}{2}x)$ is a solution of the heat equation and is never positive on $t=0$, $x=0$, or $x=2$ and hence by the maximum principle it can never be positive.
4. Proof follows as in Theorem 6.2.
5. If $v=u-\left(1-\dfrac{x}{2}-e^{-kt}\cos\dfrac{\pi x}{4}\right)$, then $v\ge0$ on $x=0$, $x=2$, $t=0$. Also $P_2v=\left(k-\dfrac{\pi^2}{16}\right)\le0.$

Hence the minimum value is attained on one of $x=0$, $x=2$, $t=0$, and hence result follows.

6. $\dfrac{4}{\pi}\displaystyle\sum_{n=1}^{\infty}\dfrac{e^{-(2n-1)^2t/4\pi^2a^2}}{(2n+1)(2n+3)}\cos(2n-1)\dfrac{\pi x}{2a}.$

Exercises 6.2

1. $-\tfrac{1}{2}\pi^{-\frac{1}{2}}(t-t')^{-\frac{1}{2}}\left\{\exp-\dfrac{(x-x')^2}{4(t-t')}+\exp-\dfrac{(x+x')^2}{4(t-t')}\right\}H(t-t').$

2. $-\dfrac{2}{a}\displaystyle\sum_{n=1}^{\infty}\exp\left\{\dfrac{(2n-1)^2}{4a^2}\pi^2(t'-t)\right\}\sin(2n-1)\dfrac{\pi x'}{2a}\sin(2n-1)\dfrac{\pi x}{2a}\ t'<t.$

Exercises 7.1

1. Put $u=x^2-x+v.$

$$v=\dfrac{16}{\pi^3}\sum_{r=0}^{\infty}\dfrac{1}{(2r+1)^3}\operatorname{sech}\{(2r+1)\pi\}\cosh\{(2r+1)\pi(y-1)\}\sin\{(2r-1)\pi x\}.$$

2. Use a sine-series representation in x for u and also represent 1 as a sine series

$$u=-\dfrac{4}{\pi}\sum_{r=1}^{\infty}\dfrac{1}{(2r-1)[1+(2r-1)^2]}\left[e^{-y}-\cos(2r-1)\pi y\right.$$
$$\left.+\dfrac{1}{(2r-1)}\sin(2r-1)\pi y\right]\sin(2r-1)\pi x.$$

3. Put

$$u=y(1-x)+\tfrac{1}{6}(x^3-3x^2+2x)+v.$$

Then

$$v(0,y)=v(1,y)=0,\qquad v(x,0)=-\tfrac{1}{6}(x^3-3x^2+2x),$$

and v can be obtained as a sine series in x.

4. $u=8\displaystyle\sum_{n=0}^{\infty}\dfrac{J_0(j_{n0}r)\cos(j_{n0}y)}{j_{n0}{}^3J_1(j_{n0})}.$

5. $u=\dfrac{2}{3}\left(\dfrac{a}{r}\right)^3 P_2(\cos\theta)+\dfrac{a}{3r}.$

Exercises 7.2

1. $\dfrac{1}{s}\exp\dfrac{x}{2a}[b-(b^2+4as)^{\frac{1}{2}}].$

2. $\left(t+\frac{1}{2}x^2\right)\mathrm{Erfc}\left(\dfrac{x}{2t^{\frac{1}{2}}}\right)-x\mathrm{e}^{-x^2/4t}\left(\dfrac{t}{\pi}\right)^{\frac{1}{2}}$.

   Use Table 7.2, (4).

5. $\dfrac{s^{-\frac{1}{2}}}{(s^2+1)}$, $\dfrac{t^{\frac{3}{2}}}{\Gamma(\frac{5}{2})}-\dfrac{t^{\frac{7}{2}}}{\Gamma(\frac{9}{2})}$.

6. $\dfrac{1}{s}\left(1-\dfrac{\sinh s^{\frac{1}{2}}(1-x)}{\sinh s^{\frac{1}{2}}}\right)$, write $\bar{u}$ as $\dfrac{1}{s}[1-\mathrm{e}^{-s^{\frac{1}{2}}x}(1-\mathrm{e}^{-2s^{\frac{1}{2}}(1-x)})$

   $\times(1-\mathrm{e}^{-2s^{\frac{1}{2}}}-\mathrm{e}^{-4s^{\frac{1}{2}}}-\cdots)]$;

invert term-by-term to give

$$u\sim 1-\mathrm{Erfc}\left(\frac{x}{2y^{\frac{1}{2}}}\right)+\mathrm{Erfc}\left(\frac{2-x}{2y^{\frac{1}{2}}}\right).$$

Exercises 7.3

1. Use exponential Fourier transform with respect to x.

$$\frac{1}{\pi}\int_{-\infty}^{\infty}\frac{\mathrm{e}^{-iwx}}{(w^2+1)}\frac{\sinh wy}{\sinh wa}\,\mathrm{d}w.$$

2. $u=\dfrac{2}{\pi}\displaystyle\int_0^{\infty}\dfrac{\cos wx}{w^2+9}\{\mathrm{e}^{-(w^2-10)t}\,\mathrm{e}^{-t}\}\,\mathrm{d}w$,

   second term is dominant as $t\to\infty$ and gives required result (use Table 7.3 (6)).

3. $\dfrac{2}{\pi}\left[\displaystyle\int_0^{\infty}\dfrac{\sin wx}{1+w^2}(w-w\mathrm{e}^{-y}\cos wy-\mathrm{e}^{-y}\sin wy)\,\mathrm{d}w\right]=\mathrm{e}^{-x}H(y-x)$

   (Integrals may be evaluated by complex variable methods or by using tables e.g. IT1, p. 8, row (11), p. 65, row. (15).)

4. $\dfrac{2}{\pi}\displaystyle\int_0^{\infty}\mathrm{e}^{-(w^2+1)^{\frac{1}{2}}y}\dfrac{w}{w^2+4}\sin wx\,\mathrm{d}w=\dfrac{1}{\pi}\displaystyle\int_{-\infty}^{\infty}\mathrm{e}^{-iwx}\dfrac{w}{w^2+4}\mathrm{e}^{-(w^2+1)^{\frac{1}{2}}y}\,\mathrm{d}w$.

   The singularity with largest negative imaginary part is $w=-i$ and Theorem 7.5 gives

$$u\sim\frac{1}{3}\left(\frac{2}{\pi}\right)^{\frac{3}{2}}yx^{-\frac{3}{2}}\mathrm{e}^{-x}.$$

5. $-\dfrac{2}{\pi}\displaystyle\int_0^{\infty}\dfrac{\mathrm{e}^{-wx}\sin wy}{w^2+1}\,\mathrm{d}w$. $u\sim-\dfrac{2}{\pi y}$ (from equation (7.38)).

Exercises 8.1

1. Prove that mean value over a sphere exceeds value at its centre and then argue as for two-dimensional case.

2. $v-u$ is harmonic and not positive on the boundary of D and hence, by the maximum principle, is not positive within D.

3. $G+\dfrac{1}{4\pi}|\mathbf{r}'-\mathbf{r}|^{-1}$ is harmonic and is positive everywhere on the boundary of V and therefore, by the minimum principle, cannot be negative within V. If the point $\mathbf{r}'=\mathbf{r}$ is excluded by a small sphere S, then, on S, G will be negative and, on the boundary of V, G is zero. Hence, by the maximum principle, G cannot have a positive maximum between S and V.

4. The function v defined by

$$\left(\frac{1}{a}-\frac{1}{b}\right)v=\left(\frac{1}{a}-\frac{1}{r}\right)M(b)+\left(\frac{1}{r}-\frac{1}{b}\right)M(a)$$

is harmonic and $v\geq u$ on $r=a$ and $r=b$ and hence $v\geq u$ for $a\leq r\leq b$.

Exercises 8.2

1. $t=3$, $t=4$, $t=4$.

2. E can be written as $\displaystyle\int_0^{c(t_0-t)}\int_C [c^{-2}u_t^2+u_r^2+r^{-2}u_\theta^2]\,\mathrm{d}s\,\mathrm{d}r$, where (r,θ) are polar coordinates with centre at (x_0,y_0) and C is the boundary of D. Using results for bounded region gives

$$\frac{\mathrm{d}E}{\mathrm{d}t}=-c\int_C [c^{-2}u_t^2+u_r^2+r^{-2}u_\theta^2]\,\mathrm{d}s+2\int_C u_t u_n\,\mathrm{d}s.$$

On C, $u_n=u_r$ and therefore

$$\frac{\mathrm{d}E}{\mathrm{d}t}=-c\int_C (c^{-1}u_t-u_r)^2\,\mathrm{d}s-c\int_C r^{-2}u_\theta^2\,\mathrm{d}s.$$

Proof then follows as in §3.3.

3. $-\dfrac{1}{2\pi}\dfrac{\cos k\{(t'-t)^2-|\mathbf{r}'-\mathbf{r}|^2\}^{\frac{1}{2}}}{\{(t'-t)^2-|\mathbf{r}'-\mathbf{r}|^2\}^{\frac{1}{2}}}H(t-t'-|\mathbf{r}'-\mathbf{r}|)$.

4. $-\dfrac{1}{4\pi}|\mathbf{r}'-\mathbf{r}|^{-1}\Big\{\delta(t-t'-|\mathbf{r}'-\mathbf{r}|)$

$\qquad\qquad -\dfrac{k\,|\mathbf{r}'-\mathbf{r}|\,J_1[k\{(t'-t)^2-(\mathbf{r}'-\mathbf{r})^2\}^{\frac{1}{2}}]}{\{(t'-t)^2-|\mathbf{r}'-\mathbf{r}|^2\}^{\frac{1}{2}}}H(t-t'-|\mathbf{r}'-\mathbf{r}|)\Big\}$

Exercises 8.3

1. Proof follows as in Theorem 6.2 when modifications indicated in §8.5 are made.

2. If $v = u - (1 - r^2)$ then

$$v_{xx} + v_{yy} - v_t = u_{xx} + u_{yy} - u_t + 4 \geq 0.$$

$v = 0$ for $t = 0$ and for $r = 1$ and hence, by exercise 1, above, it cannot exceed zero for $r < 1$, $t > 0$, and the result now follows.

3. $G = -\dfrac{1}{8\pi^{\frac{3}{2}}(t - t')^{\frac{3}{2}}} \left\{ \exp{-\dfrac{|\mathbf{r} - \mathbf{r}'|}{4(t - t')}} - \exp{-\dfrac{|\mathbf{R} - \mathbf{r}'|}{4(t - t')}} \right\} H(t - t'),$

where $\mathbf{R}$ denotes the image of $\mathbf{r}$ in the plane $z = 0$.

Exercises 9.1

1. $x^2 + y^2 > 9$.
2. $x^2 + y^2 > 2$.
3. $(cu_x + u_y) - c(cv_x + v_y) = 0$.
   $(-cu_x + u_y) - c(-cv_x + v_y) = 0$.
   On $dx/dy = v$, the first equation becomes $d(u - \frac{1}{2}v^2) = 0$.
4. $\dfrac{dy}{dx} = \tan v$, $\dfrac{dy}{dx} = -\cot v$.
   $(\cos v)u_x + (\sin v)u_y + ((\cos v)v_x + (\sin v)v_y) = 0$,
   $(\sin v)u_x - (\cos v)u_y - ((\sin v)v_x - (\cos v)v_y) = 0$,
   Equations become, on characteristics, $d(v \pm u) = 0$.

Exercises 9.2

1. Eigenvalues are 7, 2 with eigenvectors $(1, 1)^T$ and $(2, -3)^T$.

$$\mathbf{v}_y + \begin{pmatrix} 7 & 0 \\ 0 & 2 \end{pmatrix} \mathbf{v}_x = \mathbf{0},$$

$$\mathbf{u} = \begin{pmatrix} 1 & 2 \\ 1 & -3 \end{pmatrix} \mathbf{v}, \qquad \mathbf{v} = \frac{1}{5} \begin{pmatrix} 3 & 2 \\ 1 & -1 \end{pmatrix} \mathbf{u}.$$

$$\mathbf{u} = \left(\frac{-31y}{24} - \frac{7x}{24}, \frac{-37y}{8} + \frac{11x}{8} \right)^T.$$

2. Eigenvalues are 1, 2, 3 with eigenvectors $(-1, 0, 1)^T$, $(2, -1, 0)^T$, $(0, 1, -1)^T$.

$$\mathbf{v}_y + \begin{pmatrix} 1 & 0 & 0 \\ 0 & 2 & 0 \\ 0 & 0 & 3 \end{pmatrix} v_x = \mathbf{0}, \qquad \mathbf{u} = \begin{pmatrix} -1 & 2 & 0 \\ 0 & -1 & 1 \\ 1 & 0 & -1 \end{pmatrix} \mathbf{v},$$

$$\mathbf{v} = \begin{pmatrix} 1 & 2 & 2 \\ 1 & 1 & 1 \\ 1 & 2 & 1 \end{pmatrix} \mathbf{u}; \qquad \mathbf{u} = (\tfrac{1}{3}x + \tfrac{1}{3}y, \tfrac{2}{15}x - \tfrac{1}{15}y, \tfrac{2}{3}y - \tfrac{4}{3}x)^T.$$

Index

M